Sister Chromatid Exchange

Sister Chromatid Exchange

EDITED BY

SHELDON WOLFF
LABORATORY OF RADIOBIOLOGY AND ENVIRONMENTAL HEALTH, AND
DEPARTMENT OF ANATOMY
UNIVERSITY OF CALIFORNIA, SAN FRANCISCO

1807 175 YEARS OF PUBLISHING 1982

A Wiley-Interscience Publication

JOHN WILEY & SONS

New York • Chichester • Brisbane • Toronto • Singapore

Library of Congress Cataloging in Publication Data:

Main entry under title:

Sister chromatid exchange.

(A Wiley-Interscience publication)
Includes index.
1. Sister chromatid exchange. I. Wolff, Sheldon. [DNLM: 1. Crossing over—Genetics. QH 445 S623]

QH445.S57 574.87′322 81-13102
ISBN 0-471-05987-0 AACR2

Printed in the United States of America

10 9 8 7 6 5 4 3 2 1

Preface

For years geneticists debated whether or not an exchange of sister strands occurred during the process of crossing over. Largely because of data on *Drosophila,* it was commonly held that sister strand crossing over occurred in neither meiotic nor mitotic cells, although the cytological study of ring-bearing cells in maize indicated that it might occur during mitosis. In 1957, by using tritiated thymidine, J. Herbert Taylor and his colleagues succeeded in making mitotic cells whose chromosomes were composed of sister chromatids that were physically different from one another. One chromatid was radioactive and its sister was not, allowing them to be individually identified in autoradiographs. Sister chromatid exchanges (SCEs) were present in cells with such chromosomes.

The subsequent studies of SCEs in labeled chromosomes gave valuable insight into chromosome replication and structure. In fact, it is not too much to say that these studies changed the way in which cytologists ordinarily thought about chromosomes. After several fruitful years of work, however, further studies became hampered by the lack of resolution inherent in autoradiographic techniques.

In the early part of the last decade, cytologists discovered that tritiated thymidine could be replaced by halogenated thymidine analogues in experimental protocols similar to the original ones of Taylor. This procedure made the sister chromatids chemically, rather than physically, different from one another and thus they could be made to stain differentially. The resultant preparations showed SCEs with great precision and resolution.

The ease with which preparations for SCEs can be made and scored has led to an explosion of increased interest in the study of SCEs and associated phenomena. In particular, because SCEs can be readily induced by mutagenic carcinogens, they are being used not only to test for the presence of such compounds, but also to gather clues about the processes involved in chromosomal mutation and in the induction of the genetic variability that results from recombination. Interest has also been heightened by the realization that certain human diseases are characterized either by an increased level of SCE or a changed response to SCE-inducing chemicals.

Because of the large amount of work involved in investigating this relatively new cytogenetic endpoint, this book has been compiled to help biologists and other biomedical scientists understand the current position of SCE research. The chapter authors are all cytological experts who have contributed heavily to the study of SCEs and to the development of the techniques that we now use.

SHELDON WOLFF

San Francisco, California
January 1982

Contents

Chapter 1	**Sister Chromatid Exchange: Early Experiments with Autoradiography** J. HERBERT TAYLOR	**1**
Chapter 2	**Sister Chromatid Exchange: New Methods for Detection** SAMUEL A. LATT	**17**
Chapter 3	**Chromosome Aberrations, Sister Chromatid Exchanges, and the Lesions That Produce Them** SHELDON WOLFF	**41**
Chapter 4	**Sister Chromatid Exchange and Single-Gene Mutation** ANTHONY V. CARRANO AND LARRY H. THOMPSON	**59**
Chapter 5	**Induction of Sister Chromatid Exchanges by Chemical Agents** SHIN TAKEHISA	**87**
Chapter 6	**The Use of Sister Chromatid Exchange to Monitor Human Populations for Exposure to Toxicologically Harmful Agents** B. LAMBERT, A. LINDBLAD, K. HOLMBERG, AND D. FRANCESCONI	**149**

Chapter 7 **Sister Chromatid Exchanges and Disease States in Man** **183**

H. J. Evans

Chapter 8 ***In Vivo* Methods for Detecting Sister Chromatid Exchange** **229**

E. L. Schneider and D. Kram

Chapter 9 **Sister Chromatid Exchanges in Plants** **243**

B. A. Kihlman and H. C. Andersson

Chapter 10 **Sister Chromatid Exchanges in *Drosophila*** **267**

Maurizio Gatti

Index **297**

Sister Chromatid Exchange

Chapter 1

Sister Chromatid Exchange: Early Experiments with Autoradiography

J. HERBERT TAYLOR
PROFESSOR OF BIOLOGICAL SCIENCE
INSTITUTE OF MOLECULAR BIOPHYSICS
FLORIDA STATE UNIVERSITY
TALLAHASSEE, FLORIDA

I. Introduction 2
II. Semiconservative segregation of DNA in the reproduction of chromosomes 3
III. Sister chromatid exchanges 7
  A. Rationale for analysis of chromosome structure 7
  B. First experimental analyses 10
  C. Conclusions 11
IV. Models and exceptions 11
V. Some unresolved problems 13

I. INTRODUCTION

Before 1956, when Taylor et al. (1957) produced the first autoradiographs of metaphase chromosomes with [^{3}H]thymidine, the occurrence of sister chromatid exchanges was a matter of speculation. Ring chromosomes, especially small rings, usually separate freely at division, but there is a predictable frequency of double-size rings and interlocked chromosomes (McClintock, 1938). The double-sized rings may break at anaphase and reinite to produce larger and smaller rings with deletions and duplications after the ends are fused. The interlocked rings may separate by the breakage of one ring without loss of chromosome material from either cell. The mechanism of formation of the interlocks and double-sized rings was uncertain because of the lack of knowledge concerning the structure of a chromatid. A widely held view assumed that anaphase chromosomes or the two chromatids at metaphase were already double in the morphological sense because many light microscope pictures showed a visible doubleness (Sharp, 1934) and bridges formed as a result of irradiation at G_2 or prophase could be interpreted as half-chromatid bridges (e.g., see Darlington and LaCour, 1953 and Taylor, 1967a). The reunion of such double chromatids following a break in a ring could occur with many twists in the two subunits. Such a situation should give rise to very complex interlocked rings in the case of ring chromosomes at the next anaphase unless the clockwise and counterclockwise twists happen to be equal. McClintock (1938) determined that such complexities do not arise. Another possibility was that the anaphase chromosome contained a single axial unit and twists could be resolved, that is, free rotation could occur around certain bonds. Without such free rotation, if chromatids reproduce by splitting, the plane of the cleavage would have to be precise and twists would have to be eliminated in rings or numerous interlocks would occur. In summary, there was no way to distinguish between sister chromatid exchanges and some type of reproduction of chromosomes that produced the observed figures at mitosis.

The exchanges were evident in the first chromosomes examined at the second division following the S phase in which [^{3}H]thymidine had been incorporated. In fact, sister chromatid exchanges were so numerous that consistent semiconservative segregation of the DNA was sometimes difficult to demonstrate convincingly in *Vicia faba* root cells (Peacock, 1663).

A study of the frequency and pattern of sister chromatid exchanges was initiated early in 1957 following a seminar at the California Institute of Technology, in which I presented a report on work done by myself and my coworkers on the semiconservative replication of DNA in chromosomes. A small discussion group assembled the next morning to listen to Professor Max Delbruck describe the properties of the DNA helix and the restrictions on rejoining after breakage due to the difference in polarity of the two chains. Because of the difficulties in clearly demonstrating semiconservative segregation of labeled DNA in chromosomes of *Vicia* root cells without the complications introduced by sister chromatid exchanges, I had been searching for a plant with fewer large chromosomes that could be cultured for laboratory studies. I had obtained seeds of *Bellevalia romana,* a liliaceous plant with only four pairs of large chromosomes. Autoradiographs had been made to compare with results obtained with *Vicia faba,* which has chromosomes about the same size, but *Vicia* has six pairs and only the largest pair was clearly distinguishable in autoradiographs. *Bellevalia* has two larger pairs of chromosomes that are easily distinguished from each other, and two smaller pairs that are nearly the same size and have similar arm length ratios. The frequency of sister chromatid exchange appeared to be as low or lower than that in *Vicia.* In considering a suitable biological material for sister chromatid analysis, I chose *Bellevalia* even though it was more difficult to grow than *Vicia.*

The rationale for the experiment was to see if the chromatids behaved like a DNA helix in the formation of sister chromatid exchanges. Were the two subunits, which had been demonstrated in two species by that time, behaving as if the whole chromatid was a duplex of DNA chains in spite of the tremendous difference in size or diameter of the two structures? If chromatids were composed of two subunits of reversed polarity and exchanges involved only whole chromatids, the chance of producing a visible discontinuity of labeled DNA at the same site in both chromatid pairs derived from an original labeled chromosome would be considerably enhanced compared to a situation in which the two chains had identical polarity.

II. SEMICONSERVATIVE SEGREGATION OF DNA IN THE REPRODUCTION OF CHROMOSOMES

An understanding of the basis for an analysis of sister chromatid exchanges requires a brief description of the way in which semiconservative segregation of labeled chromatids was first demonstrated. We chose to use cells from roots of *Vicia faba,* which has 12 large chromosomes, because Howard and Pelc (1951) had recently determined the length of the cell cycle and the length and position of S phase over the cycle. By labeling cells for 8 hours we could be sure that some nuclei would have completed S phase in [^{3}H]thymidine, but none of these nuclei would have yet passed through division, since S phase was about 8 hours and G_2 was about 6 hours. Only cells that were labeled during the last part of the S phase would have reached mitosis. Roots were rinsed and transferred to a solution with colchicine, which blocks mitosis but does not prevent these plant cells from cycling several times. After 10 hours in colchicine some of the roots were fixed and stained by the Feulgen reaction and spread on slides for preparation of autoradiographs. Another group of roots were fixed in the same way after 34 hours in the colchicine solution and autoradiographs were prepared.

Nearly all cells examined after fixing at 10 hours had 12 chromosomes, indicating that they were blocked in the first division after labeling occurred throughout most of S phase. All these cells contained chromosomes with both chromatids labeled throughout their length. On the other hand, many of the cells in the roots fixed 34 hours after transfer to colchicine had 24 chromosomes, indicating a block of the anaphase movement of chromosomes by colchicine, but occurrence of the transition to interphase and completion of a second S phase. When chromosomes from these cells reached division, the labeled DNA synthesized during the first cycle had all segregated with one of the two chromatids of each metaphase chromosome while its sister had unlabeled DNA throughout its length. There were occasional switches of label between the two sister chromatids, indicating sister chromatid exchanges had occurred. The appearance of a few cells with 48 chromosomes indicated that some cells had cycled through c-mitosis two times during the 34 hours and had reached c-metaphase a third time. These cells contained some unlabeled chromosomes, some with only a part of one chromatid labeled, and others with one whole chromatid labeled as expected if semiconservative segregation were continuing with an occasional sister chromatid exchange.

Since it was difficult to analyze 24 or 48 chromosomes, a species with fewer large chromosomes was desirable for further studies. I obtained seeds of *Bellevalia romana*, which is a rather small liliaceous plant with only 8 (four pairs) large chromosomes, two of which are easily distinguished by their larger size and difference in position of the centromere. Instead of using colchicine for long periods I placed the roots in [^{3}H]thymidine for 6 hours and then blocked the first division by transferring roots to a solution of colchicine for 10 hours. Figure 1.1 shows a cell from a root of *Bellevalia* at the first motosis after incorporation of the [^{3}H]thymidine into DNA. Both chromatids of each chromosome are labeled. The two chromatids of each chromosome have separated in this cell, as is typical after it is blocked for several hours in c-metaphase. Yet, the sister chromatids are close enough to each other to be recognized by their position as well as their size. Alternatively, roots were rinsed and

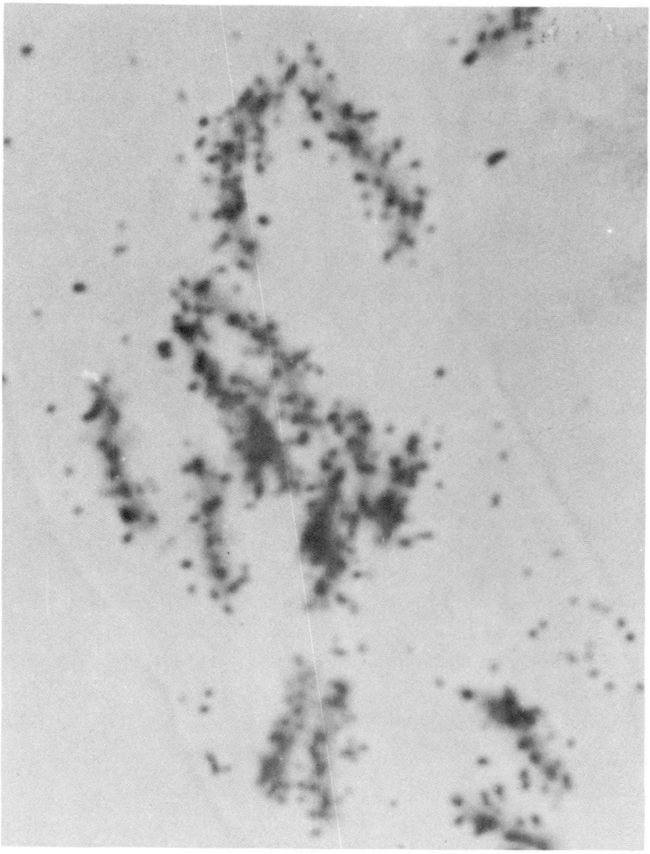

Fig. 1.1. Autoradiograph of the chromosomes in a root-tip cell of *Bellevalia* after one replication in tritium-labeled thymidine; colchicine blocked division in which the sister chromatids are separated but lie close enough together to allow the recently separated chromatids (or daughter chromosomes) to be identified. From Taylor, 1959.

time was allowed for the labeled cells to divide once before colchicine was added. For example, in one experiment roots that had been in solution with [^{3}H]thymidine for 6 hours were rinsed and grown in isotope-free solution for 12 hours and then for 12–14 hours in a solution with colchicine. Figure 1.2 shows a c-metaphase where the sister chromatids are just beginning to separate, but this one is from the roots that were free of the isotope for 12 hours before fixing. The semiconservative segregation is clear for most chromosomes in the group, with exchange sites also fairly clear.

To explain the results it was necessary to assume that chromosomes

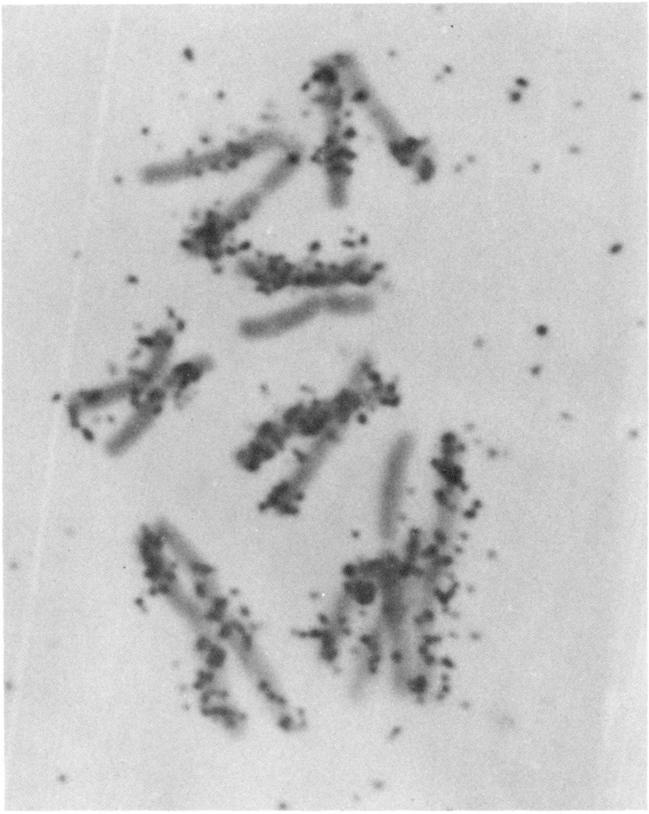

Fig. 1.2. Autoradiograph of a cell similar to that in Figure 1.1 fixed after one replication in tritium-labeled thymidine and one replication in a medium free of the labeled thymidine; colchicine blocked metaphase in which the sister chromatids are separated except in a few instances at the centromeres. From Taylor, 1959.

were composed of two subunits of DNA that extend throughout the length of a chromatid and separate at each replication or division cycle. The subunits were analogous to the two chains of the Watson-Crick model of DNA. The conceptional difficulty in accepting this interpretation without reservation was the enormous size difference between a chromatid and the DNA duplex. These very large chromatids are 1–2 microns in diameter, while the DNA helix is 20 Å, nearly a thousandfold different. It was at least conceivable that the DNA was organized by some protein structural framework in such a way that it could behave as two units with or without the separation of the chains. That possibility made further analysis of the nature and properties of the chromatid subunits highly desirable.

III. SISTER CHROMATID EXCHANGES

A. *Rationale for Analysis of Chromosome Structure*

If the chromatids were organized around a single helix of DNA or some other structure that behaved in breakage and reunion as if the subunits had a polarity, the analysis of the pattern of sister chromatid exchange should reveal this characteristic. It was clear that a chromatid did not have polarity, for reunions of all types had already been demonstrated. In the same sense a DNA helix does not have polarity—the end of any helix can reunite with the end of any other helix because its two chains have reverse polarity. To analyze these properties in chromatids, roots were placed in [^{3}H]thymidine and were allowed to incorporate the isotope into DNA for 6–8 hours; the roots were then rinsed and transferred to a solution with colchicine and were fixed at intervals until tetraploid c-metaphases could be seen. The early c-metaphases had chromatids so close together that it was sometimes very difficult or impossible to see where exchanges occurred. Therefore, the analyses were confined to chromosome groups in which the chromatids had begun to separate. Frequently sister chromatids were still very close together, as in Figures 1.1 and 1.2, but only tetraploid groups of 16 chromosomes were selected for analysis, so that all the exchanges formed in both cycles could be counted. Figure 1.3 shows a drawing of such a chromosome group. Two

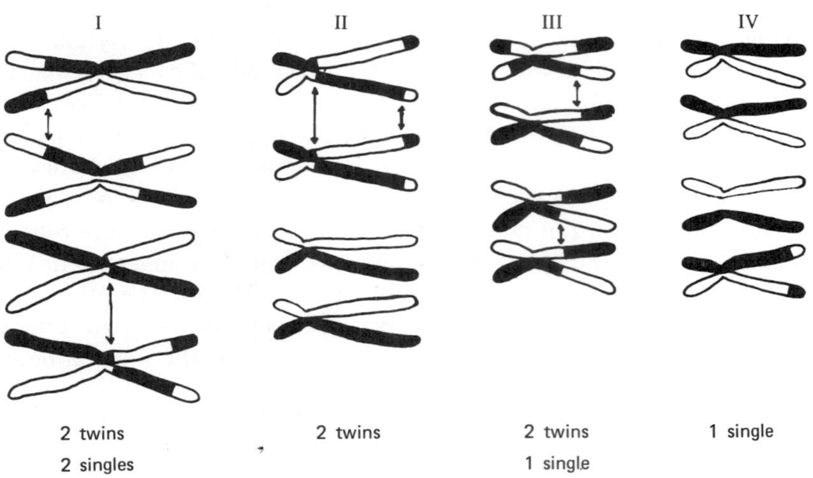

Fig. 1.3. Diagrammatic representation of a chromosome complement, from a tetraploid c-metaphase of *Bellevalia romana* at the second division after labeling with [^{3}H]thymidine, to show the position of sister chromatid exchanges; twin exchanges are indicated by arrows.

of the largest chromosomes in this group (chromosome I) have two exchanges (one in each chromosome) just to the right of the centromere. The other two have an exchange in each in the left arm at what appears to be identical positions. These exchanges are called twins. If we assume these two chromosomes are derived by replication from the same original G_1 chromosome with two subunits, any two chromosomes we would see would be composed of eight subunits, the two original unlabeled ones, two labeled ones made at the first S phase, and four unlabeled ones made at the second S phase. The paired exchanges would be the visible result of a single exchange during the first S phase when labeling occurred; these are the twin exchanges. In this instance the two chromosomes bearing twin exchanges also would have one unmatched exchange, each called a single exchange. These would occur after the chromosomes separate as a result of separate, unrelated events. Chromosome II is represented by one pair with two twin exchanges and the other pair has no exchanges. The twin exchanges occur every time an exchange occurs during the labeling cycle if there is one labeled and one unlabeled subunit in each chromatid and these subunits have opposite polarity or some property that prevents like subunits from reuniting (Figure 1.4).

Exchanges between one subunit in a chromatid and one subunit of the same polarity in its sister is theoretically possible, but if these exchanges occur, they must be repaired or converted to a complete exchange before the fork migrates far away. No unlabeled gaps, in which the sister chromatid has two times the amount of labeled DNA as adjacent segments, have been reported (Taylor, 1958). If the above half chromatid exchanges are ruled out, all exchanges during the first cycle yield twins after the second cycle if the subunits have reversed polarity (Figure 1.4). Any exchanges during the second cycle yield only unpaired or single exchanges because there is only one labeled subunit per chromatid and the exchange is visible at the division immediately after it occurs. Of course, the exchanges are mechanically or structurally like those of the first cycle and the rules of polarity are expected to apply; the difference is the one labeled subunit per two chromatids. Single exchanges occur after the chromatids of the first cycle have separated and have been replicated in the second cycle; therefore, on a per cell basis there are two times as many singles as twins if the chance for an exchange is the same at both S phases.

To present the analysis in a more quantitative way and include the possibility that the frequency of exchanges may be different in the two cycles, one may let p be the frequency per cell or per chromosome in the first cycle and q be the frequency in the second cycle (Whitehouse, 1965). The expected frequencies for the situation in which the subunits have

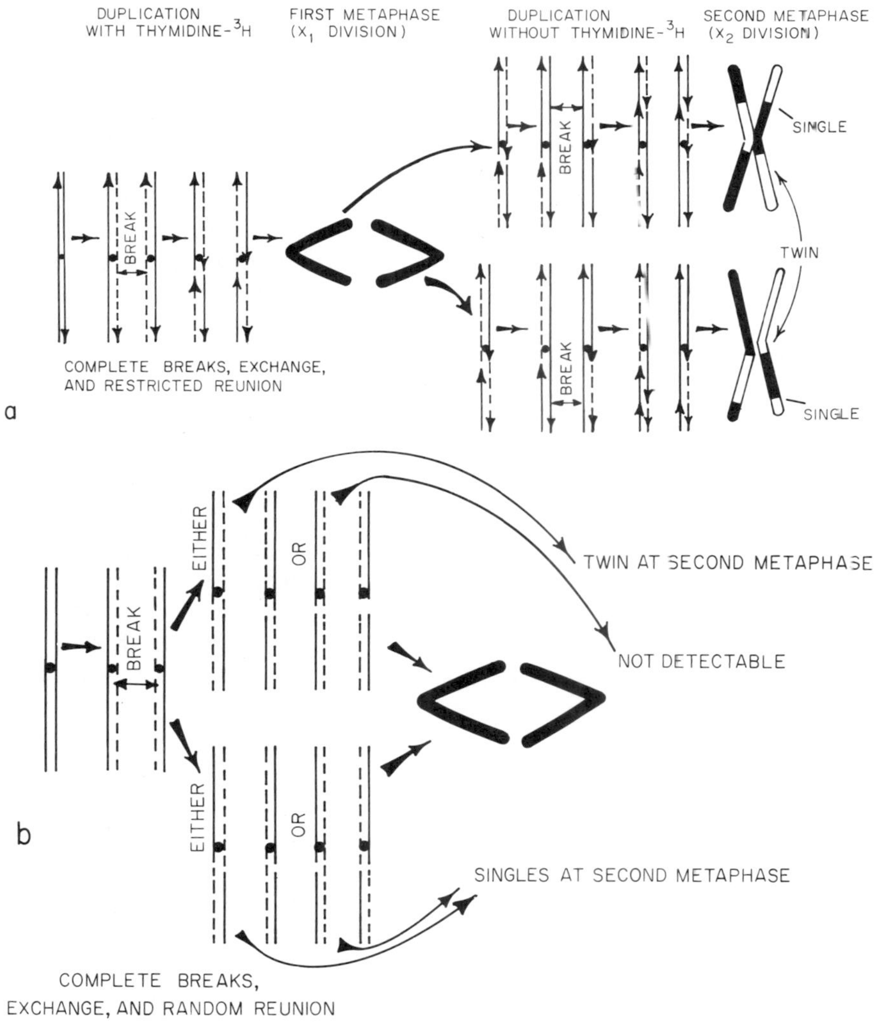

Fig. 1.4. Diagrams showing the predicted results of chromosome replication and sister chromatid exchange in which DNA was labeled during one replication with [^{3}H]thymidine; the dashed lines represent labeled subunits and the chromosomes shown in solid black represent regions showing grains in autoradiograms. The unbroken lines represent unlabeled subunits and the chromosome regions shown in outline represent regions without grains in the autoradiograms as shown in Figure 1.2. (*a*) The predicted results of exchanges when the two subunits are restricted in their reunion by a difference in sense analogous to the two chains of a DNA double helix. Exchanges are assumed to occur with an equal frequency during the first duplication with [^{3}H]thymidine and during the second duplication without [^{3}H]thymidine. All exchanges in the first duplication produce twins and the frequency of single exchanges to twin pairs is 2:1 (see text for explanation). (*b*) The predicted frequency of single and twin exchanges if the two subunits reunite at random. Since single exchanges could result from events at both the first and second duplication, the ratio of single exchanges to twins would be higher, at least 10:1. From Taylor, 1958.

opposite polarity are p twin exchanges and $2q$ single exchanges on a per cell basis since there are two times as many chromosomes undergoing exchange at the second cycle (the rate per chromosome is p for the first cycle and q for the second cycle).

The expected frequencies for the case in which the subunits composing a chromatid are alike (lack polarity) is $\frac{1}{2}p + 2q$ for singles per cell and $\frac{1}{4}p$ for twin exchanges; $\frac{1}{4}p$ produces no visible exchange because two labeled and two unlabeled subunits reunite. The $2q$ arises because there are two times as many chromosomes per cell during the second cycle. When p and q are equal:

$$\text{frequency of twins} = \tfrac{1}{4}p$$
$$\text{frequency of singles} = \tfrac{1}{2}p + 2q$$

If $\frac{1}{4}p$ is set equal to 1 and p and q are equal, $\frac{1}{2}p + 2q = 10$; the predicted ratio is 1 twin per 10 singles in tetraploid cells produced by one cycle in [^{3}H]thymidine and one cycle free of labeled thymidine where all chromosomes are retained and analyzed as a group. Even an analysis that includes some misidentifications (false twins) should distinguish between these two extremes, that is, 1 twin to 2 singles opposed to 1 twin to 10 singles.

B. *First Experimental Analyses*

Table 1.1 shows the data collected from root cells of *Bellevalia* in three separate experiments. In the first (Taylor, 1958) the ratio was found to be about two twins to each single. If the hypothesis of a difference between the two subunits were accepted from these data one would have to assume that the frequency of exchanges in the first cycle was higher than that in the second cycle ($p > q$). Two differences in the two cycles were noted. The first occurred without colchicine, while the second replication occurred in the drug. Also, the endogenous radiation per chromosome during the first division might have been two times as much as that in the second division (two subunits of four labeled in the first cycle and one subunit per four in the second cycle). Since the colchicine difference could be adjusted most easily, the second experiment was designed to have colchicine available during the first cycle when [^{3}H]thymidine was incorporated, as well as during the second cycle when it was necessary to keep all chromosomes produced in two cycles together in one cell for analysis. The ratio appeared to change in the direction expected if there were now a reduction in the frequency of exchanges in the first cycle.

TABLE 1.1
***Frequency of Single and Twin Exchanges in the Tetraploid Chromosome Sets of* Bellevalia romana**[a]

				Frequency per chromosome	
Experiment	Number of chromosomes	Twin exchanges	Single exchanges	First interphase	Second interphase
1	72 (18 cells)	36	15	1.00	0.21
2	112 (28 cells)	39	36	0.70	0.28
3	96 (24 cells)	14	26	0.29	0.27

[a]Data for only two largest chromosome pairs are given. From Taylor, 1959.

On the basis of this result a third experiment was analyzed in which colchicine was added 2 hours before the labeling began and continued throughout the experiment. Now the ratio was nearly the expected one twin to two singles and the absolute frequency of twins fell while the frequency of singles per chromosome remained the same as in the two previous experiments. On the basis of this evidence a difference in the polarity of the subunits was accepted as the most likely explanation.

C. *Conclusions*

From the analysis of these rather limited data several conclusions were drawn: (1) the two subunits of a chromatid are structurally different; effectively a difference in polarity was indicated; (2) the frequency of exchanges is probably not affected very much by the endogenous radiation of the tritium; otherwise the higher frequency of exchange in the first cycle would have been observed in all three experiments; (3) the frequency of exchanges may be reduced by prolonged treatment with colchicine and thus the reduction in the first cycle can be accounted for in the last two experiments; and (4) the sister chromatid exchanges appear to occur during replication, since the colchicine was necessary before or during the replication to give a lower frequency of exchanges.

IV. MODELS AND EXCEPTIONS

By the time these final results on the frequency of exchanges in the two cycles were known, Meselson and Stahl (1958) had demonstrated semiconservative segregation of DNA during replication in *Escherichia coli*. The separation technique, which involved the now-well-known CsCl banding technique, showed that particles or fragments of DNA as small as 7×10^6 daltons were behaving as if they were composed of two subunits. These subunits were almost certainly the two chains of the DNA double helix. This demonstration, along with the evidence that the subunits of a chromatid were of opposite polarity and segregated semiconservatively, led to a revolution in concepts of the structure of a chromosome.

For the first time there was substantial evidence that a chromatid was a single long DNA helix (Taylor, 1959) folded and coiled into the dimensions of a chromosome—up to 1 m of DNA packaged into a structure 1 or 2 microns in diameter and about 10 microns in length. Presumably interactions with proteins held the long fiber in a compact form.

The first model proposed had protein links between segments of DNA, because at that time evidence did not indicate uninterrupted polymers of such dimensions as a whole chromosome. However, as evidence accumulated on the breakage of chromosomes by DNase (Callan and McGregor, 1958; Gall, 1963), the failure of proteases to produce breaks, and the requirement for synthesis of DNA in the repair of breaks in chromosomes (Taylor et al. 1962), the idea of links was abandoned (Taylor, 1966, 1967b) and replaced by the notion that transient single-strand nicks would be sufficient to account for unwinding and separation during replication (Taylor, 1966, 1967a).

A single DNA helix was indicated more strongly when the molecular weight of a continuous double helix the size of a *Drosophila* chromosome was demonstrated by Kavenoff and Zimm (1973) in lysates of cells isolated from pupae of *D. hydei, D. virilis,* and *D. americana* and from an established cell line of *D. melanogaster* grown in culture. Although the absolute size of the DNA was sufficiently uncertain to clearly establish the principle of one molecule per chromosome, the additional evidence available makes the hypothesis the preferred one. For example, the difference in polarity of the subunits of a chromatid and the indication that chromosomes contain a significant fraction of single copy sequences (Laird, 1971; Wu et al., 1972) support the concept of one molecule of DNA per chromosome. The later demonstrations that all variations of semiconservative distributions of DNA, isolabeling, and so on, result from multiple sister chromatid exchanges (Wolff and Perry, 1974), autoradiographic image spread (Wolff, 1977), and labeling for a part of one cycle plus one more cycle with [^{3}H]thymidine or a fraction of S phase plus two cycles with BrdUrd (Wolff, 1977) made the hypothesis of one molecule per chromosome the only reasonable one. A few instances of true differential labeling for short segments of chromosomes apparently result from DNA with very asymmetric distributions of its AT base pairs, so that most of the thymine is in one chain of the DNA. Such regions produce segregation at the first division after labeling (Latt et al., 1974), as well as isolabeling after two rounds of replication in BrdUrd (Lin et al., 1974; Gibas and Limon, 1978).

V. SOME UNRESOLVED PROBLEMS

It has always been difficult to understand the mechanics of separation of very long molecules composed of two helically coiled chains. There is one twist per 10 base pairs and even with a reduction of about one turn per 200 bases to account for the structure of nucleosomes, the number of twists in a large chromosome is nearly one million. These must be unwound each time the chromosome is reproduced. The discovery of nicking and closing enzymes called topoisomerases (Wang and Liu, 1979) has simplified the conceptual problem considerably. However, there is still the problem of sorting and disentanglement during separation of chromatids, which appear to require more organization than nucleosomes and supercoiled DNA can provide. Some types of core proteins attached at intervals along the DNA with the facility for reversible binding guided by the DNA to which they are attached appear to be necessary. Studies with topoisomerases indicate that even small ring chromosomes the size of SV40 DNA do not lose all supercoils in the presence of the nicking–closing activity in solution (Wang and Liu, 1979). Some other driving force or organizing core proteins appear to be necessary for separation and disentanglement of chromatids at prophase. The observation first made by Walen (1965) that four chromatids in endoreplicated chromosomes have a specific spatial relationship to each other when they reach division may be revealing. Chromosomes in a cell sometimes undergo two successive replications without separation of the chromatids. When this happens so that [^{3}H]thymidine is available at the first replication and not at the second, the labeled chains end up in the two outer chromatids of the group of four lying in one plane. It appears that the sorting might begin at the centromere and be propagated toward the ends by side to side aggregation of what were originally called replication guides (Taylor, 1966, 1967b). Exceptions to the sorting occur when sister chromatid exchanges exist. Chains distal to an exchange sort as expected if they were a part of some polymerizing core that begins its assembly at the centromere and is propagated to the end guided by the continuity of the DNA. The chromatid is then composed of loops of DNA organized around a central core. How interlocks of the loops are prevented is a mystery. Perhaps interlocks do occur, for sister chromatids are stuck to each other along their full length in prophase, but not by adhesion of their cores. The two chromatids come apart at anaphase or shortly before anaphase. Even in cells where anaphase is blocked by colchicine the chromatids separate after being blocked at a c-metaphase for an appropriate time. It is possible that interlocks are released by the temporary opening of one of the rings. Some topoisomerase molecules bind to DNA as dimers with two cleavages made on opposite chains. Simultaneously with the cleavage, the enzyme subunits are covalently linked to the cleaved end of the DNA chains, and by a conformational change the subunits might then separate; the short segment of DNA between the nicks in opposite chains would then separate or denature. The two DNA loops would be released and then the open DNA ring would close by renaturation and perhaps binding of the two topoisomerase subunits. As the anaphase movement occurred and the two chromatids were pulled apart many rings would be passed through an opening in the interlocked ring in the sister chromatid. A model similar to this has been described by Brown and Cozzarelli (1979), who showed that supercoiling induced in small rings of DNA by gyrase is increased in steps of two supercoils per catalytic event and is decreased in the same way under conditions that relax the supercoils with gyrase. A similar enzyme would be ideal for disentanglement of DNA chains, which could be involved in separation of chromatids at anaphase.

Sister chromatid exchanges may very well be the result of malfunctioning of enzymes similar to gyrase, which operates just behind the growing fork during DNA replication. These enzymes cut both chains of the DNA at staggered sites, as do some restriction endonucleases. There appears to be some sequence specificity, but the requirement is less stringent than for the restriction nucleases. The finding that DNA breaks, presumably by double-chain cleavage, just behind the growing forks, when cells are lysed at 37°C (Guy and Taylor, 1978) lends support to the idea that such a nuclease may be active. One function of such an enzyme could be to disentangle the completed helices after replication by a mechanism similar to that proposed by Brown and Cossarelli (1979) for the action of gyrase in bacteria. One helix of DNA is passed through a break in the other without release of the ends by the gyrase molecule. The ends are then reunited before the enzyme dissociates from the DNA. Because of lesions in the DNA or a malfunction of the enzyme during the pass-through, the two chromatids might both be broken and misaligned at reunion so that a sister chromatid exchange would be produced. If the gyrase-like molecule, presumed to be present in higher plants and animals, had a sequence specificity, as bacterial gyrase is reported to have, the exchange would be equal so that no deletion or duplication would result from such a sister chromatid exchange. This appears reasonable, since few lethal events result from high rates of sister chromatid exchanges. A recent report (Mattern and Painter, 1979) indicates that a gyrase-like activity may be present in higher cells and that DNA replication is inhibited by novobiocin, which is known to react specifically with one of the subunits of the bacterial gyrase molecule and inhibits DNA replication.

REFERENCES

Brown, P. O. & Cozzarelli, N. R. (1979) A sign inversion mechanism for enzymatic supercoiling of DNA. *Science* **206,** 1081–1083.

Callan, H. G. & McGregor, H. C. (1958) Action of deoxyribonuclease on lampbrush chromosomes. *Nature* **181,** 1479–1480.

Darlington, C. D. & LaCour, L. F. (1953) The effects of irradiation on meiosis. *Heredity* **6** (Suppl.) 41–55.

Gall, J. G. (1963) Kinetics of deoxyribonuclease action on chromosomes. *Nature (Lond.)* **198,** 36–38.

Gibas, Z. & Limon, J. (1978) Isolabeling of the long arm of the human Y chromosome demonstrated by the FPG technique. *Chromosome* **69,** 113–120.

Guy, A. L. & Taylor, J. H. (1978) Actinomycin D inhibits initiation of DNA replication in mammalian cells. *Proc. Natl. Acad. Sci. U.S.* **75,** 6088–6092.

Howard, A. & Pelc, S. R. (1951) Synthesis of nucleoprotein in bean root cells. *Nature* **167,** 599–601.

Kavenoff, R. & Zimm, B. H. (1973) Chromosome-sized DNA molecules from *Drosophila. Chromosoma* **41,** 1–27.

Laird, C. D. (1971) Chromatid structure: relationship between DNA content and nucleotide sequence diversity. *Chromosoma* **32,** 378–406.

Latt, S. A., Davidson, R. L., Lin, M. S., & Gerald, P. S. (1974) Lateral asymmetry in the fluorescence of human Y chromosomes stained with 33258 Hoechst. *Exp. Cell Res.* **87,** 425–429.

Lin, M. S., Latt, S. A., & Davidson, R. L. (1978) Microfluorimetric detection of asymmetry in centromeric region of mouse chromosomes. *Exp. Cell Res.* **86,** 392–394.

McClintock, B. (1938) The production of homozygous deficient tissues with mutant characteristics by means of the aberrant mitotic behavior of ring-shaped chromosomes. *Genetics* **23,** 315–376.

Mattern, M. R. & Painter, R. B. (1979) Dependence of mammalian DNA replication on DNA supercoiling. II. Effects of novobiosin on DNA synthesis in Chinese hamster cells. *Biochim. Biophys. Acta* **563,** 306–312.

Meselson, M. S. & Stahl, F. W. (1958) The replication of DNA in *Escherichia coli. Proc. Natl. Acad. Sci. U.S.* **44,** 671–682.

Peacock, W. J. (1963) Chromosome duplication and structure as determined by autoradiography. *Proc. Natl. Acad. Sci. U.S.* **49,** 793–801.

Sharp, A. J. (1934) *Introduction to Cytology,* 3rd ed., McGraw-Hill, New York and London, p. 133.

Taylor, J. H. (1958) Sister chromatid exchanges in tritium labeled chromosomes. *Genetics* **43,** 515–529.

Taylor, J. H. (1959) The organization and duplication of genetic material. *Proc. Int. Cong. of Genetics, 10th, 1958, Montreal, Canada,* **1,** 63–78.

Taylor, J. H. (1966) The duplication of chromosomes. In *Probleme der biologischen reduplication* (P. Sitte, Ed.), Springer-Verlag, Berlin and New York.

Taylor, J. H. (1967a) Meiosis. In *Encyclopedia of Plant Physiology* (W. Ruhland, Ed.), Springer-Verlag, Berlin or Heidelberg, p. 345.

Taylor, J. H. (1967b) Patterns and mechanisms of genetic recombination. In *Molecular Genetics,* Part II (J. H. Taylor, Ed.), Academic Press, New York, pp. 95–135.

Taylor, J. H., Woods, P. S., & Hughes, W. L. (1957) The organization and duplication of chromosomes as revealed by autoradiographic studies using tritium-labeled thymidine. *Proc. Natl. Acad. Sci. U.S.* **43,** 122–128.

Taylor, J. H., Haut, W. F., & Tung, J. (1962) Effects of fluorodeoxyuridine on DNA replication, chromosome breakage and reunion. *Proc. Natl. Acad. Sci. U.S.* **48,** 190–198.

Walen, K. H. (1965) Spatial relationships in the replication of chromosomal DNA. *Genetics* **51**, 915–929.

Wang, J. C. & Liu, L. F. (1979) DNA topoisomerases: enzymes that catalyze the concerted breaking and rejoining of DNA backbone bonds. In *Molecular Genetics,* Part III (J. H. Taylor, Ed.), Academic Press, New York, pp. 65–88.

Whitehouse, H. L. K. (1965) *Towards an Understanding of the Mechanism of Heredity,* St. Martin's Press, New York, pp. 184–188.

Wolff, S. (1977) Sister chromatid exchange. *Annu. Rev. Genet.* **11,** 183–201.

Wolff, S. & Perry, P. (1974) Differential-Giemsa staining of sister chromatids and the study of sister chromatid exchange without autoradiography. *Chromosoma* **48,** 341–353.

Wu, J., Hurn, J., & Bonner, J. (1972) Size and distribution of the repetitive segments of the *Drosophila* genome. *J. Mol. Biol.,* **64,** 211–219.

Chapter 2

Sister Chromatid Exchange: New Methods for Detection

SAMUEL A. LATT
GENETICS DIVISION AND MENTAL RETARDATION CENTER
CHILDREN'S HOSPITAL MEDICAL CENTER
GENETICS DIVISION
BOSTON HOSPITAL FOR WOMEN AND
THE DEPARTMENT OF PEDIATRICS
HARVARD MEDICAL SCHOOL

I. Introduction 18
II. Nonautoradiographic methods for detecting SCEs 18
III. Basic information about SCEs 27
IV. Attempts at detecting SCEs at a molecular level 31

I. INTRODUCTION

The recent development of staining methodology for detecting DNA synthesis in cytological preparations of metaphase chromosomes has stimulated research into phenomena related to the segregation of newly synthesized DNA. The logical framework of this research was established by the pioneering efforts of J. Herbert Taylor, who used autoradiography to demonstrate the semiconservative distribution of newly replicated DNA and to detect sister chromatid exchanges (Taylor et al., 1957; Taylor, 1958). The purpose of this chapter is to review the development of newer methodology, not requiring autoradiography, for sister chromatid exchange analysis, the types of information that this methodology can provide, and the status of attempts to detect sister chromatid exchange (SCE) at a molecular level.

II. NONAUTORADIOGRAPHIC METHODS FOR DETECTING SCEs

DNA does not fluoresce appreciably at room temperature. One minor base, 5-methyl cytosine, fluoresces weakly, but at wavelengths in the ultraviolet (Gill et al., 1974), a region posing particular difficulties for fluorescence microscopy. Two different types of fluorescent base analogues, formycin and 7-deazanebularin, have been described (Ward et al., 1969a,b; Ward and Reich, 1972; Brdar and Reich, 1972). Both absorb near 300 nm and emit between 350 and 400 nm, but neither has a high quantum yield (≤ 0.01) when incorporated into DNA. Highly fluorescent derivatives of adenosine or cytosine can be prepared by reaction of these bases with chloracetaldehyde (Barrio et al., 1972), but the reaction products are modified at positions that preclude their participation in a stable double-stranded helical DNA structure. Thus, at least for the present, optical detection of DNA synthesis requires indirect approaches, based on changes in chromatin condensation, dye fluorescence or binding, or immunochemical reactivity caused by a DNA base analogue, for example, 5-bromodeoxyuridine.

The first published observation of DNA synthesis by means of incorporation of 5-bromodeoxyuridine was part of a study (Huang, 1967) of the induction of damage in rat X chromosomes by different agents. Cultured rat cells were exposed to concentrations of BrdUrd ranging from 50 to 200 μg/ml for 26 hours, an interval sufficient for somewhat more than one cycle of DNA replication. Some cells exhibited a phenomenon described as negative heteropycnosis, in which one sister chromatid stained less intensely than did the other. This occurred most often in the X chromosome, a finding that, in retrospect, probably reflected the fact that this chromosome, because of its late replication, was the chromosome most likely to undergo two cycles of BrdUrd incorporation during the 26 hour labeling period. No mechanism for the effect of BrdUrd on chromosome staining was suggested.

A second line of investigation had its origins in early observations by Stubblefield (1964) that prolonged incubation of cells in the presence of Colcemid resulted in decondensation of late replicating regions. Zakharov and Egolina (1968) exposed Chinese hamster chromosomes to this treatment and observed that heterochromatin subdivided into regions that despiralized to different extents. In a subsequent study (Zakharov and Egolina, 1972), these authors utilized 5-bromodeoxyuridine to induce chromatin decondensation. Palmer (1970) had previously shown that exposure of human peripheral lymphocytes to a terminal pulse of 75 μg/ml BrdUrd resulted in decondensation of certain regions, subsequently shown to replicate late. Zakharov and Egolina (1972) made similar observations in Chinese hamster chromosomes and did additional experiments, in which cells were exposed to 25–400 μg/ml BrdUrd during the terminal phases of two successive cycles. Chromatids that had incorporated BrdUrd for two replication cycles were decondensed, relative to their singly substituted sister chromatids. In some cases, this decondensation was evident as differential Giemsa staining, permitting detection of sister chromatid exchanges. In a related study, Zakharov and associates (Baronovskaya et al., 1972; Zakharov et al., 1974) used a terminal pulse of 200 μg/ml of BrdUrd to decondense selectively the late-replicating X chromosome of human lymphocytes. The mechanism by which a terminal pulse of high concentrations of BrdUrd decondenses chromosomes remains an unexplained clue about protein–nucleic acid interactions in chromatin.

An alternative approach for detecting DNA synthesis was based on the premise that a heavy polarizable atom such as bromine, as part of the base analogue BrdUrd, might alter the fluorescence of DNA-binding dyes (Latt, 1973), given the proper steric and electronic interactions. A related effect of BrdUrd on the phosphorescence decay of acridine dyes, when bound to synthetic polynucleotides at low temperatures, had previously been published by Galley and Purkey (1972). Without any means of predicting, *a priori*, which dye might be most affected by BrdUrd, a number of dyes, including quinacrine, ethidium, acridine orange, proflavine, and 33258 Hoechst, were surveyed (Latt, 1973). The last dye, which has been used for staining live cells (Lammler and Schutze, 1969; Hilwig, 1970) and which has been shown by Hilwig and Gropp (1972) to highlight the fluorescence of centric heterochromatin in rodent chromosomes, proved to be the most sensitive to BrdUrd substitution. Lesser effects were observed with both acridine orange and proflavine, while quinacrine and ethidium were relatively insensitive to BrdUrd (Latt, 1973, 1976).

Either BrdUrd or IdUrd, when substituted into synthetic polynucleotides, is capable of reducing the fluorescence of bound 33258 Hoechst by as much as tenfold (Latt and Wohlleb, 1975; Latt, 1976, 1977a). Quenching of dye fluorescence in cellular DNA that has incorporated BrdUrd can be as much as fivefold (Latt and Wohlleb, 1975). The latter studies also showed that BrdUrd substitution actually increases the binding affinity of the Hoechst dye to DNA and that the effect originally observed was due to a reduction in the fluorescence quantum yield of bound dye. Moreover, BrdUrd substitution can lead to a reduction in the fluorescence lifetime of bound dye (Latt, 1977a), an observation consistent with the idea that at least part of the quenching occurs at the first excited singlet state of the dye, as expected if the BrdUrd–dye interaction involves heavy-atom-promoted spin–orbit coupling (Galley and Purkey, 1972). Certain variations in solvent conditions, for example, the use of glycerol rather than water, and a reduction in pH, for example, to 4, abolish BrdUrd-dependent effects on dye fluorescence (Latt, 1974a; Latt and Wohlleb, 1975). The mechanisms of these effects has yet to be explained.

The ability of BrdUrd to reduce 33258 Hoechst fluorescence is evident both in interphase nuclei and in metaphase chromosomes. In the former, it can be detected either by fluorescence microscopy (Latt, 1974a,b) or by flow cytometry (Latt et al., 1977a), with or without fixation of cells. A number of derivatives of this bisbenzimidazole dye, provided by Dr. H. Loewe (Loewe and Urbanietz, 1974), were examined (Latt and Stetten, 1976) and were found to be sensitive to BrdUrd. It was speculated that some of these dyes, because of different ring substituents, might have greater permeability and hence utility with live cells. Arndt-Jovin and Jovin (1977) successfully employed one such derivative, 33242 Hoechst, in flow cytometric work. Other bisbenzimidazole derivatives with greater cellular permeability than 33258 have also been described (Latt, 1977b). Detection of BrdUrd substitution in metaphase chromosomes by Hoechst dyes can be used to study replication kinetics, lateral asymmetry, or sister chromatid exchange formation (Latt, 1974b; Lin et al., 1974; Latt et al., 1974; Latt, 1975; Latt, 1976; Latt, 1978). The use of this procedure for detecting segregation of DNA into sister chromatids in Figure 2.1*A*, and an example of a stained chromosome is shown in Figure 2.2*A*).

Alternative fluorescent methods for detecting BrdUrd incorporation involve the use of 4′,6-diamidino-2-phenylindole (DAPI). DAPI can stain chromosomes intensely, and, especially when slides are mounted at ele-

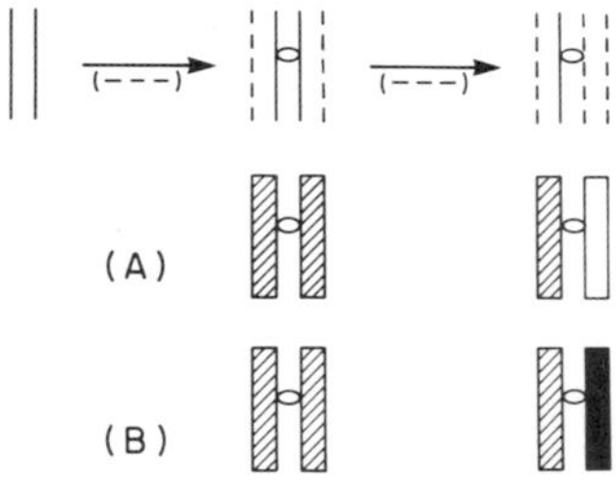

Fig. 2.1. Sister chromatid differentiation produced in metaphase chromosomes from cells that have undergone two cycles of BrdUrd incorporation. BrdUrd incorporation into DNA strands is represented by dashed lines. Rectangles represent sister chromatids containing this DNA. Shown are the results of procedures in which BrdUrd (*A*) reduces or (*B*) increases staining intensity.

vated pH, shows sister chromatid differentiation as a function of BrdUrd substitution (Lin and Alfl, 1976; Lin et al., 1977). DAPI fluorescence resembles that of 33258 Hoechst; except that the transition from BrdUrd-insensitive to BrdUrd-sensitive emission occurs with DAPI in a higher pH range.

Acridine orange was utilized by Dutrillaux et al. (1973) to accentuate contrast in late-replicating chromosomal regions, decondensed by termi-

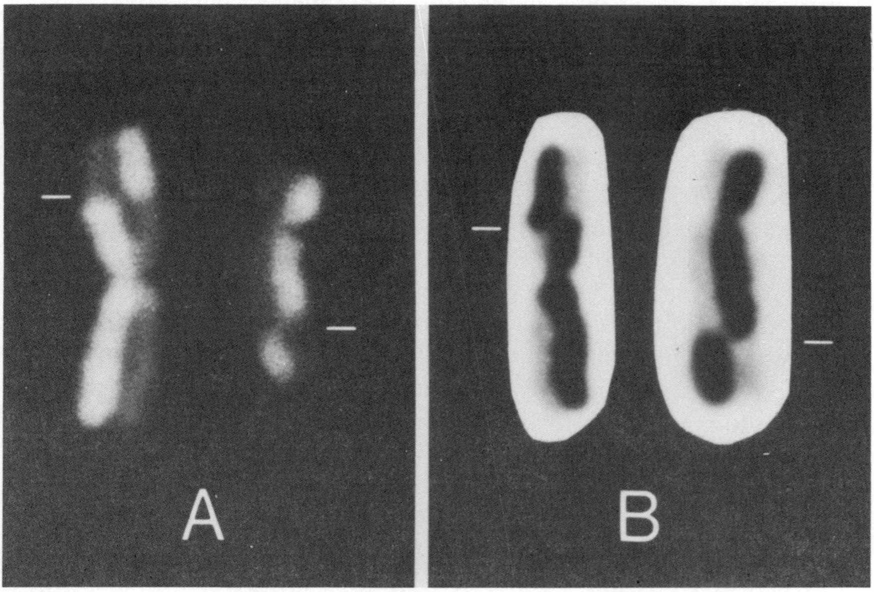

Fig. 2.2. SCEs in chromosomes from human lymphocytes that have undergone two cycles of BrdUrd incorporation. (*A*) Chromosomes stained with 33258 Hoechst and examined by fluorescence microscopy (Latt, 1973). (*B*) Chromosomes exposed to visible light while mounted in a solution containing 33258 Hoechst, incubated at 65°C in 2 × SSC, and stained with Giemsa [essentially the procedure of Perry and Wolff (1974)]. SCEs are indicated by horizontal lines.

nal pulses of BrdUrd. By analogy with later studies employing 33258 Hoechst (Willard and Latt, 1976), the effects of BrdUrd on dye fluorescence are probably distinct from those on chromosome condensation. Kato (1974), Franceschini (1974), and Perry and Wolff (1974) showed that, in cells that had incorporated BrdUrd for two cycles, sister chromatids could be differentiated after they were stained with acridine orange. This observation probably involves a number of phenomena. One, presumably, is the ability of BrdUrd to quench acridine orange fluorescence directly (Latt, 1976), an effect used recently by Daryzynkiewicz et al. (1978) in flow cytometry. A second effect appears to involve additional photochemistry, since sister chromatid differentiation with acridine orange, in contrast to that with 33258 Hoechst, increases (''burns in'') upon intense illumination. This time dependence might reflect the greater susceptibility of BrdUrd-substituted DNA to dye photosensitized single-strand nicking (Latt et al., 1977d).

Chromomycinone dyes, such as mithramycin and chromomycin A_3, can also be used for detecting DNA synthesis. BrdUrd-substituted cells stained with these dyes exhibit an approximately 25% enhanced fluorescence, detectable in flow cytometric experiments (Schwartzendruber, 1977a,b). Successful sister chromatid differentiation with these dyes requires preexposure of chromosomes to UV illumination (Buys and Osinga, 1980) (see below). Another positive signal due to BrdUrd incorporation involves fluorescent or immunoperoxidase tags following the exposure of cells or chromosomes to antibodies directed against this base analogue (Sawicki et al., 1971; Gratzner et al., 1975, 1976). This approach has been utilized to detect sister chromatid exchanges (Gratzner et al., 1976). An advantage of the immunochemical technique is its increased sensitivity to relatively short pulses of BrdUrd. Newer preparative techniques may lead to more specific and plentiful immunological reagents, which would in turn enhance the applicability of this latter methodology.

While fluorescent methods provide a vivid demonstration of BrdUrd incorporation, which is amenable to relatively definitive mechanistic analysis, and while they can be used in flow cytometry, they have the disadvantage of requiring complex optical instrumentation. Also, particularly in the case of 33258 Hoechst, the optical image fades rapidly, making prolonged or repeated observation of specimens difficult. This obstacle was overcome by the development of BrdUrd–Giemsa techniques.

Ikushima and Wolff (1974) showed that chromosomes from cells cultured in the presence of either IUdRrd or BrdUrd exhibited weak sister chromatid differentiation when stained with Giemsa. Perry and Wolff (1974) subsequently observed that contrast was greatly improved if

Giemsa staining was preceded by staining with 33258 Hoechst, illumination with ambient light, and incubation at 65°C in saline–citrate (SSC) buffer (e.g., Figure 2.2*B*). Kim's procedure (1974) is similar; the illumination occurring during fluorescence microscopy suffices as a post-Hoechst stain treatment. Good sister chromatid differentiation can also be obtained (Korenberg and Freedlender, 1974) if chromosomes, substituted with BrdUrd, are incubated at elevated pH in a high ionic strength buffer.

It has been suggested (Ikushima and Wolff, 1974; Perry and Wolff, 1974; Wolff, 1977; Wolff and Bodycote, 1977) that chromosomal proteins are responsible for the results with some BrdUrd–Giemsa techniques. It was noted that the more lightly staining, doubly substituted chromatid was more extended than was the other sister chromatid (Ikushima and Wolff, 1974). This result is reminiscent of data reported by Zakharov and Egolina (1972), although, in the latter studies, differential decondensation was seen only after *terminal* pulses; incorporation of BrdUrd for entire cycles did not produce detectable effects on chromosome condensation. The mechanism of sister chromatid differentiation by BrdUrd incorporation and Giemsa staining is not yet fully understood.

A variety of experiments have been utilized to investigate the basis of BrdUrd–Giemsa procedures. One approach focuses on the DNA, and the results indicate that photosensitized degradation and elution of BrdUrd-substituted DNA contributes to the staining contrast. A second set of studies focuses on chromosomal proteins, which can be modified, in a BrdUrd-dependent manner, during the cytological procedures. The first approaches yield results analogous to those observed with 33258 Hoechst fluorescence, with BrdUrd-substitution resulting in reduced staining. The other methods yield either similar patterns or, depending on procedural details, reverse staining patterns, in which BrdUrd substitution enhances staining.

Goto et al. (1975, 1978) showed that a variety of DNA-binding dyes could be used instead of 33258 Hoechst to photosensitize sister chromatid differentiation. All these dyes shared the feature that, in the presence of BrdUrd, and in some cases, even in its absence, they exhibited a very low fluorescence efficiency. Thus the light energy that was absorbed by the dye was not emitted and would, instead, be available for photochemical reactions. BrdUrd-substituted DNA has long been known to be sensitive to photosensitized nicking (Hutchinson, 1973), and BrdUrd-substituted cells are very sensitive to killing by light in the presence of Hoechst dye (Stetten et al., 1976, 1977). Goto and coworkers utilized Feulgen staining to show that the combination of dye, light, and SSC incubation was sufficient to remove DNA selectively from weakly straining regions. Of particular interest was their observation (Goto et al., 1978) that photosensitization by 33258 Hoechst decreased at low pH, a condition abolishing the BrdUrd-dependent quenching of 33258 Hoechst fluorescence (Latt and Wohlleb, 1975).

Similar mechanistic conclusions about BrdUrd–Giemsa staining procedures were derived from work examining metaphase chromosomes from cells grown so that the DNA was labeled with either ^{3}H or ^{14}C in either BrdUrd- or thymidine-containing strands, so that the effect of dye, light, and saline incubation on individual polynucleotide chains could be studied. Both dye and SSC incubation were needed for appreciable loss of DNA from slides or for generation of contrast in BrdUrd-substituted chromosomes (Latt et al., 1977d) (Figure 2.3). Moreover, if cells that had been incubated one cycle in the presence of ^{3}H-BrdUrd and labeled with ^{14}C in the thymidine-containing strands were exposed to dye plus light, DNA was eluted from the BrdUrd-substituted polynucleotide chain but not from its complementary chain, suggesting that the dye-induced effects occurred at the level of single DNA strands.

Suggestions that proteins are involved in the generation of sister chromatid differentiation have derived from a number of sources. One early observation (Lin and Riggs, 1972) was that BrdUrd substitution increased the affinity of DNA for the lactose operon repressor. BrdUrd-substituted DNA also has a higher affinity for 33258 Hoechst dye (Latt and Wohlleb, 1975). However, in comparing DNA in chromatin from cells with or without BrdUrd substitution, it was possible to show that the

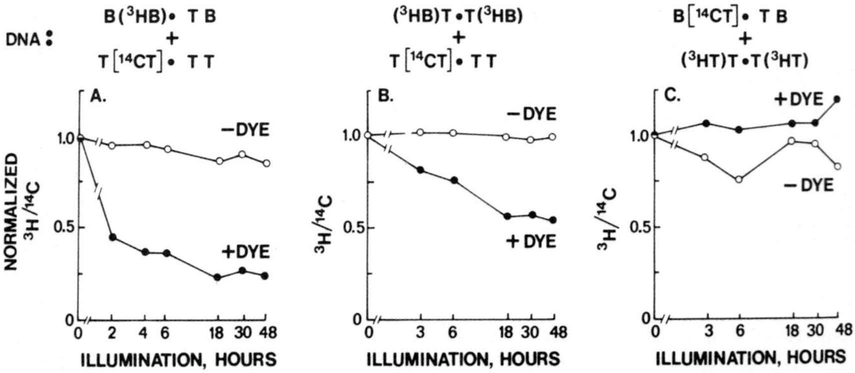

Fig. 2.3. DNA elution during a BrdUrd–dye–Giemsa procedure. Synchronized CHO cells were cultured to produce DNA substituted as shown at the top of each frame. Mixtures of Colcemid-treated cells (average mitotic index approximately 40%) were applied to cover slips, mounted at pH 7 with or without prior white lamp, for time periods indicated in the graphs, and subsequently incubated 15 min in 2 × SSC at 65°C. Relative elution of DNA species was estimated from the residual ^{3}H/^{14}C ratio (Latt et al., 1977d).

effect of protein in excluding dye binding to DNA did not seem to depend greatly on BrdUrd substitution (Latt and Wohlleb, 1975). Pathak et al. (1975) observed that exposure of human chromosomes to trypsin could generate both Giemsa banding and sister chromatid differentiation in BrdUrd-substituted chromosomes. Scheid (1976) made similar observations with *Vicia faba* chromosomes. Detailed interpretation of these results requires information about the state of the DNA in the slides before and after trypsinization.

Wolff and Bodycote (1977) reported that exposure of chromosomes to dithiothreitol or beta-mercaptoethanol, agents that can break disulfide bonds, can lead to differential sister chromatid staining with Giemsa. An indirect effect of the sulfhydryl reagents, permitting DNA rearrangement in chromatin, was postulated. Buys and Osinga (1980) reported that irradiation with near UV light led to reduced staining of BrdUrd-substituted chromosomes with either mithramycin or DAPI and suggested that BrdUrd might enhance the ability of light to break disulfide bridges within or between proteins. High resolution analysis of the wavelength dependence of this effect might provide an indication of the primary target of the incident illumination. Importantly, the conditions used by Buys and Osinga did not lead to removal of DNA from the chromosomes. Interpretation of these and related studies requires more information on the integrity and conformation of protein and DNA in fixed metaphase chromosomes, as well as on the ability of high concentrations of dye solutions to modify protein–nucleic acid interactions in partially disrupted chromosomes.

Scheid reported (1979) that when chromosomes were stained with DAPI and exposed to cysteamine, exposure to light failed to induce sister chromatid differentiation following Giemsa staining. These data were interpreted to reflect the ability of cysteamine to prevent the induction of single-strand breaks in the DNA. Such an effect would be consistent with the observation (Monkehaus, 1973) that cysteamine can act as a sink for light energy absorbed by BrdUrd-substituted DNA.

Burkholder, Scheres, and their associates (Burkholder, 1978; Burkholder and Wang, 1978; Scheres et al., 1977) developed protocols for the generation of *reverse* staining by Giemsa. That is, it is possible to stain chromosomes in which sister chromatids contain either one or two strands of DNA substituted with BrdUrd so that the doubly substituted chromatid stains more intensely than its sister (Figure 2.1*B*). Often this involves exposure of cells to low pH during staining; reexposure of slides to high pH can cause the regeneration of the standard pattern. Based on electron microscopic data, it was suggested (Burkholder and Wang, 1978; Burkholder, 1979) that, at low pH, the chromatin is much

TABLE 2.1
Protocols for Detection of BrdUrd Substitution into Metaphase Chromosomes

1. 33258 Hoechst Fluorescence (Latt, 1973; Latt et al., 1977b)
   a. *Staining*. Slides are successively dipped in PBS (0.14 M NaCl, 0.004 M KCl, 0.01 M phosphate, pH 7.0) (5′), 0.5 μg/ml 33258 Hoechst in PBS (10′), PBS (1′), PBS (5′), and H_2O (two to three changes). Stock solutions of dye (50 μg/ml) in H_2O can be stored at 4°C in the dark for at least 2 weeks. Samples of 33258 Hoechst were originally obtained from Dr. H. Loewe, Hoechst AG, Frankfurt, Germany, although the dye can now be purchased from Calbiochem.
   b. *Observation*. Microscopic observation of 33258 Hoechst fluorescence is guided by the position of the high wavelength absorption band of the dye–DNA complex (maximal near 350 nm and appreciable up to or slightly beyond 425 nm). While excitation under dark-field conditions is certainly possible, excitation with incident illumination, using a UG-1 (360 nm peak) bandpass and TK 400 (reflect $\leq$ 400 nm) dichroic mirror is especially convenient and effective. A 460 nm high wavelength pass filter in the observation pathway removes most unwanted exciting light from the fluorescence, which peaks near 475 nm.

      For optimal quenching of dye fluorescence by incorporated BrdUrd, the slides are mounted in a buffer of moderate ionic strength at or slightly above neutrality, for example, pH 7.5 McIlvaine's buffer. Primarily because of the specificity of the stain, under these conditions, for (A-T rich) DNA, and because the free dye has very weak fluorescence, slides observed as described above show little background fluorescence. However, the fluorescence of 33258 Hoechst bound to BrdUrd-substituted chromatin fades rapidly, and photography requires speed. Using a microscope with incident illumination, acceptable photographs can be obtained in 5–10 sec (e.g., Kodak Tri-X film). Reduction of the mounting medium pH shifts the dye fluorescence color from blue toward green or yellow, and the fluorescence fades less rapidly; but specific quenching due to BrdUrd substitution is decreased.
2. 33258 Hoechst plus Giemsa (adapted from Perry and Wolff, 1974; Wolff, 1981.
   a. *Staining*. Slides are stained with 10^{-4} M 33258 Hoechst in one-third strength PBS (or a comparable buffer with a pH near 7). A cover slip is applied and the slides are then placed in a Petri dish containing excess buffer to ensure that the slides remain moist, and the slides are exposed to light for a period of time that depends on the illuminating conditions. For example, approximately 6–12 hours of exposure is sufficient after the slides are positioned 6 cm from a Sylvania 20 watt cool white bulb (Latt et al., 1977b). The slides are then rinsed in H_2O, incubated at 65°C in 2 $\times$ SSC buffer (0.30 M NaCl, 0.03 M Na citrate, pH 7) for 15 min, rinsed thoroughly with H_2O, and stained with Giemsa as before. Contrast can be heightened by increasing the time during which slides are exposed to light.

TABLE 2.1
Continued

b. *Observation.* For photography, slides can be mounted in a standard embedding medium, or immersion oil can be applied directly to the slide without a cover slip. A 544 nm interference filter can be used to enhance contrast, and the optical image is recorded using, for example, High Contrast Copy film.

more diffuse, to an extent reduced by BrdUrd substitution, while in slides exposed to higher pH there was evidence for actual removal of DNA. One implication of this observation is that, if the high pH treatments had been carried out first, the low pH treatment would not have produced a reversed pattern; such studies have not been reported.

Takayama and Sakanishi observed (1977) that the exposure of chromosomes to perchloric acid or HCl could cause a reverse Giemsa staining, interpreted as reflecting increased resistance of BrdUrd-substituted DNA to acid extraction. Takayama and Sakanishi (1979) also observed a curious, time-dependent reverse staining of BrdUrd-substituted chromosomes with alkaline Giemsa that did not correlate with DNA extraction. Interpretation of these results would be assisted by quantitative analysis of staining contrast, as well as analysis of the amount and chemical status of both DNA and protein remaining in the chromosomes after treatment.

While there is a wealth of methods for differentiating between sister chromatids, a few simple techniques can suffice for most studies. Representative techniques, employing 33258 Hoechst fluorescence or 33258 Hoechst followed sequentially by illumination, SSC incubation, and Giemsa staining are detailed (Table 2.1). Procedural details for SCE analysis have also been presented by Wolff (1981). It should be noted that other techniques, related to those described in Table 2.1, can work equally well.

III. BASIC INFORMATION ABOUT SCEs

Newer techniques for sister chromatid differentiation have confirmed most of the conclusions about the overall features of SCE drawn from previous autoradiographic studies. For example, studies of third division cells (Tice et al., 1975) and of cells with diplochromosomes (Wolff and Perry, 1975) indicated that sister chromatid exchange is constrained by the polarity of the DNA helix. Third division cell analysis showed that segregation at mitosis of sister chromatids, into pairs of homologues in human lymphocytes, is random (Comings, 1970; Latt et al., 1975) (e.g.,

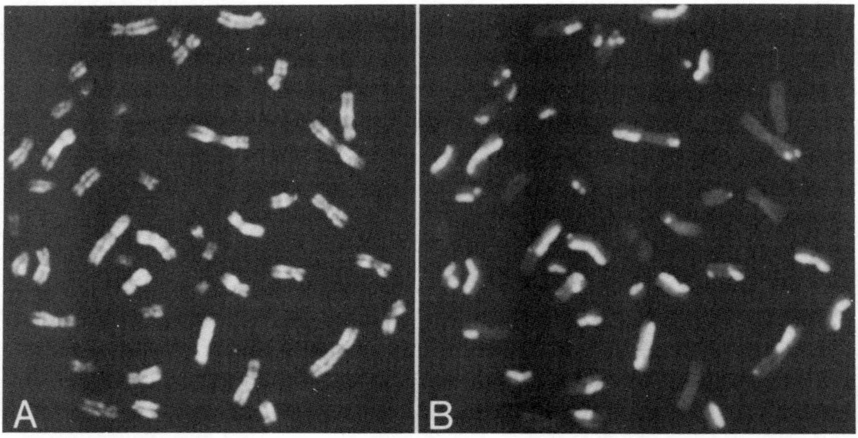

Fig. 2.4. Fluorescence of human metaphase chromosomes after three cycles of BrdUrd incorporation. Peripheral lymphocytes from a normal human female were cultured 3 days in medium containing 9×10^{-5} M BrdUrd, 4×10^{-7} M FdUrd, 6×10^{-6} M Urd. Chromosomes from the cell shown were first stained with quinacrine, photographed (*A*), destained in 3:1 methanol–acetic acid, stained with 33258 Hoechst, and rephotographed (*B*). The absence of a consistent correlation between the distribution of bright chromatids within members of a pair of homologues has been noted previously (Latt et al., 1975). Reprinted by permission of the publisher. Copyright 1975 by The Histochemical Society, Inc.

Figure 2.4) and without marked bias in chick retinal cells (Morris, 1977) and in mouse spermatogonia (Allen and Latt, 1976). However, other studies (Potten et al., 1978) suggest that chromosome segregation in certain very specialized germ cells may not be random, Analysis of endoreduplicated chromosomes indicated that apposition of newly synthesized polynucleotide chains is external to old chains with respect to centromeres (Herreros and Gianelli, 1967; Schwarzacher and Schnedl, 1965; Walen, 1965; Wolff and Perry, 1975; Latt, 1976), even in cells exhibiting sister chromatid exchange formation. Factors controlling labeling patterns in endoreduplicated chromosomes may thus be distinct, at least in most cells, from those controlling chromatid segregation at mitosis.

The greater effective resolution of BrdUrd–dye techniques has facilitated the detection of multiple, closely spaced SCEs (Kato, 1974; Latt, 1974c; Wolff and Perry, 1975). Previous claims, based on autoradiography, for the existence of isolabeling in second division metaphases, probably reflect, as previously hypothesized by Comings (1971), a methodological artifact due to poor resolution. The improved resolution of BrdUrd–dye techniques has increased the accuracy and simplicity with which SCE induction by many clastogenic agents can be quantitated

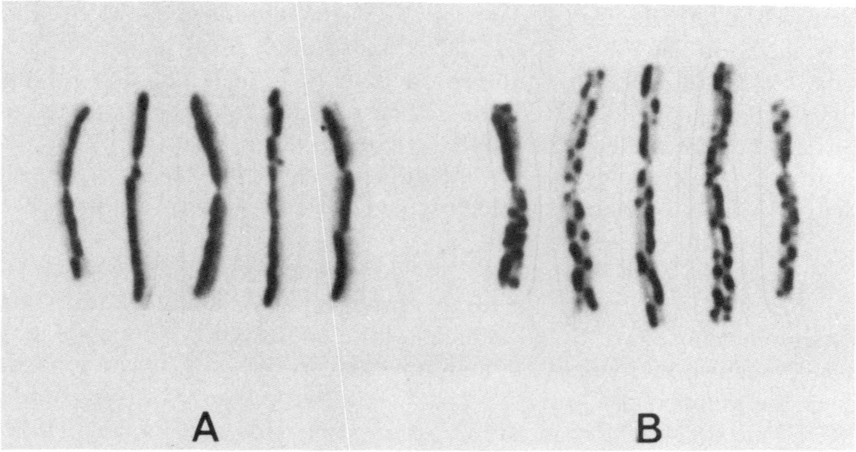

Fig. 2.5. SCE induction by 8-methoxypsoralen plus light treatment in CHO cells. Chromosomes were stained essentially as described by Perry and Wolff (1974). (*A*) Control chromosomes. (*B*) Chromosomes from cells exposed to 10^{-5} M 8-methoxypsoralen and 1.1×10^{-4} ergs/mm^2 near ultraviolet light (principally at 365 nm) before addition of BrdUrd (Latt et al., 1977c).

(Kato, 1973; Latt, 1974c; Perry and Evans, 1975; Wolff, 1977; Latt et al., 1977c) (Figure 2.5). Modified Giemsa methods have also provided material that is amenable to automated SCE analysis; preliminary efforts in this direction have been described (Zack et al, 1976, 1977). BrdUrd itself, like [^{3}H]dT (Brewen and Peacock, 1969; Gibson and Prescott, 1972; Marin and Prescott, 1964), induces SCEs (Gatti et al., 1979; Kato, 1974; Latt, 1974c; Mazrimas and Stetka, 1978; Wolff and Perry, 1974) and may be responsible for most of the baseline SCEs observed in the absence of additional clastogens. In this regard, Gatti et al. (1979) achieved sister chromatid differentiation in *Drosophila* somatic cells at 1 μg/ml BrdUrd and observed no SCEs in 361 cells. Such results with *Drosophila* are equivalent, on a DNA basis, to a spontaneous SCE rate of $\leq \frac{1}{20}$ SCE per human cell.

In spite of BrdUrd-induced baseline levels, increments in SCEs can easily be scored and the extent of SCE induction, at least by mitomycin C, does not seem to be very sensitive to BrdUrd levels to which cells are exposed (Ishii and Bender, 1978; Davidson et al., in press). Effects due to BrdUrd appear to be at least twofold. Thus an increase in BrdUrd incorporation is associated with increased SCE formation, but at very high BrdUrd concentrations, BrdUrd incorporation levels off, while SCE induction increases sharply (Latt, 1974c). Davidson et al. (1980) have recently resolved effects of BrdUrd in CHO cells into those due to BrdUrd concentration and those due to DNA substitution by BrdUrd. While a small effect of DNA substitution on both SCE and mutation (at the HGPRT locus) was observed, much larger effects could be attributed to BrdUrd concentration. These latter effects could be reversed by dC, an approach initially suggested by Meuth and Green (1974) for reversing BrdUrd effects on ribonucleotide reductase, which in turn can influence nucleotide pool balance in cells.

The position of SCEs detected by fluorescence or Giemsa can be reasonably well localized relative to chromosome banding patterns. In human chromosomes, SCEs induced either by BrdUrd or by alkylating agents occur preferentially, though not exclusively, in Q-negative bands or at the junction of Q-positive and Q-negative regions (Crossen et al., 1977; Latt, 1974d; Morgan and Crossen, 1977; Hoo and Parslow, 1979; Hagland and Zech, 1979) (e.g., Figure 2.6). SCEs that appear to be in Q-positive bands might conceivably occur in Q-negative subbands, which would be observable only in highly extended chromosomes. SCEs are also very rare in homogeneously staining regions of human neuroblastoma cells (Balaban-Malenbaum and Gilbert, 1978).

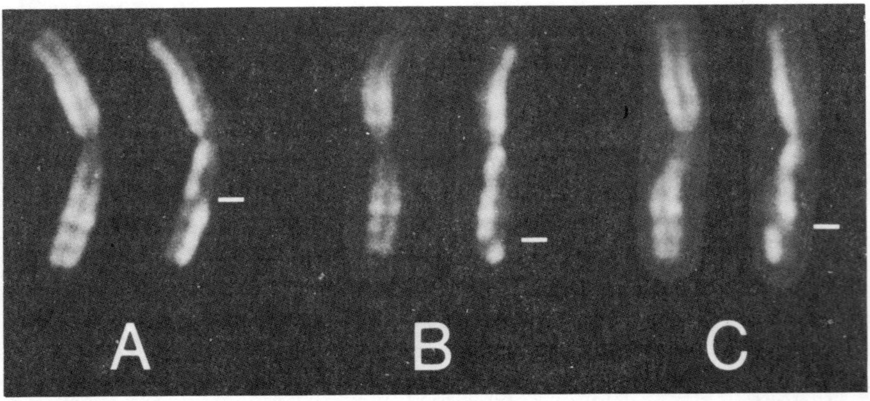

Fig. 2.6. Localization of sister chromatid exchanges in human chromosomes. Chromosomes were prepared from peripheral leukocytes obtained from normal human subjects and grown 70–72 hours in medium containing 0.01 or 0.02 mM BrdUrd. Chromosomes were stained with quinacrine (left-hand member of each pair), destained, and restained with 33258 Hoechst (right hand member of each pair) (Latt, 1974d). Sister chromatid exchanges (indicated by horizontal lines) are evident as abrupt reciprocal alternations in 33258 Hoechst fluorescence along chromatids. The exchanges shown are located at or near junctions between quinacrine-bright and quinacrine-dull bands. Copyright 1974 by the American Association for the Advancement of Science.

Additional studies have detected a clustering of SCE at junctions between heterochromatic and euchromatic regions in muntjac (Carrano and Wolff, 1975), kangaroo rat (Bostock and Christie, 1976), microtus, and hamster chromosomes (Hsu and Pathak, 1976). The significance of these "junctional" regions is as yet unknown. Although Dolfini (1978) failed to observe SCE formation in heterochromatin of cultured *Drosophila* lines, Gatti et al. (1979) found a preponderance of SCEs in *Drosophila* cells (virtually all BrdUrd induced) at heterochromatin–euchromatin junctions, but he also found them to occur with heterochromatin. Lin and Alfi (1976) observed, in one mouse marker chromosome, that both BrdUrd-induced and mitomycin C plus BrdUrd-induced SCEs were *more* frequent in centric heterochromatin than in the chromosome arms. Similarly, Shubert et al. (1979) found SCE hotspots near (though not necessarily within) heterochromatin in chromosomes from *Vicia* treated with alkylating agents, while Vosa (1977) described the approximate location of SCEs in *Vicia* to be sparse in the vicinity of at least certain heterochromatic regions. It might prove especially interesting if the sites of SCE formation were somehow related to special classes of DNA (e.g., repetitive DNA). In fact, reports of unusually high or low SCE frequencies within heterochromatin suggest that SCE formation is somehow sensitive to DNA sequence repetition, a result potentially compatible with suggestions (Smith, 1976) that SCE formation may, by unequal crossing over, mediate (Kurnit, 1979) alterations in the amount of repeated DNA in individual chromatids.

IV. ATTEMPTS AT DETECTING SCEs AT A MOLECULAR LEVEL

Current methodology for detecting SCEs is restricted to analysis of metaphase or prometaphase chromosomes. It would be useful, however, to be able to detect SCEs in interphase cells, at the DNA level, as well. Such a capability would circumvent the sensitivity of SCE analysis to selective effects on the progression of cells to metaphase, and it would provide a system for investigating the molecular features of SCE formation. Unfortunately, in spite of claims to the contrary, no method for unambiguous analysis of SCEs at the DNA level has been published.

Two types of experimental approaches have been used to search for DNA interchange corresponding to sister chromatid exchange. Both utilize cells that have incorporated BrdUrd for less than one cycle and thus contain DNA substituted in only one strand. Following sister-strand exchange, junctions of substituted and unsubstituted polynucleotide should result and appear as material of intermediate density in alkaline

CsCl gradients [Rommelaere and Miller-Faures (1975)]. The Holliday (1964) model for DNA recombination also predicts that segments of double-substituted heavy-heavy (HH) and light-light (LL) DNA exist at interchange sites in neutral CsCl gradients.

Rommelaere and Miller-Faures (1975) reported the detection of Chinese hamster DNA with intermediate density in alkaline CsCl gradients, but most of this DNA appeared to be cross-linked. When DNA from the Chinese hamster cells was centrifuged in neutral CsCl, approximately 0.1% of the material exhibited density greater than that of hybrid, heavy-light (HL) DNA and was considered to contain segments of bifilarly substituted, heavy-heavy (HH) DNA. The amount of this DNA was increased by UV irradiation prior to BrdUrd incorporation, but the overall yield was more than 10 times that expected from the number of SCEs in these cells.

Moore and Holliday (1976) similarly detected 0.1–0.2% HH DNA from rapidly growing CHO cells cultured not quite one cycle in medium containing BrdUrd. Mitomycin C, when administered at a very high concentration (1 μg/ml) 5 hours prior to harvest, appeared to increase both HH DNA and the SCE frequency. Again, the amount of HH DNA was much more than expected for the number of SCE observed. A recent paper by Resnick and Moore (1979), which in part deals with HH DNA, reiterates these conclusions. An increased amount (six- to tenfold over a baseline of 0.04%) of HH DNA in Bloom's syndrome fibroblasts, relative to normal human fibroblasts, has also been observed (Waters et al., 1978) and hypothesized to be associated with the elevated SCE frequency in this condition.

In further experiments, in which synchronized CHO cells were employed, a small amount of unusually dense DNA was observed, but it had some properties not expected for HH DNA (Loveday and Latt, 1978, 1979). Cells that had incorporated BrdUrd for one cycle exhibited 0.4 ± 0.2% of DNA with a density expected for HH DNA, but this amount was not increased by addition of sufficient mitomycin C (0.03 μg/ml, at the start of S) to more than triple the SCE frequency. Significantly, the dense DNA persisted after a subsequent round of replication in the absence of BrdUrd (calling its bifilar substitution into question), and material with a similar density shift from the main band DNA was seen in cells that incorporated [^{3}H]dT (but not BrdUrd). Moreover, the "heavy" DNA interacts differently with netropsin or actinomycin D, A-T, and G-C specific dyes, respectively, than does main band DNA (Loveday and Latt, 1979), in a manner expected for a DNA species containing clusters of G-C base pairs. Displacement of this DNA by actinomycin D left a virtually undetectable residuum of true HH DNA that might have been associated with SCE formation. While these data do not rule out the existence of the HH DNA predicted by the Holliday model, they suggest that the biochemical evidence thus far claimed for this DNA may be weak and that additional experiments are necessary to detect SCE formation at a molecular level and to clarify the chemical events associated with DNA interchange during SCE formation. In view of the small number of DNA fragments containing interchange points expected, even in cells with 100 SCEs, such studies may well require amplification of specific DNA segments by recombinant DNA techniques.

A more definite evaluation of the role of Holliday-type DNA interchange during SCE formation is essential. In addition to other factors, it might be noted that the Holliday model, which envisions recombinational events to involve staggered DNA interchange, may be more appropriate for meiotic recombination, which occurs *after* semiconservative DNA synthesis is complete, than for SCEs, which presumably occur during DNA synthesis, perhaps at the growing point of semiconservative DNA replication. A requirement for special types of DNA sequence repeats at the site of DNA damage in order for this damage response to cause an SCE might in part account for the low ratio of SCEs to DNA alkylating products (Cassel and Latt, 1980). Such a constraint would be consistent with SCE localization data (cited earlier). Elucidation of the type of sequences involved in SCE formation might also ultimately provide information about the details of DNA interchange at the junction of a sister chromatid exchange event. Improved understanding of the molecular details of sister chromatid exchange formation might in turn provide an indication of the biological significance of this phylogenetically widespread process.

References

Allen, J. W. & Latt, S. A. (1976) *In vivo* BrdU-33258 Hoechst analysis of DNA replication kinetics and sister chromatid exchange formations in mouse somatic and meiotic cells. *Chromosoma* **58,** 325–340.

Arndt-Jovin, D. J. & Jovin, T. M. (1977) Analysis and sorting of living cells according to deoxyribonucleic acid content. *J. Histochem. Cytochem.* **25,** 585–589.

Balaban-Malenbaum, G. & Gilbert, F. (1978) Origin of double minute chromosomes in human neuroblastoma cells. *Amer. J. Hum. Genet.* **29,** 21A.

Baranovskaya, L. I., Zakharov, A. F., Dutrillaux, B., Carpentier, S., Prieur, M., & Lejeune, J. (1972) Differenciation des chromosomes *X* par les methodes de despiralisation au 5-bromodeoxyuridine (BUDR) et de denaturation thermique menagee. *Ann. Genet.* **15,** 271–274.

Barrio, J. R., Secrist, J. A., & Leonard, N. J. (1972) Fluorescent adenosine and cytidine derivatives. *Biochem. Biophys. Res. Commun.*, **46,** 597–604.

Bostock, C. J., & Christie, S. (1976) Analysis of the frequency of sister chromatid exchange in different regions of chromosomes of the kangaroo rat. *Chromosoma* **56,** 275–287.

Brdar, B. & Reich, E. (1972) 7-Deazanebularin metabolism in cultures of mouse fibroblasts and incorporation into cellular and viral nucleic acids. *J. Biol. Chem.* **247,** 725–730.

Brewen, J. G. & Peacock, W. J. (1969) The effect of tritiated thymidine on sister chromatid exchange in a ring chromosome. *Mutat. Res.* **7,** 433–440.

Burkholder, G. D. (1978) Reciprocal Giemsa staining of late DNA replicating regions produced by low and high pH sodium phosphate. *Exp. Cell Res.* **111,** 489–492.

Burkholder, G D. (1979) An investigation of the mechanism of the reciprocal differential staining of BUdR-substituted and unsubstituted chromosome regions. *Exp. Cell Res.* **121,** 209–219.

Burkholder, G. D. & Wang, H. C. (1978) Electron microscopy of differentially BrdU-substituted sister chromatids treated with sodium phosphate. *Chromosoma* **70,** 101–107.

Buys, C. H. C. M. & Osinga, J. (1980) The mechanism of differential sister chromatid fluorescence as studied with the GC-specific DNA-ligand mithramycin. *Exp. Cell Res.* **125,** 105–109.

Carrano, A. V. & Wolff, S. (1975) Distribution of sister chromatid exchanges in the euchromatin and heterochromatin of the Indian muntjac. *Chromosoma* **53,** 361–369.

Cassel, D. M. & Latt, S. A. (1980) Relationship between DNA adduct formation and sister chromatid exchange induction by [^{3}H]8-methoxypsoralen in Chinese hamster ovary cells. *Exp. Cell Res.* **128,** 15–22.

Comings, D. E. (1970) The distribution of sister chromatids at mitosis in Chinese hamster cells. *Chromosoma* **29,** 428–433.

Comings, D. E. (1971) Isolabelling not compatible with single stranded model. *Nature, New Biol.* **229,** 24–25.

Crossen, P. E., Drets, M. E., Arrighi, F. E., & Johnston, D. A. (1977) Analysis of the frequency and distribution of sister chromatid exchanges in cultured human lymphocytes. *Hum. Genet.* **35,** 345–352.

Darzynkiewicz, A., Andreeff, M., Traganos, F., Sharpless, T., & Melamed, M. R. (1978) Discrimination of cycling and non-cycling lymphocytes by BUdR-suppressed acridine orange fluorescence in a flow cytometric system. *Exp. Cell Res.* **115,** 31–36.

Davidson, R. L., Kaufman, E. R., Dougherty, C. P., Ouellette, A. M., DiFolco, C. M., & Latt, S. A. (1980) Induction of sister chromatid exchanges by BUdR is largely independent of the BUdR content of DNA. *Nature* **284,** 74–76.

Dolfini, S. F. (1978) Sister chromatid exchanges in *Drosophila melanogaster* cell lines *in vitro*. *Chromosoma* **69,** 339–347.

Dutrillaux, B., Laurent, C., Couturier, J, & Lejeune, J. (1973) Coloration des chromosomes humains par l'acridine orange apres traitement par 5-bromodeoxyuridine. *C. R. Acad. Sci. (D) Paris* **276,** 3179–3181.

Franceschini, P. (1974) Semiconservative DNA duplication in human chromosomes treated with BUdR and stained with acridine orange. *Exp. Cell Res.* **89,** 420–421.

Galley, W. C. & Purkey, R. M. (1972) Spin-orbital probes of biomolecular structure. A model DNA–acridine system. *Proc. Natl. Acad. Sci. U.S.* **69,** 2198–2202.

Gatti, M., Pimpinelli, S., Santini, G., & Olivieri, G. (1979) Lack of spontaneous sister chromatid exchange (SCE) in somatic cells of *Drosophila melanogaster*. *Genetics* **91,** 255–274.

Gibson, D. A. & Prescott, D. M. (1972) Induction of sister chromatid exchanges in chromosomes of rat kangaroo cells by tritium incorporated into DNA. *Exp. Cell Res.* **74,** 397–402.

Gill, J. E., Mazrimas, J. A., & Bishop, C. C. (1974) Physical studies on synthetic DNAs containing 5-methyl-cytosine. *Biochim. Biophys. Acta* **335,** 330–348.

Goto, K., Akematsu, T., Shimazu, H., & Sugiyama, T. (1975) Simple differential Giemsa staining of sister chromatids after treatment with photosensitive dyes and exposure to light and the mechanism of staining. *Chromosoma* **53,** 223–230.

Goto, K., Maeda, S., Kano, Y., & Sugiyama, T. (1978) Factors involved in differential Giemsa staining of sister chromatids. *Chromosoma* **66,** 351–359.

Gratzner, H. G., Leif, R. C., Ingram, O. J., & Castro, A. (1975) The use of antibody specific for bromodeoxyuridine for the immunofluorescent determination of DNA replication in single cells and chromosomes. *Exp. Cell Res.* **95,** 88–94.

Gratzner, H. G., Pollack, A., Ingram, D. J., & Leif, R. C. (1976) Deoxyribonucleic acid replication in single cells and chromosomes by immunologic techniques. *J. Histochem. Cytochem.* **24,** 34–39.

Haglund, U. & Zech, L. (1979) Simultaneous staining of sister chromatid exchanges and Q-bands in human chromosomes after treatment with methyl methane sulphonate, quinacrine mustard, and quinacrine. *Hum. Genet.* **49,** 307–317.

Herreros, B. & Giannelli, F. (1967) Spatial distribution of old and new chromatid subunits and frequency of chromatid exchange in induced human lymphocyte endoreduplications. *Nature* **182,** 286–288.

Hilwig, I. (1970) On the influence of a benzimidazol derivative (fluorochrome) on cell lines in tissue culture. *Z. Zellforsch.* **104,** 127–137.

Hilwig, I. & Gropp, A. (1972) Staining of constitutive heterochromatin in mammalian chromosomes with a new fluorochrome. *Exp. Cell Res.* **75,** 122–126.

Holliday, R. (1964) A mechanism for gene conversion in fungi. *Genet. Res.* **5,** 282–304.

Hoo, J. J. & Parslow, M. I. (1979) Relation between the SCE points and the DNA replication bands. *Chromosoma (Berl.)* **73,** 67–74.

Hsu, T. C. & Pathak, S. (1976) Differential rates of sister chromatid exchanges between euchromatin and heterochromatin. *Chromosoma* **58,** 269–273.

Huang, C. C. (1967) Induction of a high incidence of damage to the X chromosomes of rattus (Mastomys) natalensis by base analogues, viruses and carcinogens. *Chromosoma* **23,** 162–179.

Hutchinson, F. (1973) The lesions produced by ultraviolet light in DNA containing 5-bromouracil. *Q. Rev. Biophys.* **6,** 201–246.

Ikushima, T. & Wolff, S. (1974) Sister chromatid exchanges induced by light-flashes to 5-bromodeoxyuridine and 5-iododeoxyuridine-substituted Chinese hamster chromosomes. *Exp. Cell Res.* **87,** 15–19.

Ishii, Y. & Bender, M. (1978) Factors influencing the frequency of mitomycin C-induced sister chromatid exchanges in 5-bromodeoxyuridine substituted human lymphocytes in culture. *Mutat. Res.* **51,** 411–418.

Kato, H. (1973) Induction of sister chromatid exchanges by UV light and its inhibition by caffeine. *Exp. Cell Res.* **82,** 382–390.

Kato, H. (1974) Spontaneous sister chromatid exchanges detected by BudR-labelling method. *Nature* **251,** 70–72.

Kato, H. (1974) Is isolabelling a false image? *Exp. Cell Res.* **89,** 416–420.

Kim, M. A. (1974) Chromatid austausch und heterochromatin veränderungen menschlicher chromosomen nach BrdU-markierung. *Humangenetik* **25,** 179–188.

Korenberg, J. R. & Freedlender, E. (1974) Giemsa technique for the detection of sister chromatid exchanges. *Chromosoma* **48,** 355–360.

Kurnit, D. M. (1979) Satellite DNA and heterochromatin variations. The case for unequal mitotic crossovers. *Hum. Genet.* **47,** 169–186.

Lammler, G. & Schutze, H. R. (1969) Vital-fluorochromierunt tierischer zellkerne mit einem neuen fluorochrom. *Naturwissenschaften* **56,** 286.

Latt, S. A. (1973) Microfluorometric detection of deoxyribonucleic acid replication in human metaphase chromosomes. *Proc. Natl. Acad. Sci. U.S.* **70,** 3395–3399.

Latt, S. A. (1974a) Detection of DNA synthesis in interphase nuclei by fluorescence microscopy. *J. Cell Biol.* **62,** 546–550.

Latt, S. A. (1974b) Microfluorometric analysis of DNA replication and sister chromatid exchanges in human chromosomes. *J. Histochem. Cytochem.* **22,** 478–491.

Latt, S. A. (1974c) Sister chromatid exchanges, indices of human chromosome damage and repair: detection by fluorescence and induction by mitomycin C. *Proc. Natl. Acad. Sci. U.S.* **71,** 3162–3166.

Latt, S. A. (1974d) Localization of sister chromatid exchanges in human chromosomes. *Science* **185,** 74–76.

Latt, S. A. (1975) Fluorescence analysis of late DNA replication in human metaphase chromosomes. *Somat. Cell Genet.* **1,** 293–321.

Latt, S. A. (1976) Longitudinal and lateral differentiation of metaphase chromosomes based on the detection of DNA synthesis by fluorescence microscopy. In *Chromosomes Today*, Vol. 5 (P. L. Pearson and K. R. Lewis, Eds.), New York, Wiley, pp. 367–394.

Latt, S. A. (1977a) Fluorescent probes of chromosome structure and replication. *Can. J. Genet. Cytol.* **19,** 603–623.

Latt, S. A. (1977b) Fluorometric detection of deoxyribonucleic acid synthesis: possibilities for interfacing bromodeoxyuridine dye techniques with flow fluorometry. *J. Histochem. Cytochem.* **25,** 913–926.

Latt, S. A. (1978) Analysis of sister chromatid exchange and chromosome replication kinetics using BrdU–dye techniques. *Virchows Arch. B Cell Pathol.* **29,** 19–27.

Latt, S. A. & Stetten, G. (1976) Spectral studies on 33258 Hoechst and related bisbenzimidazole dye, useful for fluorescent detection of deoxyribonucleic acid synthesis. *J. Histochem. Cytochem.* **24,** 24–33.

Latt, S. A. & Wohlleb, J. C. (1975) Optical studies of the interaction of 33258 Hoechst with DNA, chromatin, and metaphase chromosomes. *Chromosoma* **52,** 297–316.

Latt, S. A., Davidson, R. L., Lin, M. S., & Gerald, P. S. (1974) Lateral asymmetry in the fluorescence of human Y chromosomes stained with 33258 Hoechst. *Exp. Cell Res.* **87,** 425–429.

Latt, S. A., Stetten, G., Juergens, L. A., Willard, H. F., & Scher, C. D. (1975) Recent developments in the detection of deoxyribonucleic acid synthesis by 33258 Hoechst fluorescence. *J. Histochem. Cytochem.* **23,** 493–505.

Latt, S. A., George, Y. S. & Gray, J. W. (1977a) Flow cytometric analysis of BrdU-substituted cells stained with 33258 Hoechst. *J. Histochem. Cytochem.* **25,** 927–934.

Latt, S. A., Allen, J. W., Rogers, W. E., & Juergens, L. A. (1977b) *In vitro* and *in vivo* analysis of sister chromatid exchange formation. In *Handbook of Mutagenicity Test Procedures* (B. Kilbey, M. Legator, W. Nichols, and C. Ramel, Eds.), Elsevier/North-Holland Biomedical Press, Amsterdam, pp. 275–291.

Latt, S. A., Allen, J. W., & Stetten, G. (1977c) *In vitro* and *in vivo* analysis of chromosome structure, replication, and repair using BrdU-33258 Hoechst techniques. *International Cell Biology* (K. R. Porter and W. Brinkler, Eds.), Rockefeller University, New York.

Latt, S. A., Munroe, S. H., Disteche, C., Rogers, W. E., & Cassell, D. M. (1977d) Uses of fluorescent dyes to study chromosome structure and replication. In *Chromosomes Today*, Vol. 6 (A. de La Chapelle and M. Sorsa, Eds.), Elsevier, Amsterdam, pp. 27–36.

Lin, M. S. & Alfl, O. S. (1976) Detection of sister chromatid exchanges by 4′-6-diamidino-2-phenylindole fluorescence. *Chromosoma* **57,** 219–225.

Lin, S. Y. & Riggs, A. D. (1972) Lac operator analogues: bromodeoxyuridine substitution in the lac operator affects the rate of dissociation of the lac repressor. *Proc. Natl. Acad. Sci. U.S.* **69,** 2574–2576.

Lin, M. S., Latt, S. A., & Davidson, R. L. (1974) Microfluorometric detection of asymmetry in the centromeric region of mouse chromosomes. *Exp. Cell Res.* **86,** 392–394.

Lin, M. S., Comings, D. E., & Alfi, O. S. (1977) Optical studies of the interaction of 4′-6-diamidino-2-phenylindole with DNA and metaphase chromosomes. *Chromosoma* **60,** 15–25.

Loewe, H. & Urbanietz, J. (1974) Basisch substitutiente 2,6-bisbenzimidazol-derivate, eine neue chemotherapeutisch active Korperklasse. *Arzneim. Forsch.* **24,** 1927–1933.

Loveday, K. S. & Latt, S. A. (1978) Search for DNA interchange corresponding to sister chromatid exchanges in Chinese hamster ovary cells. *Nucleic Acids Res.* **5,** 4087–4104.

Loveday, K. S. & Latt, S. A. (1979) Is there biochemical evidence for sister chromatid exchange formation? *Am. J. Hum. Genet.* **31,** 103A.

Marin, G. & Prescott, D. M. (1964) The frequency of sister chromatid exchanges following exposure to varying doses of ^{3}H-thymidine or X-rays. *J. Cell Biol.* **21,** 159–167.

Mazrimas, J. A. & Stetka, D. G. (1978) Direct evidence for the role of incorporated BudR in the induction of sister chromatid exchanges. *Exp. Cell Res.* **117,** 23–30.

Meuth, M. & Green, H. (1974) Induction of a deoxycytidineless state in cultured mammalian cells by bromodeoxyuridine. *Cell* **2,** 109–112.

Monkehaus, F. (1973) Influence of cysteamine on intramolecular energy transfer in 5-bromouracil-substituted phage-DNA. *Int. J. Radiat. Biol.* **24,** 517–524.

Moore, P. D. & Holliday, R. (1976) Evidence for the formation of hybrid DNA during mitotic recombination in Chinese hamster cells. *Cell* **8,** 573–579.

Morgan, W. F. & Crossen, P. E. (1977) The frequency and distribution of sister chromatid exchanges in human chromosomes. *Hum. Genet.* **38,** 271–278.

Morris, V. B. (1977) Random segregation of sister chromatids in developing chick retinal cells demonstrated *in vivo* using the fluorescence plus Giemsa technique. *Chromosoma (Berl.)* **60,** 139–145.

Palmer, C. G. (1970) 5-Bromodeoxyuridine-induced constrictions in human chromosomes. *Can. J. Genet. Cytol.* **12,** 816–830.

Pathak, S., Stock, A. D., & Lusby, A. (1975) A combination of sister chromatid differential staining and Giemsa banding. *Experienta* **31,** 916–917.

Perry, P. & Evans, H. J. (1975) Cytological detection of mutagen-carcinogen exposure by sister chromatid exchange. *Nature* **258,** 121–124.

Perry, P. & Wolff, S. (1974) New Giemsa method for differential staining of sister chromatids. *Nature* **261,** 156–158.

Potten, C. S., Hume, W. J., Reid, P., & Cairns, J. (1978) The segregation of DNA in epithelial stem cells. *Cell* **15,** 899–906.

Resnick, M. A. & Moore, P. D. (1979) Molecular recombination and the repair of DNA double-stranded breaks in CHO cells. *Nucleic Acids Res.* **6,** 3145–3160.

Rommelaere, J. & Miller-Faures, A. (1975) Detection by density equilibrium centrifugation of recombinantlike DNA molecules in somatic mammalian cells. *J. Mol. Biol.* **98,** 195–218.

Sawicki, D. L., Erlanger, B. F., & Beiser, S. M. (1971) Immunochemical detection of minor bases in nucleic acids. *Science* **174,** 70–72.

Scheid, W. (1976) Mechanism of differential staining of BUdR-substituted *Vicia faba* chromosomes. *Exp. Cell Res.* **101,** 55–58.

Scheid, W. (1979) Influence of cysteamine on differential staining of BUdR-substituted human chromosomes. *Can. J. Genet. Cytol.* **21,** 145–149.

Scheres, J. M. J. C., Hustinx, T. W. J., Ruttem, F. J., & Merkx, G. F. M. (1977) "Reverse" differential staining of sister chromatids. *Exp. Cell Res.* **109,** 466–468.

Schubert, I., Sturelid, S., Dobel, P., & Rieger, R. (1979) Intra-chromosomal distribution patterns of mutagen-induced SCEs and chromatid aberrations in reconstructed karyotypes of *Vicia faba*. *Mutat. Res.* **59,** 27–38.

Schwarzacher, H. G. & Schnedl, W. (1965) Endoreduplication in human fibroblast cultures. *Cytogenetics* **4,** 1–18.

Smith, G. P. (1976) Evolution of repeated DNA sequences by unequal crossover. *Science* **191,** 528–535.

Stetten, G., Latt, S. A., & Davidson, R. L. (1976) 33258 Hoechst enhancement of the photosensitivity of bromodeoxyuridine-substituted cells. *Somat. Cell Genet.* **2,** 285–290.

Stetten, G., Davidson, R. L., & Latt, S. A. (1977) 33258 Hoechst enhances the selectivity of the bromodeoxyuridine-light method of isolating conditional lethal mutants. *Exp. Cell Res.* **108,** 447–452.

Stubblefield, E. (1964) DNA synthesis and chromosomal morphology of Chinese hamster cells cultured in media containing *N*-deacetyl-*N*-methylcolchicine (colcemid). In *Symposium of the International Society of Cell Biology 3, Cytogenetics of Cells in Culture* (R. J. C. Harris, Ed.), Academic Press, New York, pp. 223–248.

Swartzendruber, D. G. (1977a) Microfluorometric analysis of cellular DNA following incorporation of bromodeoxyuridine. *J. Cell Phys.* **90,** 445–454.

Swartzendruber, D. G. (1977b) A bromodeoxyuridine (BudR)-mithramycin technique for detecting cycling and non-cycling cells by flow microfluorometry. *Exp. Cell Res.* **109,** 439–443.

Takayama, S. & Sakanishi, S. (1977) Differential Giemsa staining of sister chromatids after extraction with acids. *Chromosoma* **64,** 109–115.

Takayama, S. & Sakanishi, S. (1979) Sister chromatid differential staining by direct staining in Na_2HPO_4–Giemsa solution and the mechanism involved. *Chromosoma (Berl.)* **75,** 37–44.

Taylor, J. H. (1958) Sister chromatid exchanges in tritium-labelled chromosomes. *Genetics* **43,** 515–529.

Taylor, J. H., Woods, P. S., & Hughes, W. L. (1957) The organization and duplication of chromosomes as revealed by autoradiographic studies using tritium-labelled thymidine. *Proc. Natl. Acad. Sci. U.S.* **43,** 122–128.

Tice, R., Chaillet, J., & Schneider, E. L. (1975) Evidence derived from sister chromatid exchanges of restricted rejoining of chromatid sub-units. *Nature* **256,** 642–644.

Vosa, C. G. (1977) Sister chromatid exchange bias in *Vicia faba* chromosomes. In *Current Chromosome Research* (K. Jones & P. E. Brandham, Eds.), Elsevier/North-Holland Biomedical Press, Amsterdam, pp. 105–114.

Walen, K. H. (1965) Spatial relationships in the replication of chromosomal DNA. *Genetics* **51,** 915–929.

Ward, D. C. & Reich, E. (1972) Fluorescence studies of nucleotides and polynucleotides. II. 7-deazanebularin: coding ambiguity in transcription with base pairs containing fewer than two hydrogen bonds. *J. Biol. Chem.* **247,** 705–719.

Ward, D. C., Cerami, A., Reich, E., Acs, G., & Altwerger, L. (1969a) Biochemical studies of the nucleoside analogue, formycin. *J. Biol. Chem.* **244,** 3243–3250.

Ward, D. C., Reich, E., & Stryer, L. (1969b) Fluorescence studies of nucleotides and polynucleotides. I. Formycin, 2-aminopurine riboside, 2,6-diaminopurine riboside, and their derivatives. *J. Biol. Chem.* **244,** 1228–1237.

Waters, R., Regan, J. D., & German, J. (1978) Increased amounts of hybrid (heavy/heavy) DNA in Bloom's syndrome fibroblasts. *Biochem. Biophys. Res. Commun.* **83,** 536–541.

Willard, H. F. & Latt, S. A. (1976) Analysis of deoxyribonucleic acid replication in human X chromosomes by fluorescence microscopy. *Am. J. Hum. Genet.* **28,** 213–237.

Wolff, S. (1977) Sister chromatid exchanges. *Annu. Rev. Genet.* **11,** 183–201.

Wolff, S. (1981) Measurement of sister chromatid exchange in mammalian cells. In DNA Repair: A Laboratory Manual of Research Procedures, Vol. 1, Part B (E. C. Friedberg, & P. C. Hanawalt, Eds.), Dekker, New York.

Wolff, S. & Bodycote, J. (1977) The production of harlequin chromosomes by chemical and physical agents that disrupt protein structure. In *Molecular Human Cytogenetics; ICN-UCLA Symposia on Molecular and Cellular Biology VII* (R. S. Sparkes, D. E. Comings, & C. F. Fox, Eds.), Academic Press, New York, pp. 335–340.

Wolff, S. & Perry, P. (1974) Differential Giemsa staining of sister chromatids and the study of sister chromatid exchanges without autoradiography. *Chromosoma* **48,** 341–353.

Wolff, S. & Perry, P. (1975) Insights of chromatid structure from sister chromatid exchange ratios and the lack of both isolabelling and heterolabelling as determined by the FPG technique. *Exp. Cell Res.* **93,** 23–30.

Zack, G. W., Rogers, W. E. & Latt, S. A. (1977) Automatic measurement of sister chromatid exchange frequency. *J. Histochem. Cytochem.* **25,** 741–753.

Zack, G. W., Spriet, J. A., Latt, S. A., Granlund, G. H. & Young, I. T. (1976) Automatic detection and localization of sister chromatid exchanges. *J. Histochem. Cytochem.* **24,** 168–177.

Zakharov, A. F. & Egolina, N. A. (1968) Asynchrony of DNA replication and mitotic spiralization along heterochromatic portions of Chinese hamster chromosomes. *Chromosoma* **23,** 365–385.

Zakharov, A. F. & Egolina, N. A. (1972) Differential spiralization along mammalian mitotic chromosomes. I. BUdR-revealed differentiates in Chinese hamster chromosomes. *Chromosoma* **38,** 341–355.

Zakharov, A. F., Baranovskaya, L. I., Ibraimov, A. I., Benjusch, V. A., Demintseva, V. S., & Oblapenko, N. G. (1974) Differential spiralization along mammalian mitotic chromosomes. II. 5-bromodeoxyuridine and 5-bromodeoxycytidine-revealed differentiation in human chromosomes. *Chromosoma (Berl.)* **44,** 343–359.

Chapter 3

Chromosome Aberrations, Sister Chromatid Exchanges, and the Lesions That Produce Them

SHELDON WOLFF
LABORATORY OF RADIOBIOLOGY AND DEPARTMENT OF ANATOMY
UNIVERSITY OF CALIFORNIA, SAN FRANCISCO
SAN FRANCISCO, CALIFORNIA

I.	Introduction	42
II.	Chromosome aberrations	42
III.	Sister chromatid exchange	44
IV.	Lesions that lead to aberrations	45
V.	Possible lesions for the induction of SCEs	46
VI.	Do SCEs actually represent exchanges of chromatids?	51
VII.	Conclusion	54

Work performed under the auspices of the U.S. Department of Energy.

I. INTRODUCTION

Historically, genetic damage could be measured as an increase in mutations detected after treatment of an organism with physical or chemical mutagens. The mutations were divided into two groups: those that were intragenic or true point mutations, and those that were intergenic or the result of chromosomal rearrangements and losses. We now know that the two classes are not mutually exclusive but blend into one another as the size of the intergenic mutation decreases. Thus deletions of but a single base pair in DNA can lead to frameshift mutations that ordinarily would be classified as intragenic, but larger deletions that affect more DNA than encompasses the gene would be considered, by definition, to be intergenic.

Many of the intergenic mutations are large enough to be visible when the chromosomes are examined microscopically. Because the total genome is approximately a million times as large as a structural gene, the frequency of cytologically observable chromosomal effects per cell is high enough to make cytogenetic analysis of chromosome aberrations a favored way to study mutagens.

The advent of the new methods to visualize sister chromatid exchanges (SCEs) provided another cytological endpoint that could be used to measure the effects of mutagens (Fig. 3.1). In fact, SCEs provide a far more sensitive method for the detection of many chemical mutagens than do chromosome aberrations. It should be noted, however, that although conventional chromosome aberrations and SCEs are both cytogenetic manifestations of damage to the genome, they are seemingly unrelated and arise by very different mechanisms (Wolff, 1978a).

II. CHROMOSOME ABERRATIONS

Chromosome aberrations are induced readily both by physical mutagens, such as ionizing radiation, and by chemical mutagens. The types of aberrations formed and the sensitive stages of the cell cycle vary for different types of mutagens. For instance, ionizing radiations, which can induce strand breaks as well as base damage in DNA, are effective at all stages. Thus if a cell is irradiated in G_1 before DNA replication, the chromosomes react to radiation as though they were single stranded and the unit of breakage is the whole unreplicated G_1 chromosome. This is true even though the chromosomes consist of a DNA double helix and its associated protein. These breaks can undergo repair by either rejoining with another broken end nearby to form an aberration or rejoining in the

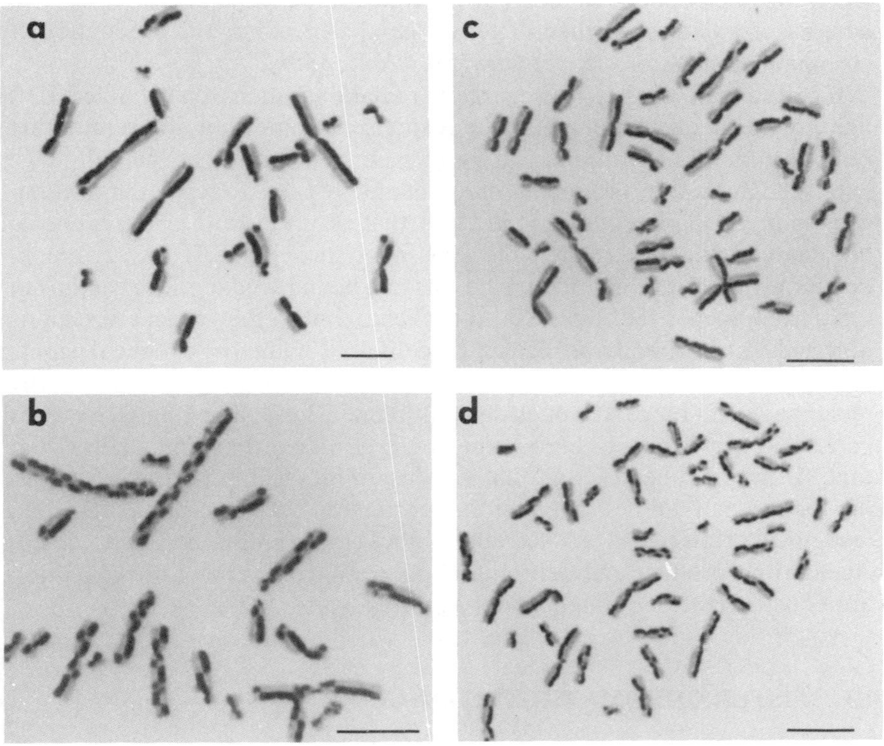

Fig. 3.1. Harlequin (FPG) stained cells showing SCEs. (*a*) CHO cell. (*b*) CHO cell after exposure to cyclophosphamide plus S-9 mix. (*c*) Human lymphocyte. (*d*) Human lymphocyte exposed to mitomycin C. The bars are 10 μm long.

original configuration (restituting) and not forming an aberration. The breaks can also remain open or unrepaired, which results in a terminal deletion. When the cells enter S and the chromosomes replicate so that they then contain two daughter chromatids, the aberrant chromosomes have both chromatids similarly affected. Since the whole chromosome is involved, the aberrations induced in G_1 are termed chromosome aberrations, which distinguishes them from those aberrations found when cells are irradiated in S or G_2; at these stages the chromosomes react to radiation as though they had already doubled and the chromatid is the unit of breakage and reunion. The broken ends of chromatids can undergo the same types of repair as occur in G_1, so that analogous types of chromatid and chromosome aberrations are formed. Thus, in S or G_2 cells, as in G_1 cells, translocations, inversions, dicentrics, rings, interstitital deletions, terminal deletions, and so on can be induced. The only difference is that when the aberration is induced in G_1, it is a full chromosome aberration, whereas irradiation at other stages of the cycle leads to the production of chromatid aberrations (for review see Wolff, 1981).

If cells are exposed to less energetic radiations, such as ultraviolet (UV) light, or to chemical mutagens, the response is somewhat different in that only chromatid aberrations can be formed irrespective of the cell stage exposed. Instead of producing chromosome breaks directly, these mutagens induce long-lived lesions in DNA that can lead to the production of chromatid aberrations when the cells pass through S. If the treatment occurs within S, chromatid aberrations can be induced in those regions of the chromosomes that have yet to replicate, but in the replicated regions of S cells, or in already-replicated G_2 cells, UV and most chemical agents do not induce aberrations that can be observed at the first succeeding metaphase. If, however, such lesions remain unrepaired until the cells enter another S phase, chromatid aberrations can be produced at that time. It has also been found that any lesion induced in G_1 that was both unrepaired and did not give rise to a chromatid break in the first S could lead to aberrations at subsequent replications (Evans & Scott, 1969). Agents that produce only chromatid aberrations by errors of replication are considered to be S-dependent agents (Evans, 1977).

III. SISTER CHROMATID EXCHANGE

SCEs, like chromosome aberrations, affect both chromatids, but because they are produced during replication (Wolff et al., 1974) they are more analogous to isolocus chromatid aberrations in which the sister chromatids are translocated. In fact, at one time it was thought that SCEs arose by breakage and reunion of sister chromatids and were merely one class of ordinary chromosomal aberrations that now could be visualized because the two chromatids stained differentially (Evans, 1977). Many differences exist between aberrations and SCEs, however, and the preponderance of the evidence now indicates that the two cytogenetic endpoints are produced by different lesions and/or cellular processes (Wolff, 1978a; Carrano et al., 1979). For instance, as is indicated above, chromosome breakage and rejoining commonly occur after X-irradiation. SCEs, however, are relatively insensitive to induction by ionizing radiation and, furthermore, are not induced throughout the whole of the cell cycle. Contrary to what is found for ordinary chromosome aberrations, irradiation in G_1 or G_2 produces either no increase in SCEs (Wolff et al., 1974; Morgan & Crossen, 1980; Galloway, 1977; Littlefield et al., 1979) or at the very most only a slight increase even at high doses that produce large numbers of broken chromosomes. Even when cells are irradiated in S, where somewhat more SCEs can be induced, the increase is minor compared to that for aberrations (Perry & Evans, 1975).

IV. LESIONS THAT LEAD TO ABERRATIONS

Experiments on the induction of the two types of effects indicate that they are induced by different lesions. The most important lesions leading to the production of chromosome breaks are double-strand breaks in DNA (Wolff, 1972; Evans, 1977), which is not the case for SCEs. Five types of experimental results relate double-strand breaks to chromosome aberrations. The first comes from experiments on the efficiency of various types of ionizing radiations to produce aberrations, in which it was found that densely ionizing high-LET radiations are more efficient than sparsely ionizing low-LET radiations (see Wolff, 1961; Evans, 1962). High-LET radiations induce a higher ratio of double-strand breaks in DNA to other types of damage, such as single-strand breaks and base damage, than do low-LET radiations, thus establishing a relation between the numbers of aberrations induced and the numbers of double-strand breaks. The second line of evidence comes from experiments in which it was found that X rays, which can produce double-strand breaks in DNA throughout the whole cell cycle, can also induce aberrations throughout the whole cell cycle, whereas most chemical mutagens and UV light, which cannot produce double-strand breaks, cannot induce aberrations in G_1 or G_2, but can only produce lesions that lead to errors in replication at the time of S (Griggs & Bender, 1973). Another argument can be made from experiments carried out in *Vicia faba* in which excision repair of base damage could not be detected after X-irradiation (Painter & Wolff, 1973). In *Vicia* as in other organisms, X rays can induce aberrations in G_1 and G_2, indicating that excision repair, and thus base damage, is not involved in the formation of the aberrations (Wolff & Scott, 1969). Another indication of this is that UV and most chemical mutagens, which do not produce double-strand breaks, can induce an excision repair response in G_1 and G_2 cells but do not induce aberrations in these stages. Lastly, X rays, unlike UV and chemicals, do not produce long-lived lesions that can lead to aberrations in subsequent cell cycles. It should also be noted that the other common lesion induced by ionizing radiation, the single-strand break in DNA, cannot be related to aberration yields. Thus single-strand breaks have been found to be repaired much faster than chromosome breaks, and treatments that are known to modify the rejoining of single-strand breaks do not seem to affect the repair of chromosome aberrations (see Wolff, 1972).

The response of SCEs is very different. They are not readily induced by ionizing radiation at all, but are only produced in S (Wolff et al., 1974). Some chemical mutagens, such as bleomycin, which can induce double-strand breaks in DNA (Fujiwara & Kondo, 1973), can effectively induce aberrations but not SCEs (Perry & Evans, 1975; Gebhart & Kappauf, 1978). Furthermore, unlike X rays, chemicals that produce base damage produce long-lived lesions that lead to the formation of SCEs at subsequent generations. Thus it has been found that if Chinese hamster ovary (CHO) cells are exposed to a variety of chemical agents, including acetoxyacetylaminofluorene, which has a half-life of only $7\frac{1}{2}$ min in water (Miller, 1970), SCEs can be induced not only in the first round of replication after exposure, but also in the second. In this line of cells the lesions did not last long enough to give SCEs in the third cycle after treatment (Wolff, 1978b).

V. POSSIBLE LESIONS FOR THE INDUCTION OF SCEs

When chromosome aberrations are induced with UV light, not only are long-lived lesions induced that can produce aberrations in subsequent cell cycles, but the most likely UV lesion to lead to this effect is the cyclobutane thymine dimer. In experiments with *Xenopus* cells, which contain the photoreactivating enzyme that can monomerize dimers, Griggs & Bender (1973) showed that photoreactivating light could remove the lesions responsible for the aberrations. With SCEs, however, the data are somewhat equivocal. In 1974 Kato reported that in his experiments with *Potorous tridactylis* cells the lesions leading to SCEs after UV treatment were readily photoreactivable. A series of experiments carried out with xeroderma pigmentosum (XP) lines of varying excision repair capacity, however, showed that although XP lines were generally more sensitive to UV than the normal cells, the number of SCEs could not be related to the excision repair capacity of the lines (Figure 3.2) and thus could not be related to the removal of thymine dimers in any simple fashion (de Weerd-Kastelein et al., 1977). In addition, experiments with Chinese hamster cell lines of differing repair capacities had shown they were equally sensitive to the induction of SCEs by UV light (Wolff et al., 1974). Wolff and Bodycote (in Wolff, 1978a) have reported that in experiments with chick cells, which contain the photoreactivating enzyme, there was no photoreactivation effect on SCE yields after low doses of UV. With the

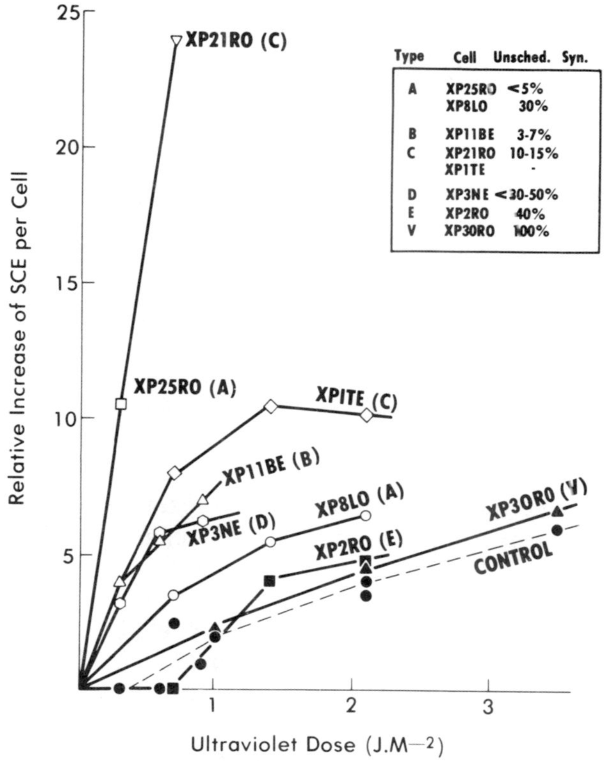

Fig. 3.2. UV-light-induced SCEs in various xeroderma pigmentosum lines. From data of de Weerd-Kastelein et al., 1977.

same system, however, Natarajan et al. (1980) found a photoreactivation effect that could easily be masked by the fact that the photoreactivating light itself can induce SCEs. At low doses they obtained results that did not differ from those of Wolff and Bodycote. Nevertheless, Ishizaki et al. (1980), working with the newer methods for SCE detection, have obtained results in *Potorous* that are as striking as those reported by Kato (1974). Thus it appears that there is a great discrepancy among the results of different types of experiments obtained in different laboratories regarding whether or not there is a direct relation between the numbers of thymine dimers and the yield of SCEs.

Another series of experiments indicating that the lesions responsible for aberrations and SCEs are different have been carried out in the excision repair-defective strain of human cells, XP12RO, which come from a group A xeroderma pigmentosum patient (Wolff et al., 1977). This line of cells has less than 5% of the normal repair capability of normal cells and is unable to repair damage induced by UV radiation or a group of chemicals such as 4-nitroquinoline 1-oxide that have been termed "UV-like" (Cleaver, 1973). That is, treatment with these agents does not elicit an excision repair response in these cells and the cells are sensitive to these agents (Sasaki, 1973). This sensitivity is manifested both in high chromosome aberration yields and in a high degree of killing. With X rays and a group of chemicals that have been called X-ray-like, however, these cells exhibit a normal response (Sasaki, 1973; Stich et al., 1973); that is, excision repair occurs and the aberration and survival levels are equivalent to those in normal cells. When SCEs were studied in the same line (Wolff et al., 1977), however, it was found that the yield of SCEs was high for both X-ray-like and UV-like chemicals, and it was independent of whether or not the chemical elicited an excision repair response. It thus appears that SCEs are induced by unexcised lesions and that these are such a small fraction of the total number of lesions that they do not affect the relatively crude measurement of excision repair. The results also indicate that the major lesions responsible for the excision repair response, aberration formation, and cell killing are not the same lesions responsible for the formation of SCEs.

The exact nature of the minor lesion that could be responsible for the production of SCEs, however, is as yet unknown. The xeroderma pigmentosum results indicate that the lesion formed at the major site of alkylation in DNA that leads to excision repair—the N-7 position of guanine—is not the lesion that would lead to SCEs. In fact, in experiments with the same XP12RO cells, Goth-Goldstein (1977) was able to measure the repair of alkylations at the O-6 position of guanine after treatment with ethylnitrosourea, which produces 40% of its alkylguanine at this position. She found that O-6 alkylguanine is removed much more slowly than N-7 alkylguanine. This is contrary to the results obtained with normal cells, in which the O-6 alkylguanine was removed as fast as N-7 alkylguanine or even faster. Since O-6 alkylguanine is ordinarily a minor alkylation product of DNA, the possibility exists that it is the minor product responsible for SCE production.

This conclusion is consistent with the results obtained with the Mer$^-$ strain of cells isolated from human tumors or from SV40-transformed normal fibroblasts (Day et al., 1980). The Mer$^-$ phenotype makes the cells sensitive to *N*-methyl-*N'*-nitro-*N*-nitrosoguanidine with respect to survival and SCE induction, and this sensitivity has now been linked to a defect in removing O-6 methyl guanine from the treated cells.

Another lesion that has been considered as a likely candidate to give rise to SCEs is the interstrand DNA cross-link of the type that can be induced by mitomycin C, which is a potent inducer of SCEs. However, Fanconi's anemia cells, which seem to be defective in cross-link repair and have more aberrations after treatment with mitomycin C than normal cells, have been found to have a reduced mitomycin C response for SCEs (Latt et al., 1975); this behavior indicates that the cross-link might not be the lesion responsible for SCE formation (Wolff, 1978a). Carrano et al. (1979) have recently carried out experiments in which they used not only mitomycin C, which can form cross-links as well as large numbers of monoadducts, but also decarbamoyl mitomycin C, which cannot form cross-links. It was found that decarbamoyl mitomycin C was as effective as ordinary mitomycin C in inducing SCEs or more so, again indicating that the cross-link does not lead to the formation of the SCE.

This conclusion was also reached by Bradley et al. (1979), who carried out a series of experiments in Chinese hamster V79 cells with agents that would induce various types of lesions. All the agents induced SCEs, leading the authors to conclude that single-strand DNA breaks, DNA–DNA cross-links, and DNA–protein cross-links are not necessary for SCE formation.

Experiments similar to those of Carrano et al. (1979) have been carried out by Linnainma and Wolff (in preparation), who used not only mitomycin C and decarbamoyl mitomycin C, but also 8-methoxypsoralen and its noncrosslinking derivative, angelicin. As was the case with the mitomycins, the noncrosslinking agent could induce SCEs, which implicates the monoadduct rather than the cross-link as the lesion formed by these compounds that leads to SCE formation. As noted by Singer (1976), however, DNA can be alkylated at many different positions. It is still not known which of these, if any, could be responsible for the misreplication that leads to the production of an SCE in S.

Although it is known that SCEs are formed in S, the exact mechanism by which they are brought about is as yet unknown. Several models have been proposed that invoke strand breakage and exchange at replication forks (Painter, 1980; Ishii and Bender, 1980; Kato, 1977; Evans, 1977; Taylor, 1981; Sasaki, 1980; Ikushima, 1980) and in some cases invoke an enzymatic attack by topoisomerases (Ishii and Bender, 1980; Taylor, 1981). So far these models are representational and have not lent themselves to critical testing.

A further representational model has been proposed to account for the induction of SCEs by cross-linking agents such as mitomycin C (Shafer, 1977). In this model the cross-link is bypassed at the time of replication with a switch of strands used as the template. It has been pointed out (Stetka, 1979) that in such a model, because the lesion remains *in situ*, no twin exchanges would be induced by events that took place in the first round of replication. Ordinarily, an SCE produced in the first cycle affects both sister chromatids identically, so that after the next round of replication both daughter chromosomes have the same SCE. This is usually noted when the spindle is inhibited to prevent the dispersal of the sister chromatids into separate cells (Taylor, 1958), that is, if the cells are made tetraploid after the first cycle in bromodeoxyuridine (BrdUrd). If, however, one of the sister chromatids still had the lesion at the point of exchange, then the bypass mechanism would unmake the SCE in that chromosome at the second round of replication so that a twin would be converted to a single that is observed as an SCE in only one of the two daughter chromosomes.

Linnainma and Wolff's data obtained with CHO cells (Table 3.1) show that contrary to expectations from the bypass model, both mitomycin C and 8-methoxypsoralen, which can induce cross-links, produce twin SCEs. Interestingly enough, the data also show that the compounds are relatively ineffective at inducing single SCEs; that is, the lesions induce SCEs in the first cycle after treatment, but not the second. As is noted earlier, the noncrosslinking derivatives of the two compounds, decarbamoyl mitomycin C and angelicin, give similar results.

The results obtained with these compounds differ from those obtained with several other compounds in experiments designed to determine the lifetimes of the SCE-inducing lesions (Wolff, 1981). If cells are exposed to a chemical that produces lesions that last only one cycle, then only twins are induced. If the lesions last two cycles, then both twins and singles are formed. Furthermore, because cells that have replicated three times in the presence of BrdUrd have half of their chromatin in differentially stained sister chromatids, SCEs found in the third cycle can also be detected. In CHO cells it was found that ethyl methanesulfonate, methyl methanesulfonate, 4-nitroquinoline 1-oxide, and acetoxyacetylaminofluorene (AAAF) all induced both twins and singles, that is, the lesions lasted until the second round of replication. In additional experiments with AAAF, however, the lesions did not last to produce SCEs in the third cycle. The lifetimes of the lesions were highly dependent on the excision repair capacity of the cells, because similar experiments performed with B14FAF Chinese hamster cells showed that in this line, which is highly deficient in DNA repair, AAAF could induce SCEs three cycles after it was administered. Similarly, the compound could induce SCEs for three cycles in repair-defective human XP fibroblasts, but not in normal fibroblasts or in normal human lymphocytes. In fact, in normal human cells the repair was so effective that most of the lesions were removed even before the first round of DNA replication.

VI. DO SCEs ACTUALLY REPRESENT EXCHANGES OF CHROMATIDS?

Stahl (1964) pointed out that the labeling patterns attributed to sister chromatid exchange could also be obtained if sister chromatid exchange did not occur, but if the DNA within a chromatid broke and the polynucleotide strands rejoined randomly within a chromatid, that is, rejoined as if polarity did not obtain and actually no exchange occurred between the two sisters. Since this alternative derived with assumed identity of chromosomal subunits mimics sister chromatid exchange, an independent demonstration that sister chromatid exchange did indeed involve exchange was needed. This was obtained in two independent experiments.

The first of these (reported in Peacock, 1971), was an analysis of the exchange pattern noted in endoreduplicated chromosomes performed by Brewen and Peacock. They labeled cells with tritiated thymidine in the first round of replication, let the cells divide, and then induced endoreduplication. The endoreduplicated chromosomes thus contained one labeled chromatid and three unlabeled chromatids. If a true sister chromatid exchange had occurred during the first or second round of replication, there would be an exchange of label involving each of the chromosomes making up the diplochromosomes, whereas an exchange in the third round of replication would yield one unlabeled chromosome and one chromosome with an SCE. The ratio of the two types would be 2:1. On the other hand, if no actual exchanges had occurred but there was random subunit reunion within chromatids, only a first-division reunion would lead to an exchange of label between the two chromosomes of the diplochromosome, and only a second-division reunion would lead to one unlabeled chromosome and one with an SCE. A third-division reunion would not lead to label exchange. Thus the ratio of the two types would be 1:1. Brewen and Peacock's data are consistent with expectations derived from the occurrence of true sister chromatid exchange.

A further clear indication that sister chromatid exchange actually does occur was obtained in experiments by Wolff et al. (1976), who induced ring chromosomes by X-irradiating synchronized G_1 CHO cells. Ordinarily, after two rounds of replication in the presence of BrdUrd, two monocentric rings would be found in cells made tetraploid by the addition of Colcemid before the first mitosis. These rings would each have one dark-staining and one light-staining chromatid when processed by the fluorescence-plus-Giemsa technique to produce harlequin chromosomes. If no true sister chromatid exchange had occurred, but only an exchange of polynucleotide strands within a chromatid, the cells would still contain a pair of monocentric rings. If, however, a true sister chromatid exchange

TABLE 3.1

SCEs Induced in the Large Acrocentric Chromosome of CHO Cells by Bifunctional Cross-linking Agents and Their Monofunctional Derivatives

		First cycle		Second cycle		
Treatment (60 min in G_1)	SCEs/cell in 20 cells[a] (mean ± SD)	#Twin SCEs/50 chromosome pairs	SCEs/pair	#Single SCEs/100 chromosomes	SCEs/ chromosome	Ratio of second cycle to first cycle SCEs
Control	19.2 ± 1.0	19	0.38	38	0.38	2.00
Mitomycin C						
10^{-8} M	25.3 ± 1.12	24	0.48	30	0.30	1.25
5×10^{-8} M	31.5 ± 1.25	24	0.38	31	0.31	1.29
10^{-7} M	42.0 ± 1.5	45	0.90	36	0.36	0.80
5×10^{-7} M	75.8 ± 1.9	94	1.88	68	0.68	0.72
10^{-6} M	116.6 ± 2.4	145	2.90	68	0.68	0.47
Decarbamoyl mitomycin C						
10^{-8} M	29.9 ± 1.2	28	0.56	32	0.32	1.14
5×10^{-8} M	37.9 ± 1.4	31	0.62	34	0.34	1.09
10^{-7} M	57.9 ± 1.7	44	0.88	41	0.41	0.93
6×10^{-7} M	125.7 ± 2.5	108	2.16	87	0.87	0.80
10^{-6} M	151.7 ± 2.8	136	2.72	85	0.85	0.63

Control (dark)	16.1 ± 1.2	18	0.36	34	0.34	1.89
Control (light)[b]	19.3 ± 1.3	19	0.38	35	0.35	1.84
Psoralen + light[b]						
5×10^{-7} M	35.0 ± 2.2	37	0.74	39	0.39	1.05
10^{-6} M	72.2 ± 2.2	122	2.44	51	0.51	0.42
Angelicin + light[b]						
2.5×10^{-5} M	35.2 ± 2.0	40	0.80	61	0.61	1.52
5×10^{-5} M	51.5 ± 2.7	72	1.44	59	0.59	0.82
10^{-4} M	69.3 ± 3.1	78	1.56	60	0.60	0.77

[a] Entire genome scored for SCEs.

[b] 560 J/m^2 of 360 nm black light.

had occurred, then the monocentric rings would be converted to double-sized dicentric rings. The results showed that approximately 20% of the rings that should have formed two monocentrics after the second round of replication were converted to dicentric rings that could come about only if true sister chromatid exchange took place.

These two experiments provide convincing evidence that sister chromatid exchange is a real phenomenon and does not merely represent an interpretation of the labeling and staining patterns obtained in chromosomes.

VII. CONCLUSION

Treatment of cells with mutagenic carcinogens can lead to cytogenetic damage manifested as either chromosome aberrations or SCEs. The two types of damage are caused by different lesions. Aberrations are most likely the result of double-strand breaks in DNA, whereas SCEs come from unremoved adducts.

Experiments on SCE induction are consistent with the notion that O-6 alkylation of guanine could be the responsible lesion, but lesions at other sites of alkylation and other lesions such as thymine dimers could also be involved. Our uncertainty is a reflection of the fact that we still do not know exactly how SCEs are formed. We do know, however, that SCEs represent actual exchanges from one chromatid to another and are not merely reflections of a differential staining pattern brought about by other mechanisms, such as the breakage and random rejoining of polynucleotide strands within a single chromatid. Because the exchange between the two sister chromatids takes place in S, most models of SCE induction involve putative events that occur at the replication fork.

REFERENCES

Bradley, M. O., Hsu, I. C., & Harris, C. C. (1979) Relationships between sister chromatid exchange and mutagenicity, toxicity and DNA damage. *Nature (Lond.)* **282,** 318–320.

Carrano, A. V., Thompson, L. H., Stetka, D. G., Minkler, J. L., Mazrimas, J. A., & Fong, S. (1979) DNA crosslinking sister-chromatid exchange and specific-locus mutations. *Mutat. Res.* **63,** 175–188.

Cleaver, J. E. (1973) DNA repair with purines and pyrimidines in radiation- and carcinogen-damaged normal and xeroderma pigmentosum human cells. *Cancer Res.* **33,** 362–369.

Day, R. S., III, Ziolkowski, C. H. J., Scudiero, D. A., Meyer, S. A., Lubiniecki, A. S., Girardi, A. J., Galloway, S. M., & Bynum, G. D. (1980) Defective repair of alkylated

DNA by human tumour and SV40-transformed human cell strains. *Nature (Lond.)* **288,** 724–727.

de Weerd-Kastelein, E. A., Keijzer, W., Rainaldi, G., and Bootsma, D. (1977) Induction of sister chromatid exchanges in xeroderma pigmentosum cells after exposure to ultraviolet light. *Mut. Res.* **45,** 253–261.

Evans, H. J. (1962) Chromosome aberrations induced by ionizing radiations. *Int. Rev. Cytol.* **13,** 221–321.

Evans, H. J. (1977) Molecular mechanisms in the induction of chromosome aberrations. In *Progress in Genetic Toxicology*, Vol. 2 (D. Scott, B. A. Bridges, & F. H. Sobels, Eds.), Elsevier/North-Holland, Amsterdam, pp. 57–74.

Evans, H. J. & Scott, D. (1969) The induction of chromosome aberrations by nitrogen mustard and its dependence on DNA synthesis. *Proc. R. Soc. Lond. B.* **173,** 491–512.

Fujiwara, Y. & Kondo, T. (1973) Strand-scission of HeLa cell deoxyribonucleic acid by bleomycin *in vitro* and *in vivo*. *Biochem. Pharmacol.* **22,** 323–333.

Galloway, S. M. (1977). Ataxia telangiectasia: The effects of chemical mutagens and X-rays on sister chromatid exchanges in blood lymphocytes. *Mutat. Res.* **45,** 343–349.

Gebhart, E. & Kappauf, H. (1978) Bleomycin and sister-chromatid exchange in human lymphocyte chromosomes. *Mutat. Res.* **58,** 121–124.

Goth-Goldstein, R. (1977) Repair of DNA damaged by alkylating carcinogens is defective in xeroderma pigmentosum derived fibroblasts. *Nature (Lond.)* **267,** 81–82.

Griggs, H. G. & Bender, M. A. (1973) Photoreactivation of ultraviolet-induced chromosomal aberrations. *Science* **179,** 86–88.

Ikushima, T. (1980) Sister chromatid exchange and its significance. *Heredity (Tokyo)* **34,** 34–53. In Japanese.

Ishii, Y. & Bender, M. A. (1980) Effects of inhibitors of DNA synthesis on spontaneous and ultraviolet light-induced sister-chromatid exchanges in Chinese hamster cells. *Mutat. Res.* **79,** 19–32.

Ishizaki, K., Nikaido, O., & Takebe, H. (1980) Photoreactivation of ultraviolet light-induced sister chromatid exchanges in potorous cells. *Photochem. Photobiol.* **31,** 277–279.

Kato, H. (1974) Photoreactivation of sister chromatid exchanges induced by ultraviolet irradiation. *Nature (Lond.)* **249,** 552–553.

Kato, H. (1977) Mechanisms for sister chromatid exchanges and their relation to the production of chromosomal aberrations. *Chromosoma (Berl.)* **59,** 179–191.

Latt, S. A., Stetten, G., Juergens, L. A., Buchanan, G. R., & Gerald, P. S. (1975) Induction by alkylating agents of sister chromatid exchanges and chromatid breaks in Fanconi's anemia. *Proc. Natl. Acad. Sci. U.S.* **72,** 4066–4070.

Linnainma, K. & Wolff, S., in preparation.

Littlefield, L. G., Colver, S. P., Joiner, E. E., & DuFrain, R. J. (1979) Sister chromatid exchanges in human lymphocytes exposed to ionizing radiation during G_0. *Radiat. Res.* **78,** 514–521.

Miller, J. A. (1970) Carcinogenesis by chemicals: An overview—G. H. A. Clowes Memorial Lecture. *Cancer Res.* **30,** 559–576.

Morgan, W. F. and Crossen, P. E. (1980). X irradiation and sister chromatid exchange in cultured human lymphocytes. *Environ. Mutagenesis* **2,** 149–155.

Natarajan, A. T., van Zeeland, A. A., Verdegaal-Immerzeel, E. A. M., & Filon, A. R. (1980) Studies on the influence of photoreactivation on the frequencies of UV-induced chromosomal aberrations, sister-chromatid exchanges and pyrimidine dimers in chicken embryonic fibroblasts. *Mutat. Res.* **69,** 307–317.

Painter, R. B. (1980). A replication model for sister-chromatid exchange. *Mutat. Res.* **70,** 337–341.

Painter, R. B. & Wolff, S. (1973) Apparent absence of repair replication in *Vicia faba* after X-irradiation. *Mutat. Res.* **19,** 133–137.

Peacock, W. J. (1971) Cytogenetic aspects of the mechanism of recombination in higher organisms. In *Stadler Genetics Symposium*, Vols. 1 and 2 (G. Kimber & G. P. Redei, Eds.), The University of Missouri Agricultural Experiment Station, Columbia, Missouri, pp. 123–152.

Perry, P. & Evans, H. J. (1975) Cytological detection of mutagen-carcinogen exposure by sister chromatid exchange. *Nature (Lond.)* **258,** 121–125.

Sasaki, M. S. (1973) DNA repair capacity and susceptibility to chromosome breakage in xeroderma pigmentosum cells. *Mutat. Res.* **20,** 291–293.

Sasaki, M. S. (1980) Chromosomal aberration formation and sister chromatid exchange in relation to DNA repair in human cells. In *DNA Repair and Mutagenesis in Eukaryotes* (W. M. Generoso, M. D. Shelby, & F. J. de Serres, Eds.), Plenum Press, New York, pp. 285–314.

Shafer, D. A. (1977) Replication bypass model of sister chromatid exchanges and implications for Bloom's syndrome and Fanconi's anemia. *Hum. Genet.* **39,** 177–190.

Singer, B. (1976) All oxygens in nucleic acids react with carcinogenic ethylating agents. *Nature (Lond.)* **264,** 333–339.

Stahl, F. W. (1964). In *The Mechanisms of Inheritance*, 1st ed., Prentice-Hall, Englewood Cliffs, N.J., p. 80.

Stetka, D. G., Jr. (1979) Further analysis of the replication bypass model for sister chromatid exchange. *Hum. Genet.* **49,** 63–69.

Stich, H. F., Stich, W. and San, R. H. C. (1973) Chromosome aberrations in xeroderma pigmentosum cells exposed to the carcinogens, 4-nitroquinoline-1-oxide, and *N*-methyl-*N'*-nitronitrosoguanidine. *Proc. Soc. Exp. Biol. Med.* **142,** 1141–1144.

Taylor, J. H. (1958) Sister chromatid exchanges in tritium-labeled chromosomes. *Genetics* **43,** 515–529.

Taylor, J. H. (1981) Sister chromatid exchange: Early experiments with autoradiography. In *Sister Chromatid Exchange* (S. Wolff, Ed.), Wiley-Interscience, New York, pp. 1–16.

Wolff, S. (1961) Radiation genetics. In *Mechanisms in Radiobiology*, Vol. 1 (M. Errera & A. Forssberg, Eds.), Academic Press, New York, pp. 419–475.

Wolff, S. (1972) Genetic effects and radiation-induced cell death. In *Frontiers of Radiation Therapy and Oncology,* Vol. 6, Karger, Basel, and University Park Press, Baltimore, pp. 459–469.

Wolff, S. (1978a) Relation between DNA repair, chromosome aberrations, and sister chromatid exchanges. In *DNA Repair Mechanisms* (P. C. Hanawalt, E. C. Friedberg, & C. F. Fox, Eds.), Academic Press, New York, pp. 751–760.

Wolff, S. (1978b) Chromosomal effects of mutagenic carcinogens and the nature of the lesions leading to sister chromatid exchange. In *Mutagen-Induced Chromosome Damage in Man* (H. J. Evans & D. C. Lloyd, Eds.), Edinburgh University Press, pp. 208–215.

Wolff, S. (1981) Induced chromosome variation. In *Chromasomes Today*, Vol. 7 (M. D. Bennett, M. Bobrow, & G. M. Hewitt, Eds.), George Allen & Unwin, London, pp. 226–241.

Wolff, S. & Scott, D. (1969) Repair of radiation-induced damage to chromosomes: Independence of known dark repair mechanisms. *Exp. Cell Res.* **55**, 9–16.

Wolff, S., Bodycote, J., & Painter, R. B. (1974) Sister chromatid exchanges induced in Chinese hamster cells by UV irradiation of different stages of the cell cycle: The necessity for cells to pass through S. *Mutat. Res.* **25**, 73–81.

Wolff, S., Lindsley, D. L., & Peacock, W. J. (1976) Cytological evidence for switches in polarity of chromosomal DNA. *Proc. Natl. Acad. Sci. U.S.* **73**, 877–881.

Wolff, S., Rodin, B., & Cleaver, J. E. (1977) Sister chromatid exchanges induced by mutagenic carcinogens in normal and xeroderma pigmentosum cells. *Nature (Lond.)* **265**, 347–349.

Chapter 4

Sister Chromatid Exchange and Single-Gene Mutation

ANTHONY V. CARRANO AND LARRY H. THOMPSON
LAWRENCE LIVERMORE NATIONAL LABORATORY
BIOMEDICAL SCIENCES DIVISION
UNIVERSITY OF CALIFORNIA
LIVERMORE, CALIFORNIA

I. Introduction 60
II. The nature of alterations producing sister chromatid exchange 60
III. The nature of alterations producing mutant phenotypes 63
IV. Methodology for comparing sister chromatid exchange and single-gene mutation 66
  A. Rationale for the choice of a cell system and loci for mutation assay 66
  B. Treatment and assay of cells 66
  C. The role of bromodeoxyuridine 68
  D. Limitations on data interpretation 69
V. Experimental evidence relating sister chromatid exchange and single-gene mutation 70
  A. Quantitative dose-response studies in CHO cells—our own data 70
  B. SCE–mutation studies of other investigators 76
VI. Genetic approaches to determining the mechanism of SCE 78
VII. Concluding remarks 79

Work performed under the auspices of the U.S. Department of Energy by the Lawrence Livermore National Laboratory under contract number W-7405-ENG-48.

I. INTRODUCTION

The biological response of a cell to insult is governed by two factors: the nature of the initial lesions and the manner in which the cell converts the lesions to biologically significant events. Physical and chemical agents induce a broad spectrum of cellular damage, cytoplasmic and nuclear. This chapter deals with two kinds of events that result from lesions in the genetic material (DNA): sister chromatid exchanges (SCEs) and single-gene mutations.

The lesions induced in DNA by physical and chemical agents generally fall into two broad classes: (1) those that directly break the DNA backbone to form either single-strand or double-strand DNA scissions and (2) those that modify the backbone by producing adducts or otherwise distort the existing structure. Ionizing radiation, an example of an agent of the first type, is a relatively poor inducer of both SCE and single-gene mutation when compared to most agents of the second type at equivalent levels of survival (Perry and Evans, 1975; Littlefield et al., 1979; see also data below). The fate of the initial lesions in the DNA also depends on the ability of the cell to repair or otherwise process the damage. A DNA lesion may be left unmodified and, if consistent with viability, may persist without any genetic consequences or produce an SCE or mutation. A DNA lesion that is modified by a repair mechanism may also have no genetic consequences, may form an SCE or mutation, or may become lethal through misrepair.

Thus there is no *a priori* reason to expect that either a unique DNA lesion or a single repair event might be responsible for either SCE or mutation. Our initial approach to understanding the biological significance of the SCE has been to compare quantitatively the efficiency of a variety of DNA lesions for inducing both SCE and single-gene mutation in the same cell system, Chinese hamster ovary (CHO) cells. Before we discuss data relevant to this comparison, we review what is known about the nature of SCEs and single-gene mutations.

II. THE NATURE OF ALTERATIONS PRODUCING SISTER CHROMATID EXCHANGE

Four observations suggest a close correlation between SCE and single-gene mutation for many agents: (1) the S-phase dependence for the production of both endpoints; (2) the induction of SCEs by known mutagens and carcinogens; (3) the detectability of both endpoints at doses that produce reasonably high cell survival; and (4) the similarity in spontaneous frequencies over the entire genome.

The necessity for a cell to pass through S-phase in order to form an SCE was demonstrated by Kato (1974a) for visible light, by Wolff et al. (1974) for UV light in Chinese hamster cells, and by Kihlman et al. (1978) for fluorescent light in *Vicia faba*. Schvartzman et al. (1979a) later demonstrated an S-dependent process for the induction of SCEs with visible light in BrdUrd (bromodeoxyuridine) substituted chromosomes of *Allium cepa* but could not rule out a post replicational recombination mechanism closely associated with semiconservative DNA replication. In fact, a model for SCE induction based on a recombination event at the junction between a completely duplicated replicon cluster and a partially duplicated cluster has been postulated (Painter, 1980). This model accounts for the relatively poor SCE-inducing ability of agents that inhibit replicon initiation (e.g., X rays) and for the relatively potent SCE induction by agents that inhibit replication fork progression (e.g., mitomycin C). Such S-phase dependence requires that the SCE-forming lesions induced in the DNA be present at the time of semiconservative replication. In this regard, the mechanism of SCE formation is analogous to the process of mutation fixation, that is, a premutational lesion can be converted to a mutation during DNA synthesis.

Strong evidence for the similarity of mutational and SCE lesions derives from the many studies demonstrating that SCEs can be readily induced by known mutagens and carcinogens (Latt, 1974; Perry and Evans, 1975; Abe and Sasaki, 1977). SCE induction by an indirect-acting mutagen also shows a requirement for metabolic activation (Natarajan et al., 1976; Stetka and Wolff, 1976). Most of these investigations demonstrated a linear dose response for SCE induction, suggesting a single hit mechanism for this event. Again, at moderate to low toxicity, mutational events in mammalian cells also exhibit linear induction kinetics (Hsie et al., 1978) in most cases. Chemicals that produce cell transformation *in vitro* were also examined for their ability to induce SCEs (Benedict et al., 1978; Banerjee and Benedict, 1979). Eight of the 11 chemicals tested produced both transformation and SCE induction, while the remaining 3 chemicals did not produce significant increases for either endpoint. The doses necessary to elicit detectable SCEs were substantially lower than those required to produce transformation. Thus, to the extent that transformation *in vitro* is a measure of carcinogenic potential *in vivo*, the measurement of SCEs appears to be a sensitive method for detecting presumptive carcinogenic compounds.

Our initial study with four chemicals demonstrated a linear relation for the induction of SCEs and single-gene mutation (Carrano et al., 1978). It was also evident from these results that the concentration range necessary to produce readily detectable SCE and mutation was the same. With some agents both genetic effects could be detected at low doses that produced little or no cellular toxicity. This behavior is in sharp contrast to the induction of chromosomal aberrations, which, for the most part, are a direct measure of genetic toxicity (Dewey et al., 1970; Carrano, 1973) and require much higher doses for a detectable effect (Perry and Evans, 1975).

If the formation of SCE and single-gene mutation do share one or more common processes, one might expect their baseline frequencies to be similar. A comparison can probably best be made using the data from mammalian cells in culture. For SCEs the baseline frequency depends on the concentration of BrdUrd. Wolff and Perry (1974) titrated the amount of BrdUrd in the Chinese hamster cell system and found the baseline SCE frequency to vary from 0.35 to 0.70/chromosome. Back extrapolation of their data to zero BrdUrd dose suggests a "spontaneous" level of 0.30/chromosome. This value is consistent with other reported frequencies for hamster cells *in vitro* (Stetka and Carrano, 1977; Kato, 1974b). The frequency of 0.30/chromosome, or approximately 6.0/cell, is obtained after two cycles of DNA synthesis and, therefore, must be halved to obtain the baseline rate of 3.0 SCE/cell/generation. The baseline or spontaneous mutation rate for a hemizygous or heterozygous gene is obtained by fluctuation analysis (Luria and Debruck, 1943). For CHO cells this spontaneous rate has been determined for the *aprt* (adenine phosphoribosyltransferase) locus to be 8.6×10^{-7}/cell/generation (Thompson et al., 1980b). Estimates for the *hprt* (hypoxanthine phosphoribosyltransferase) locus fall into the same range (Thompson and Baker, 1973). Since the SCEs are integrated over all genes in the cell, for comparison, SCE and single-gene mutation rates can be expressed as the frequency or probability of occurrence per DNA base pair. Assuming the mammalian genome contains about 6×10^9 base pairs and that an average locus is about 2×10^3 base pairs, then the baseline rates are 5×10^{-10} SCE/base pair and 4.3×10^{-10} mutations/base pair, using the above value for the *aprt* locus. Without a further understanding of the molecular nature of the lesions we can only regard this similarity in rates as fortuitous, but provocative. It seems likely that many SCEs, if they represent exact recombination events, will be genetically neutral, and also that most single-gene mutations will result from nonrecombinational events, such as base substitutions. In addition, the spatial distribution of SCE and mutation along each chromosome may be dissimilar.

In an elegant series of experiments Gatti et al. (1980) examined the relation among the processes of meiotic recombination, mitotic recombination (symmetrical chromatid interchange formation), and SCE. Using

Drosophila melanogaster cultured cells derived from two mutant stocks defective in meiotic recombination and in postreplication repair, they found that the frequency of SCE in these cells was either unchanged or slightly reduced, but that the frequency of chromatid interchanges was markedly reduced. These results suggest that the molecular processes involved in meiotic recombination are shared in the formation of symmetrical chromatid interchanges but are different from those that produce SCEs. In one of these mutant stocks, but not the other, the spontaneous mutation frequency was also increased. Thus there appears to be no clear relation among the mechanisms of meiotic recombination, SCE, and single-gene mutation, although this does not rule out a recombination process unique to somatic cells.

An interesting aspect of any correlation between SCE and mutation is the apparent nonrandom distribution of SCEs along a chromosome. Baseline SCEs were reported to be preferentially increased at euchromatin–heterochromatin junctions in three species: the Indian muntjac (Carrano and Wolff, 1975); the kangaroo rat, *Dipodomys ordii* (Bostock and Christie, 1976); and *D. melanogaster* (Gatti et al., 1979). In man the SCE frequency appears to be higher at the interface between light and dark G- or Q-bands (Latt, 1974; Hoo and Parslow, 1979). It is interesting to speculate that these junctions may represent regions between two replicons that differ in their rate of fork progression, regions that may be susceptible to SCE induction (Painter, 1980). Little is known, however, concerning the spatial distribution of mutations in mammalian cells and the influence of adjacent heterochromatin.

Before we explore the quantitative relation between SCE and single-gene mutation, it is important to review our current understanding of the nature of the mutagenic alterations that occur in mammalian cells *in vitro*.

III. THE NATURE OF ALTERATIONS PRODUCING MUTANT PHENOTYPES

Historically, a major concern has been that a variant phenotype might owe its origin to a change in DNA expression not involving an actual change in the structure and coding properties of DNA, that is, mutation. However, there is little direct evidence actually indicating that particular phenotypes arise from nonmutational changes (DeMars, 1974; Hsie et al., 1979), or even from mutational changes in regulatory genes that might control the expression of the structural genes under consideration (Skaper et al., 1977). On the other hand, there are numerous examples in which the evidence strongly suggests that mutations have occurred. The need for such evidence is greatest with enzyme-deficiency markers, in which a variant phenotype might arise because of the absence of *production* of the normal protein. This class of markers includes the commonly studied *hprt, aprt,* and *tk* (thymidine kinase) loci, which code for nonessential, salavage-pathway enzymes.

Among the gene loci used for quantitative assays of induced mutations, the *hprt* locus is by far the best characterized in terms of the kinds of molecular events that occur at the protein level to give rise to the mutant phenotypes (see review by Caskey and Kruh, 1979). In the earlier studies, induced mutations at this locus were usually detected in selective medium containing 8-azaguanine (AG) (Chu and Malling, 1968), but many investigators now prefer 6-thioguanine (TG) because it selects more faithfully for mutants that have little or no HPRT activity (Sharp et al., 1973; Hsie et al., 1978; Thompson et al., 1980a). TG also has fewer associated problems arising from dialyzable serum factors because the affinity of HPRT for TG is ~300-fold higher than that for AG (van Diggelen et al., 1979). Among those mutants that retain residual HPRT activity, many have been characterized as having alterations that are indicative of structural gene changes. Alterations in kinetic properties of the enzyme have been shown for mutants isolated in L cells (Sharp et al., 1973), V79 hamster (Fenwick et al., 1977a), and CHO hamster cells (Chasin and Urlaub, 1976). Mutants possessing thermosensitive HPRTs (Sharp et al., 1973; Fenwick and Caskey, 1975; and Wahl et al., 1975) or HPRTs with altered isoelectric points (Chasin and Urlaub, 1976; Milman et al., 1976) further suggest that point mutations of the missense type have occurred. In experiments using highly specific antisera to the wild-type HPRT protein, mutants that lack enzyme activity have been analyzed to determine whether they possess protein that cross reacts with the antisera, that is, CRM-positive material. Among V79 hamster mutants selected with AG, only 3 of 35 $HPRT^-$ clones induced by alkylating agents contained CRM-positive material; in L cells the fraction of MNNG (*N*-methyl-*N′*-nitro-*N*-nitrosoguanidine) induced $HPRT^-$ mutants that were CRM positive was $\frac{43}{105}$, while that of the HeLa cell mutants induced with MNNG and EMS (ethyl methanesulfonate) was only $\frac{1}{24}$ (Milman et al., 1976). Thus in all three systems the majority of the $HPRT^-$ mutants are CRM^-, which is consistent with an alteration occurring elsewhere than in the structural gene. However, in one study (Fenwick et al., 1977b) in which revertants were obtained from a CRM^- $HPRT^-$ mutant the revertants differed in their antigenicity from the wild-type cells, as well as among themselves, indicating the original mutant contained a structural gene mutation.

In several instances peptide analysis of the HPRT protein in mutant cells provided further insights concerning the nature of the mutations.

Two mutants of V79 cells that have reduced subunit molecular weights (Fenwick et al., 1977a) show alterations in specific peaks in peptide profiles and appear to contain a defect in HPRT precursor processing in one case and a small deletion in the structural gene in the other (Caskey et al., 1979; Fenwick, 1979). A HeLa cell mutant with an altered isoelectric point on two-dimensional gels was shown to have an altered lysine-containing peptide (Milman et al., 1977). One of the L-cell mutants that is HPRT$^-$ has the properties of a nonsense mutation; it has an altered carboxyterminal peptide and a phenotype that is correctable by ochre-suppressor tRNA from yeast or bacteria (Capecchi et al., 1977). All these results together with those mentioned above indicate that many thioguanine-resistant (TG^r) and azaguanine-resistant (AG^r) clones carry mutations in the structural gene but leave open the possibility that other kinds of changes may also be involved. In the case of TG^r mutants of human fibroblasts induced by ionizing radiation, it appears that as many as 40% of the mutants are the result of cytologically visible chromosomal exchanges and deletions involving the X chromosome (Cox and Masson, 1978). In summary, the *hprt* locus appears to allow detection of a variety of point mutations, as well as changes at the chromosomal level for certain agents.

In CHO cells both the *aprt* and *tk* loci can be used to measure mutation induction in the appropriate heterozygous lines, although these markers are less well characterized than the *hprt* locus. Azaadenine resistance (AA^r) is generally associated with low or missing APRT activity (Jones and Sargent, 1974; Thompson et al., 1980a; Carver et al., 1980), and some induced mutants have an APRT$^-$ CRM$^+$ phenotype, while others are APRT$^-$ CRM$^-$ (Taylor et al., 1979). Variants at the *tk* locus detected using trifluorothymidine resistance in CHO cells occur at an inordinately high spontaneous mutation rate (Adair et al., 1980), that is approximately 2×10^{-4}, suggesting the involvement of additional genetic loci or possibly epigenetic events. In mouse lymphoma L5178Y cells, genetic changes leading to the TK$^-$ phenotype appear to involve both point mutations and chromosomal changes (Moore-Brown and Clive, 1979; Clive et al., 1979; Clive and Moore-Brown, 1979).

Mutations involving a more restricted class of genetic alterations can be measured using markers such as ouabain resistance (OUA^r) (Baker et al., 1974; Baker, 1979), which is a dominant mutation, and emetine resistance (Gupta and Siminovitch, 1977), a recessive mutation affecting the 40 S ribosomal subunit, which can be readily detected in CHO cells as a result of functional hemizygosity. Because mutations at these loci require a functional but altered protein having reduced drug binding, only DNA base changes compatible with these constraints can be observed. Thus these genetic markers are expected not to respond to certain mutagens, and this has been found to be true for OUA^r following treatment with ionizing radiations or the frameshift mutagen ICR191 (Baker, 1979). In addition, mutations to OUA^r are approximately tenfold lower in frequency than mutations to TG^r, which means that treated cultures must be proportionately greater in cell number to allow accurate determination of induced mutation frequency. For these reasons OUA^r is not a suitable marker for studying relationships between SCE and mutation induction in terms of broad spectrum of mutagens. Under certain conditions, however, OUA^r may be useful in helping to establish whether mutations induced at different loci correlate in the same manner with SCEs.

IV. METHODOLOGY FOR COMPARING SISTER CHROMATID EXCHANGE AND SINGLE-GENE MUTATION

A. *Rationale for the Choice of a Cell System and Loci for Mutation Assay*

For quantitatively comparing induced SCEs and mutations a cell system is desired that combines good growth properties, validated conditions for measuring single-gene mutations, and a karyotype favorable for cytogenetic analysis. The Chinese hamster ovary (CHO) line satisfies these criteria and has been used extensively for the assay of induced mutations (Hsie et al., 1978; O'Neill and Hsie, 1977; Thompson et al., 1980a) and of SCEs (Wolff and Perry, 1974; Perry and Evans, 1975; Stetka and Wolff, 1976; Abe and Sasaki, 1977), as well as of the two endpoints in combination (Carrano et al., 1978; Carrano et al., 1979). Clones of CHO cells are available in which enzyme-deficiency mutations can be scored for both TG^r and AA^r (Jones and Sargent, 1974; Thompson et al., 1980a, Carver et al., 1980), so that correlations between mutation and SCE can be determined for more than one genetic marker involving enzyme deficiency. Because CHO cells have been widely used for genetic and biochemical studies, a relatively large amount of information is available about the system. These cells offer the additional advantage that a variety of DNA repair-deficient mutant strains are now available (Thompson et al., 1980c; Busch et al., 1980; Adair and Clarkson, 1980), providing an important dimension in studying how mutation and SCE may be mechanistically related.

B. *Treatment and Assay of Cells*

For comparisons of induced mutations and SCEs, we recommend using cells from common treated cultures to minimize the influence of experimental errors. Because the accuracy of mutation frequency measurements has often been limited by the size of the cultures used (Irr and Snee, 1979), it is important to carry enough cells through each stage of the procedure that fluctuations associated with sampling remain small. For studies of TG^r and AA^r in CHO cells we have used 200 ml suspension cultures that contain a minimum of 5×10^6 cells at the time of treatment and after subculture dilutions, that is, a minimum concentration of 2.5×10^4 cells/ml. Thus, for a representative spontaneous mutant frequency of 1×10^{-5}, the average minimum number of mutants in the culture would be 50, resulting in a Poisson standard error of approximately 15% at each dilution. Mutagen treatments are performed on cultures in strict exponential growth to ensure a reproducible response. When treating in suspension, the cell concentration at the time of mutagen addition is usually in the range $1.0–1.5 \times 10^5$/ml; cells can also be treated in monolayer. A short exposure, such as 1 hour, is preferable, since it may be more reproducible than long treatments, and many mutagens have short half-lives. Both cultures and test chemicals should be protected from exposure to regular fluorescent light, which contains wavelengths that have been shown to be mutagenic (Bradley and Sharkey, 1977; Hsie et al., 1977; Jostes et al., 1977). In addition, incorporated BrdUrd sensitizes the DNA to photolysis by UV light (Kaufman and Davidson, 1978). Cultures should be handled in the dark or under gold fluorescent lights (Burki and Lam, 1978) at all steps until the cells are fixed or placed in a dark incubator.

When cultures are exposed to a DNA-damaging agent, cells in all phases of the cycle receive DNA damage, but these lesions must be fixed in order to be expressed as either mutations or SCEs. As is stated earlier, SCEs are produced only during S phase, and it is probable that most mutations also arise during S phase from replication errors, although little information is available on this point. SCEs arise in the second, as well as the first, S phase following exposure; this might also occur for mutations. Observation of SCEs is made after allowance for two complete S phases after exposure, while mutations cannot be assayed until the genotypic change has been converted into a biochemically altered phenotype. In practice, each culture treated with test agent is divided so that one portion is used for SCE analysis and the other for mutation. Sizable fluctuations have been shown to occur in the SCE frequency determined at different times after treatment (Carrano et al., 1979). Thus cells for SCE are further divided into replicate cultures, BrdUrd is added, and the cultures are incubated. The replicate cultures are harvested sequentially, usually at 4 hour intervals, to collect essentially all the second-division cells. The induced SCE frequency is calculated as a mean value weighted according to the frequency of second-division mitotic figures in each sample.

For measurement of induced mutations, cultures are maintained in exponential growth in suspension by diluting at 2-day intervals as needed. In our CHO system the expression of TG^r is completed by 5 days or less, regardless of the mutagen or level of survival (Thompson et al., 1980a); for AA^r 3 days appear to be adequate. Because the induced frequencies are stable—a necessary condition for obtaining quantitative results—plating for both drug markers can be done at day 5 or later.

C. *The Role of Bromodeoxyuridine*

Of potential concern in determining dose–response relationships for SCE are the synergistic effects that might arise between the test agent and the incorporated BrdUrd that is needed to produce visible SCEs. An example of such synergy has been described by Stetka and Carrano (1977) for the DNA-binding dye Hoechst 33258. These nonadditive effects can be minimized by maintaining a relatively low BrdUrd concentration while at the same time avoiding artifacts that can arise when the ratio of BrdUrd molecules to the number of cells is too low (Stetka and Carrano, 1977).

In addition to potential synergism, there exists the problem that BrdUrd itself can induce SCE (Kato, 1974b; Latt, 1974; Wolff and Perry, 1974; Schvartzman et al., 1979b) and mutation (Burki and Aebersold, 1978; Kaufman and Davidson, 1978). The important question has been raised as to whether the amount of BrdUrd incorporated into the DNA or the concentration in the medium is responsible for these increases. Stetka and Carrano (1977) demonstrated that at low to moderate concentrations of BrdUrd, the SCE frequency was dependent on the concentration of BrdUrd, the volume of medium, and the number of cells, that is, the SCE frequency depended on the number of BrdUrd molecules per base pair (BrdUrd/b.p.) over the range studied. Later, Mazrimas and Stetka (1978) showed that the SCE frequency was linearly related to the amount of BrdUrd incorporated into the DNA. For both of these studies Chinese hamster cells were exposed to BrdUrd at a concentration of approximately 3–50 molecules/b.p. pair. Recently it was proposed that the induction of SCE by BrdUrd in Chinese hamster cells is largely independent of the BrdUrd content of DNA (Davidson et al., 1980). These investigators showed that when the BrdUrd concentration ranged from 12 to 300 μM, almost a fourfold increase in SCEs occurred and the level of SCEs increased more than twofold at approximately constant levels of BrdUrd substitution in DNA (92.3–96.0%). Using the data from Figure 1 of their report, we calculated the BrdUrd concentration in terms of molecules per base pair for comparison to other studies. The results of the three experiments from the two laboratories are shown in Figure 4.1. There is only one point of overlap of the data from the two laboratories, namely, at 14 BrdUrd molecules/b.p., which represents about 75% DNA substitution

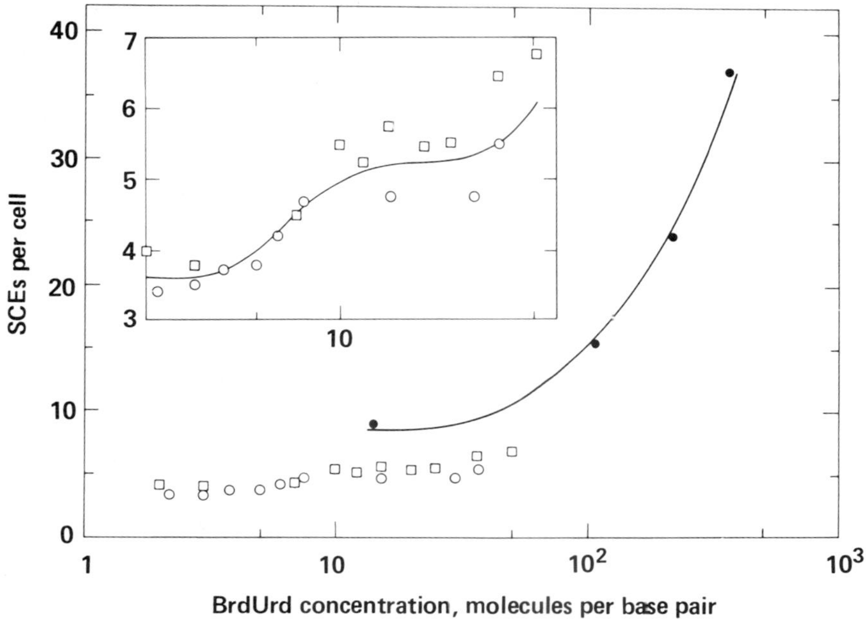

Fig. 4.1. SCE induction in CHO cells as a function of BrdUrd concentration. The abscissa is a log scale. Data redrawn from (○) Stetka and Carrano (1977); (□) Mazrimas and Stetka (1978); and (●) Davidson et al. (1980). The inset, redrawn on a larger scale, illustrates the increase in SCEs at moderate BrdUrd concentrations.

with BrdUrd. From about 3 to 10 molecules/b.p. there is about a 50% rise in SCE frequency (inset, Figure 4.1) that is followed by the much larger increase appearing at a concentration of about 50 molecules/b.p. when the DNA is more than 90% substituted. Thus there appear to be both BrdUrd-substitution-dependent and BrdUrd-concentration-dependent increases in SCE frequency. Since the majority of SCE studies are performed at BrdUrd/b.p. ratios under 50, the primary concern should be with the smaller increase due to DNA substitution. Under these conditions frequencies may change by 25–100%, depending on the cell systems. These variations in the baseline frequency could limit the ability to detect agents that are weak inducers of SCE. For these reasons we choose to use a BrdUrd/b.p. ratio between 10 and 20, a range over which the SCE frequency is relatively constant.

D. *Limitations on Data Interpretation*

Two other aspects underlying the measurement of SCE-mutation ratios that place interpretational constraints should be kept in mind. The two genetic endpoints are not necessarily being measured in the same population of cells whenever cell killing occurs. This is because only surviving cells can produce events scored as mutations, whereas SCEs are scorable in both viable cells and any second-division cells that are incapable of forming colonies. To the extent that SCEs are higher in nonviable progeny, bias is introduced into correlations with induced mutations. A second limiting aspect of such quantitative correlations is the fact that the measured mutation frequencies may reflect only a fraction of the actual mutational changes within the given locus. For enzyme-deficiency markers, only mutations that lead to inactivation of the protein are detected. Thus, unlike the SCE frequencies, the absolute numbers of mutations cannot be determined and are undoubtedly underestimated. Furthermore, different loci respond differently to a mutagen, depending on the time in S phase when the gene replicates and other factors.

V. EXPERIMENTAL EVIDENCE RELATING SISTER CHROMATID EXCHANGE AND SINGLE-GENE MUTATION

Over the past 3 years a considerable number of physical and chemical agents have been examined for the induction of both SCE and gene mutation. In this section we discuss first the investigations conducted in our own laboratory and then experiments performed by other investigators.

A. *Quantitative Dose-Response Studies in CHO Cells—Our Own Data*

As is indicated in the preceding section, our approach to the study of the relation between SCE and mutation has been to apply the test agent and then subdivide the cultures to perform each genetic assay. We have investigated the dose-response characteristics for seven chemicals: ethylnitrosourea (ENU), ethyl methanesulfonate (EMS), mitomycin C (MMC), decarbamoyl mitomycin C (DCMMC), 7-bromomethylbenz(*a*)anthracene (BMBA), ICR-191, and proflavine (PRO). In addition, we examined two physical agents: ultraviolet light and 4 MeV electrons. Since these agents produce different lesions in the DNA, they were chosen so that we might obtain information on the efficiency of each type of lesion for inducing the two genetic endpoints. EMS and ENU are both monofunctional ethylating agents, but ENU ethylates the O^6 position of guanine to a greater extent (Lawley, 1974; Singer, 1976). This has implications for mutagenesis, as the alkylation of the O^6 position in guanine has been suggested and demonstrated to be a premutagenic lesion (Loveless, 1969; Gerchman and Ludlum, 1973). In addition, its removal from DNA is defective in at least some cultured cells derived from patients with xeroderma pigmentosum (XP) (Goth-Goldstein, 1977; Bodell et al., 1979; Altamirano-Dimas et al., 1979). MMC, a DNA cross-linking agent, and DCMMC, a monofunctional alkylating agent, are structurally similar except that the latter lacks the carbamoyl moiety from the C-10 position and does not cross-link DNA (Otsuji and Murayama, 1972; Fujiwara et al., 1977; Carrano et al., 1979). A site of DNA alkylation for MMC (and presumably DCMMC) is suggested to be the O^6 position of guanine, and the ratio of mono- to bifunctional alkylations for MMC to be about 10:1 (Lipsett and Weissbach, 1965; Tomasz et al., 1974). BMBA, a polycyclic aromatic hydrocarbon, has been shown to preferentially and monofunctionally alkylate the extracyclic amino groups in the purines (Dipple et al., 1971). In Chinese hamster and HeLa cells a large portion of these adducts persist through DNA replication and are consistent with high viability (Dipple and Roberts, 1977). The two chemicals PRO and ICR-191 are both intercalating agents but ICR-191 also alkylates DNA (Creech et al., 1972). The actions of the last two agents, UV light and 4 MeV e^-, on DNA are well characterized; the former produces pyrimidine dimers as the major photoproduct, while the latter results primarily in direct DNA strand breakage. With this battery of agents we examined whether the dose-response kinetics for induced SCE and mutation would be quantitatively similar for a single agent, as well as among the agents, with the hope that some agents might produce unique lesions that would uncouple the genetic endpoints.

The dose-response curves for SCE and mutation at the *hprt* locus are shown in Figure 4.2. The toxicity dependence of both endpoints is given in Figure 4.3. Several general statements can be made concerning the data in these two figures. First, for all agents, and for both SCE and mutation, the dose responses when plotted on a linear scale (not shown) are linear at low to moderate doses, that is, in the absence of high toxicity. Second, with the exception of MMC and DCMMC, and to a lesser extent ICR-191, the SCE and mutation curves for any specific agent have similar slopes for their dose-response curves. Among agents, however, the slopes can vary widely. This behavior suggests that the induction kinetics for both SCE and mutation are lesion dependent and that for a single agent both genetic endpoints may possess some similar mechanism for their production. Third, both SCE and mutation can be induced at no detectable loss of survival, unlike chromosomal aberrations. Fourth, with the exception of MMC and DCMMC, both mutation and SCE are induced with similar kinetics over the same range of survival. At any given level of survival, however, ENU, EMS, MMC, and DCMMC are more potent inducers of SCE than the remaining five agents. Therefore, the lesions produced by

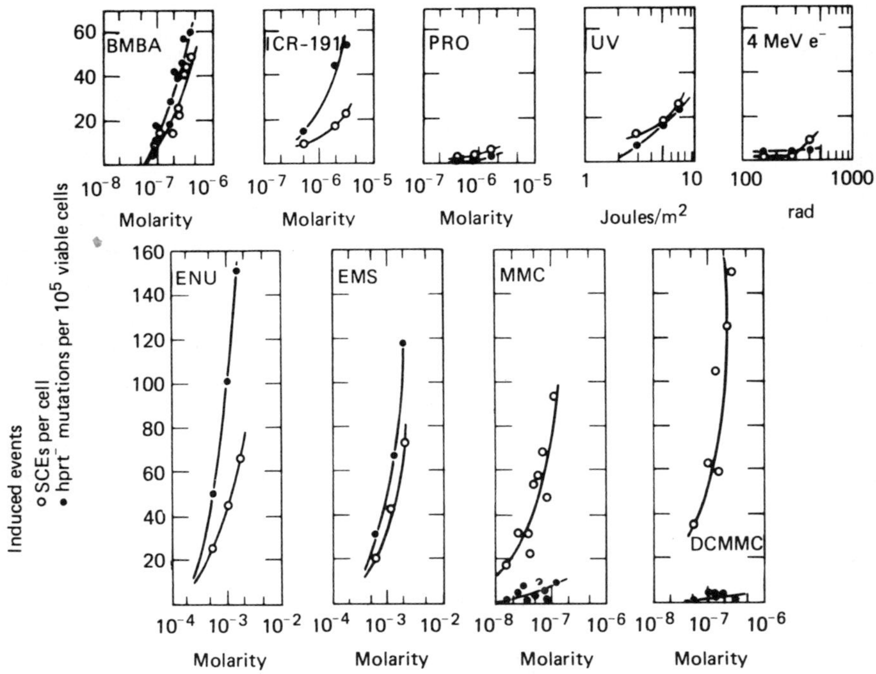

Fig. 4.2. Dose-response curves for the induction of (○) SCE and (●) *hprt*⁻ mutations in CHO cells. The abscissa is a log scale. Note the change in the ordinate scale between the top and bottom panels.

BMBA, ICR-191, PRO, UV, and 4 MeV e^- have a relatively greater probability of being processed as a lethal lesion than an SCE.

Because of the disparity between mutation and SCE for MMC and DCMMC, we postulated that the DNA cross-link is not likely the lesion responsible for either the induction of SCE or mutation but may represent a lethal lesion in these cells (Carrano et al., 1979). This has also been suggested by Wolff (1978) for SCE and by Zwelling et al. (1979) for mutation, although studies in yeast using psoralen and UV light may not support this view (Grant et al., 1979). The exact nature of the MMC-induced lesion responsible for the nonlethal endpoints is yet to be elucidated.

If the O^6-ethylation of guanine were important in producing SCE or mutation, one might expect a significant difference in response between EMS and ENU. In fact, at the same levels of survival ENU is only a slightly more potent inducer of SCE and mutation than EMS. Thus other ethylation products may play a role in these endpoints. Alkylation of the purine or pyrimidine ring is, in itself, not a requirement for SCE or mutation induction. This is demonstrated by both the chemical agent BMBA

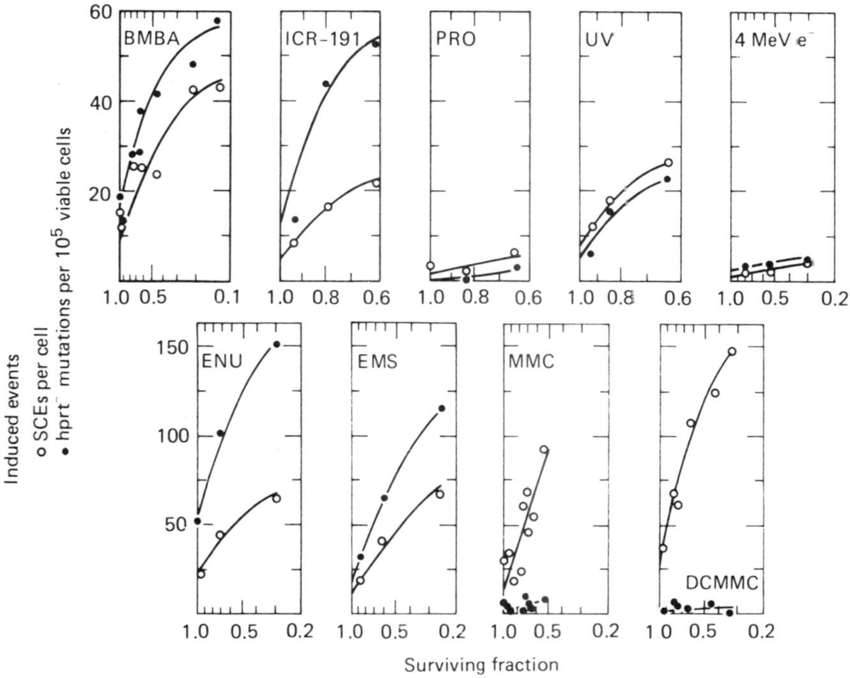

Fig. 4.3. The frequency of (○) SCE and (●) *hprt*⁻ mutations induced in CHO cells as a function of survival. The abscissa is a log scale. Note the change in the ordinate scale between the top and bottom panels.

and the physical agent UV light. BMBA is known to react with the extracyclic amino groups of the nucleic acid bases (Dipple et al., 1971), yielding products that are chemically stable under physiological conditions (Lieberman and Dipple, 1972). In both V79 Chinese hamster cells and HeLa cells the BMBA-modified adenine residues are excised two to four times more rapidly than the modified guanine residues, and sizable percentages of both of these modified bases are present even after 30 hours. It would be interesting to compare the persistence of SCE-forming lesions with the half-lives of the different purine adducts.

The thymine dimer, a major photoproduct of UV light and demonstrated mutagenic lesion in prokaryotes, is probably responsible for mutations also in animal cells (Freed et al., 1978). It is not surprising, therefore, that this lesion is suspected to play a role in the formation of SCE. The experiments to demonstrate this, however, have been somewhat contradictory. Kato (1974c) reported that UV-induced SCEs could be photoreactivated in *Potorous* cells, thus implicating the thymine dimer as the inducing lesion. This result was confirmed and extended by Ishizaki et al. (1980), who showed that the photoreactivation of SCEs was both a time- and temperature-dependent process. Photoreactivation of UV-induced SCEs was also demonstrated in chicken embryonic fibroblasts (Natarajan et al., 1980). In a study using CHO cells, the number of UV-endonuclease-sensitive sites induced by both far and near UV light was linearly related to the frequency of induced SCE, further suggesting that the pyrimidine dimer is the responsible lesion. The ratio of SCEs to dimers, however, was about 1:20,000 (Reynolds et al., 1979). On the other hand, it is difficult to reconcile these experiments with studies on excision repair defective cells. UV experiments carried out with Chinese hamster cells demonstrated a similar induction of SCEs in cells of apparently different excision repair capacities (Wolff et al., 1974). In xeroderma pigmentosum cells the frequency of UV-induced SCEs was not clearly related to the excision repair capability of different XP strains, although XP cells were always more sensitive than normal fibroblasts (de Weerd-Kastelein et al., 1977). Finally, Wolff (1978) has indicated that he was unable to produce photoreactivation of UV-induced SCEs either in cells of *Potorous* or chicken embryos. It is evident that further data are needed to assess critically the role of pyrimidine dimers in SCE formation.

The agents ICR-191, PRO, and 4 MeV e^- form an interesting group. PRO, an intercalating agent and frameshift mutagen in prokaryotes, induces relatively few SCEs and mutations in CHO cells compared to the other chemicals. The results are similar for 4 MeV e^-. When the induction of these endpoints is related to cell survival (Figure 4.3) it becomes obvious that the lesions induced by these agents are more likely converted to lethal lesions. Chromosomal aberrations, cytological indicators of genetic toxicity, are readily produced by 4 MeV e^- and, over the same concentration range, PRO induces a greater increase in chromosomal aberrations than SCEs (Kato, 1974d). Thus the direct DNA strand breakage that results in aberration formation is an inefficient inducer of both SCE or mutation. The alkylating activity of ICR-191 may account for the increased ability of this chemical, relative to PRO, to induce both SCE and mutation.

A most striking comparison of SCE and mutation induction for these nine agents is shown in Figure 4.4. In this graph of induced SCE versus induced mutation, the relation is linear, that is, for any increment in SCE frequency there is a fixed increment in mutation frequency. These results suggest that SCEs might be used as an estimator for predicting mutation frequencies. This approach has obvious implications to human population monitoring because there is not yet any simple measure of mutation frequency in an "exposed" individual, whereas SCEs can be readily measured in stimulated human lymphocytes. Further, since each agent appears to have a unique slope, the estimation of mutation frequency from

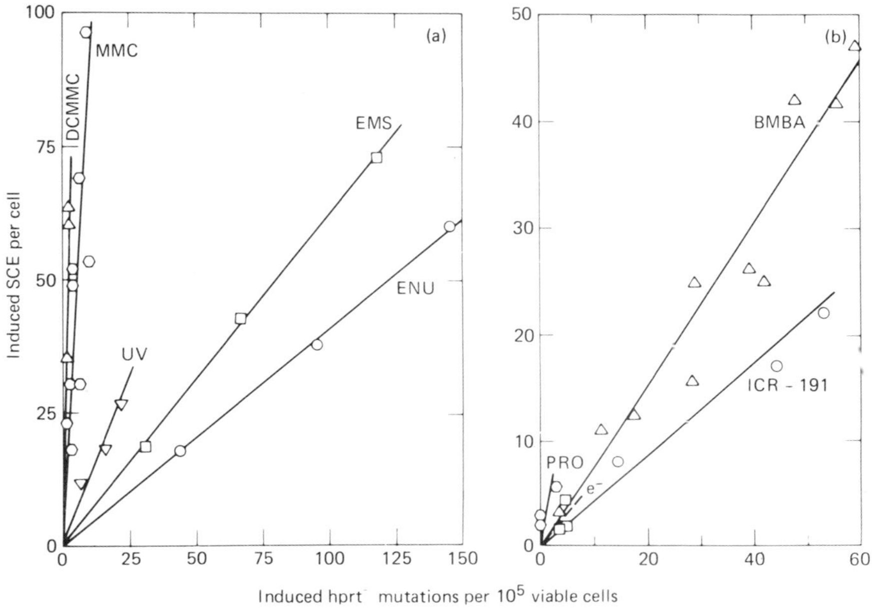

Fig. 4.4. The relation between induced SCEs per cell and induced *hprt*⁻ mutations per 10^5 viable cells. The agents tested are indicated adjacent to their curves. The ordinate and abscissa scales differ for panels *a* and *b*.

SCE frequency is only valid if a "calibration" curve exists, the identity of the agent is known, and cellular differences in DNA repair between hamster and human do not change the ratio. The differences in slope indicate to us that the induction of either an SCE or mutation depends on both the types of lesions produced and the processing of the lesions by the cell. In summary, MMC and DCMMC are potent inducers of SCE but relatively weak mutagens, while EMS, ENU, BMBA, and ICR-191 are more efficient mutagens relative to their SCE-inducing ability. PRO, UV, and 4 MeV e^- are intermediate in their ability to induce both endpoints.

It might be argued that the SCE–mutation correlation obtained for our agents might not hold if mutation were measured at a different locus. Thus we quantified both SCE and mutation at the *aprt* locus for five of the nine agents: MMC, DCMMC, ICR-191, UV, and 4 MeV e^-. The SCE–mutation relation was linear for this locus and, with the exception of the 4 MeV e^-, the slopes obtained using the *aprt* locus did not differ considerably compared to the *hprt* locus. The differences in slope are shown in Figure 4.5. In addition, the slopes obtained by using the two alternative selecting agents, AG and TG, for mutations at the *hprt* locus also are shown.

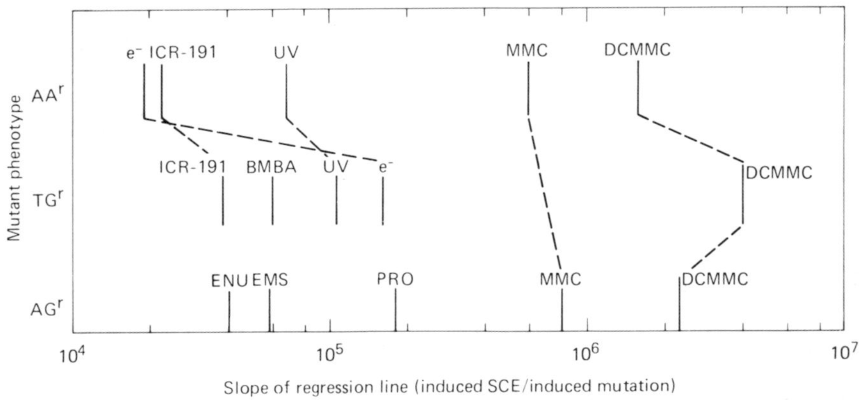

Fig. 4.5. A comparison of the slopes of the regression lines for induced SCE versus induced mutations per 10^5 viable cells. The slopes are calculated for AA^r as a measure of (*top row*) *aprt*⁻ mutants and (*middle row*) TG^r or (*bottom row*) AG^r as measures of *hprt*⁻ mutants.

We have examined the ability of the phorbol ester, TPA, to induce both SCE and single-gene mutation to TG^r, AA^r, and OUA^r. This study was stimulated by reports that TPA induced SCEs (Kinsella and Radman, 1978; Nagasawa and Little, 1979) but not mutations (Trosko et al., 1977). Our results failed to provide any evidence that TPA could significantly induce either SCE or mutation (Thompson et al., 1980b). The inability of TPA to induce SCEs was also reported by Loveday and Latt (1979).

B. SCE–Mutation Studies of Other Investigators

Several investigations have been conducted that extend our initial results in CHO cells to additional cell systems, agents, and mutational endpoints. We limit our discussion to those studies that measured SCE and mutation.

The ability of the antitumor drug *cis*-platinum (II) diamine dichloride to produce SCE, *hprt*⁻ mutations, and cytotoxicity was examined in V79 Chinese hamster cells (Turnbull et al., 1979). This chemical is a potent inducer of both SCE and mutation. If the frequency of induced SCE is compared to the induced TG^r mutant frequency for two doses, a ratio of approximately 15×10^5 SCE per mutation is obtained. This is the middle of the range obtained from our own studies with the other agents (Figure 4.5).

Connell (1979) examined SCE and *hprt*⁻ mutation induction in V79 Chinese hamster cells after treatment with BMBA and three ultimate metabolites of benzo(*a*)pyrene: diol-epoxide I; its geometric isomer, diol-epoxide II; and k-epoxide. If the data are related to survival, the relative mutagenic potency of the four chemicals decreases as follows:

diol-epoxide I > BMBA > diol-epoxide II ⩾ k-epoxide

The relative SCE inducing ability decreases as follows:

diol-epoxide I > BMBA > k-epoxide > diol-epoxide II

Although the author concludes that SCE may not be *directly* related to mutagenesis, the data for the four chemicals do suggest that each has a different SCE/mutation ratio, in analogy with our own results above. The diol-epoxide I was also found to be a more potent inducer of SCE and OUA^r mutations compared to diol-epoxide II in V79 cells (Hsu et al., 1979).

Sirianni and Huang (1980) examined the relation among SCE, AG^r, and OUA^r mutants induced in V79 cells in diffusion chambers in mice treated with cyclophosphamide, ifosfamide, and 1-(pyridyl-3)-3,3-dimethyltriazene (PyDT). All three chemicals produced dose-dependent increases in SCE and mutation at both loci. When the frequency of induced SCE was plotted against the induced mutation for PyDT, a positive linear correlation was observed for each locus. The slope of the SCE-OUA^r regression line was considerably greater than the SCE-AG^r regression, which presumably reflects the restricted nature of mutations that are detectable using OUA^r.

The ability of BrdUrd at high concentrations to induce mutation and SCE was thoroughly examined by Davidson et al. (1980) over a concentration range of 12–300 μM in CHO cells. SCEs and mutation were both increased above 90 μM in a dose-dependent manner. This increase could be demonstrated without any change in the level of substitution of deoxythymidine by BrdUrd in the DNA, suggesting a mechanism of both SCE and mutation induction that may be mediated metabolically, possibly by means of perturbation of deoxycytidine metabolism. It is interesting that both endpoints showed a similar response even when the presumptive mechanism did not directly involve DNA. Furthermore, the induction of both genetic events could be largely reversed by the addition of deoxycytidine.

Finally, the relationships among SCE, mutagenicity, and toxicity were examined for seven agents in V79 cells by Bradley et al. (1979). It is difficult to draw specific conclusions from their data, however, since the SCE and mutation studies were apparently independent and neither dose-response nor toxicity information was reported for the SCE study. For *cis*-Pt (II) diamine dichloride, an unfiltered sunlamp, and an unfiltered cool white fluorescent lamp Bradley et al. detected an increase in SCE and mutations at the *hprt* locus. Four of the seven agents [trans-Pt (II) diamine dichloride, a sunlamp plus a *p*-aminobenzoic acid filter, a cool white fluorescent lamp plus the same filter, and hydrogen peroxide] produced an increase in SCE but no detectable increase in *hprt*$^-$ mutations. The *p*-aminobenzoic acid filter blocks wavelengths below 345 nm and hence would indicate that SCEs are being produced at wavelengths greater than this, a result that is inconsistent with results reported for *Potorous* cells (Ishizaki et al., 1980). For the fluorescent lamp with filter and the hydrogen peroxide, 29 and 75% increases in SCE were detected, respectively. Without associated toxicity and dose-response information it is premature to classify these as positive SCE responses. In these cases, which demonstrated discordance between the SCE and mutation endpoints, the SCE was always the apparent positive endpoint while mutation was the apparent negative.

VI. GENETIC APPROACHES TO DETERMINING THE MECHANISM OF SCE

As is discussed above, one approach to gain insight into the events by which SCEs are produced is to relate the efficiency of SCE production with the kinds of DNA lesions known to result from various physical and chemical agents and also to correlate SCEs with single-gene mutations produced by these agents. There are major limitations in this approach because virtually all agents produce a variety of DNA lesions. Also, because the number of induced SCEs per cell is very small compared to the total number of DNA lesions of any particular type, there will remain considerable uncertainty as to a causal relationship.

The most powerful approach to discerning the mechanism(s) of SCE production is likely to be through the use of specific mutant cells that are altered in their spontaneous and/or induced SCE levels. Mutants should exist that are defective in the gene products that mediate the molecular events involved in SCE formation. For example, cells from Bloom's syndrome (BS) are noted for their highly elevated baseline SCEs (Chaganti et al., 1974). Although BS individuals are at a high risk for cancer and show chromosome instability in the form of homologous chromatid interchanges (Hecht and McCaw, 1977), it is not known whether the underlying defect involves DNA repair, DNA replication, or some other aspect of DNA metabolism.

Mutants that are defective in the repair of DNA damage will likely exhibit elevated SCE frequencies when the lack of repair increases the number of available (unexcised) lesions that can give rise to SCEs when the DNA is replicated. In this sense, reduced repair would have essentially a dose-modifying effect on SCE production. Examples of this class of mutants probably include certain strains of XP that show hypersensitivity for SCE induction in response to UV radiation (de Weerd-Kastelein et al., 1977; Cheng et al., 1978) or alkylating agents (Wolff et al., 1977). Studies in our laboratory using UV-sensitive DNA repair-deficient mutants of CHO cells (Thompson et al., 1980c) indicate an analogous behavior. In these mutants the hypersensitivity for SCE induction closely parallels their increased sensitivity to cell killing and to mutation induction at the *hprt* and *aprt* loci (L. H. Thompson and A. V. Carrano, unpublished work). An understanding of the specific types of damage whose repair is deficient may lead to some understanding of which classes of lesions cause SCEs.

In addition to dose-modifying effects, DNA repair processes may also contribute to SCE formation in a more direct manner. The enzymatic steps presumed to be involved in the formation of SCEs could be an integral part of one or more repair pathways. If this were the case, a mutant deficient in such a repair step would be expected to have lesser induction of SCEs than the repair-proficient cells. To our knowledge none of the human repair-deficient mutants have exhibited this kind of behavior with regard to SCE induction. However, in experiments in our laboratory using the direct acting carcinogen BMBA, a repair-deficient mutant of CHO cells has shown substantially fewer induced SCEs than the parental cells (L. H. Thompson and A. V. Carrano, unpublished work). It is remarkable that the induction of TG^r mutations and OUA^r mutations in this mutant strain is also substantially reduced, which suggests that the repair process in wild-type CHO cells that acts on BMBA lesions somehow contributes to both mutations and SCEs. We expect that an extension of these studies with a variety of CHO strains that are hypersensitive to mutagens will enhance our understanding of the origin of SCEs and their relation to specific gene mutations.

VII. CONCLUDING REMARKS

Simply because SCE and single-gene mutation are linearly correlated for some agents, all agents do not necessarily demonstrate such a relation, nor are the lesions that produce an SCE necessarily identical to those that produce mutation. Such correlations may only imply similarity in either the inducing lesions or the cellular processing of the lesion to the genetic endpoint, or both. Thus not every agent that induces SCE may be mutagenic and vice versa. For reasons we discuss earlier, any attempts to demonstrate quantitative correlations between the two endpoints should meet three criteria: (1) the treatment of the cells should be common to both the SCE and mutation assays; (2) cellular toxicity should be used as a measure of effective dose; and (3) both endpoints should be measured at moderate to high survivals (≥50%) to eliminate potential uncoupling of SCE and mutation due to nongenetic toxicity and/or measurement of SCE in cells that would not be viable for the mutant selection. Only through more precise correlations that seek to identify specific lesions potentially responsible for either or both endpoints can we hope to understand the relation of SCE to mutation.

References

Abe, S. & Sasaki, M. (1977) Chromosome aberrations and sister chromatid exchanges in Chinese hamster cells exposed to various chemicals. *J. Natl. Cancer Inst.* **58,** 1635–1641.

Adair, G. M. & Clarkson, J. M. (1980) DNA repair and mutagenesis in UV light-sensitive mutants of CHO-AT3-2 cells. Abstracts 11th Annual Meeting of the Environmental Mutagen Society, p. 87.

Adair, G. M., Carver, J. H. & Wandres, D. L. (1980) Mutagenicity testing in mammalian cells. I. Derivation of a Chinese hamster ovary cell line heterozygous for the adenine phosphoribosyltransferase and thymidine kinase loci. *Mutat. Res.,* **72,** 187–205.

Altamirano-Dimas, M., Sklar, R., & Strauss, B. (1979) Selectivity of the excision of alkylation products in a xeroderma pigmentosum-derived lymphoblastoid line. *Mutat. Res.* **60,** 197–206.

Baker, R. M. (1979) Nature and use of ouabain-resistant mutants. In *Banbury Report 2, Mammalian Cell Mutagenesis: The Maturation of Test Systems* (A. W. Hsie, J. P. O'Neill, & V. K. McElheny, Eds.), Cold Spring Harbor Laboratory, pp. 237–244.

Baker, R. M., Brunette, D. M., Mankovitz, R., Thompson, L. H., Whitmore, G. F., Siminovitch, L., & Till, J. E. (1974) Ouabain-resistant mutants of mouse and hamster cells in culture. *Cell* **1,** 9–21.

Banerjee, A. & Benedict, W. F. (1979) Production of sister chromatid exchanges by various cancer chemotherapeutic agents. *Cancer Res.* **39,** 797–799.

Benedict, W. F., Banerjee, A., & Venkatesan, N. (1978) Cyclophosphamide-induced oncogenic transformation, chromosomal breakage, and sister chromatid exchange following microsomal activation. *Cancer Res.* **38,** 2922–2924.

Bodell, W. J., Singer, B., Thomas, G. H., & Cleaver, J. E. (1979) Evidence for removal at different rates of *O*-ethyl pyrimidines and ethylphosphotriesters in two human fibroblast cell lines. *Nucleic Acids Res.* **6,** 2819–2829.

Bostock, C. J. & Christie, S. (1976) Analysis of the frequency of sister chromatid exchange in different regions of chromosomes of the kangaroo rat (*Dipodomys ordii*). *Chromosoma (Berl.)* **56,** 275–287.

Bradley, M. O. &. Sharkey, N. A. (1977) Mutagenicity and toxicity of visible fluorescent light to cultured mammalian cells. *Nature (Lond.)* **266,** 724–726.

Bradley, M. O., Hsu, I. C., & Harris, C. C. (1979) Relationships between sister chromatid exchange and mutagenicity, toxicity and DNA damage. *Nature (Lond.)* **282,** 318–320.

Burki, H. J. & Aebersold, P. M. (1978) Bromodeoxyuridine-induced mutations in synchronous Chinese hamster cells: Temporal induction of 6-thioguanine and ouabain resistance during DNA replication. *Genetics* **90**, 311–321.

Burki, H. J. & Lam, C. K. (1978) Comparison of the lethal and mutagenic effects of gold and white fluorescent lights on cultured mammalian cells. *Mutat. Res.* **54**, 373–377.

Busch, D. B., Cleaver, J. E., & Glaser, D. A. (1980) Large-scale isolation of UV-sensitive clones of CHO cells. *Somat. Cell Genet.* **6**, 407–418.

Capecchi, M. R., Vonder Haar, R. A., Capecchi, N. E., & Sveda, M. M. (1977) The isolation of a suppressible nonsense mutant in mammalian cells. *Cell* **12**, 371–381.

Carrano, A. V. (1973) Chromosome aberrations and radiation-induced cell death II. Predicated and observed cell survival. *Mutat. Res.* **17**, 355–366.

Carrano, A. V. & Wolff, S. (1975) Distribution of sister chromatid exchanges in the euchromatin and heterochromatin of the Indian muntjac. *Chromosoma (Berl.)* **53**, 361–369.

Carrano, A. V., Thompson, L. H., Lindl, P. A., & Minkler, J. L. (1978) Sister chromatid exchanges as an indicator of mutagenesis. *Nature (Lond.)* **271**, 551–553.

Carrano, A. V., Thompson, L. H., Stetka, D. G., Minkler, J. L., Mazrimas, J. R., & Fong, S. (1979) DNA crosslinking, sister chromatid exchange and specific-locus mutations. *Mutat. Res.* **63**, 175–188.

Carver, J. H., Adair, G. M., & Wandres, D. L. (1980) Mutagenicity testing in mammalian cells. II. Validation of multiple drug-resistance markers having practical application for screening potential mutagens. *Mutat. Res.*, **72**, 207–230.

Caskey, C. T. & Kruh, G. D. (1979) The HPRT locus. *Cell* **16**, 1–9.

Caskey, C. T., Fenwick, R. G., Jr., & Kruh, G. D. (1979), HGPRT mutants in Chinese hamster V79 cells. In *Banbury Report 2, Mammalian Cell Mutagenesis: The Maturation of Test Systems* (A. W. Hsie, J. P. O'Neill, & V. K. McElheny, Eds.), Cold Spring Harbor Laboratory, pp. 23–33.

Chaganti, R. S. K., Schonberg, S., & German, J. (1974) A manyfold increase in sister chromatid exchange in Bloom's syndrome lymphocytes. *Proc. Natl. Acad. Sci. U.S.* **71**, 4508–4512.

Chasin, L. A. & Urlaub, G. (1976) Mutant alleles for hypoxanthine phosphoribosyltransferase: codominant expression, complementation, and segregation in hybrid Chinese hamster cells. *Somat. Cell Genet.* **2**, 453–467.

Cheng, W.-S., Tarone, R. E., Andrews, A. D., Whang-Peng, J. S., & Robbins, J. H. (1978) Ultraviolet light-induced sister chromatid exchanges in xeroderma pigmentosum and in Cockayne's syndrome lymphocyte cell lines. *Cancer Res.* **38**, 1601–1609.

Chu, E. H. Y. & Malling, H. V. (1968) Mammalian cell genetics II. Chemical induction of specific locus mutations in Chinese hamster cells *in vitro*. *Proc. Natl. Acad. Sci. U.S.* **61**, 1306–1312.

Clive, D. & Moore-Brown, M. M. (1979) The L5178Y/TK$^{+/-}$ mutagen assay system: mutant analysis. In *Banbury Report 2, Mammalian Cell Mutagenesis: The Maturation of Test Systems* (A. W. Hsie, J. P. O'Neill, & V. K. McElheny, Eds.), Cold Spring Harbor Laboratory, pp. 421–429.

Clive, D., Johnson, K. O., Spector, J. F. S., Batson, A. G., & Brown, M. M. M. (1979) Validation and characterization of the L5178Y/TK$^{+/-}$ mouse lymphoma mutagen assay system. *Mutat. Res.* **59**, 61–108.

Connell, J. R. (1979) The relationship between sister chromatid exchange, chromosome aberration and gene mutation induction by several reactive polycyclic hydrocarbon metabolites in cultured mammalian cells. *Int. J. Cancer.* **24,** 485–489.

Cox, R. & Masson, W. K. (1978) Do radiation-induced thioguanaine-resistant mutants of cultured mammalian cells arise by HGPRT gene mutation or X-chromosome rearrangement? *Nature (Lond.)* **276,** 629–630.

Creech, H. J., Preston, R. K., Peck, R. M., O'Connell, A. P., & Ames, B. N. (1972) Antitumor and mutagenic properties of a variety of heterocyclic nitrogen and sulfur mustards. *J. Med. Chem.* **15,** 739–746.

Davidson, R. L., Kaufman, E. R., Dougherty, C. P., Ouellette, A. M., DiFolco, C. M., & Latt, S. A. (1980) Induction of sister chromatid exchanges by BUdR is largely independent of the BUdR content of DNA. *Nature (Lond.)* **284,** 74–76.

DeMars, R. (1974) Resistance of cultured human fibroblasts and other cells to purine and pyrimidine analogues in relation to mutagenesis detection. *Mutat. Res.* **24,** 335–364.

de Weerd-Kastelein, E. A., Keijzer, W., Rainaldi, G. & Bootsma, D. (1977) Induction of sister chromatid exchanges in xeroderma pigmentosum cells after exposure to ultraviolet light. *Mutat. Res.* **45,** 253–261.

Dewey, W. C., Furman, S. C., & Miller, H. H. (1970) Comparison of lethality and chromosomal damage induced by x-rays in synchronized Chinese hamster cells *in vitro*. *Radiat. Res.* **43,** 561–581.

Dipple, A. & Roberts, J. J. (1977) Excision of 7-bromomethylbenz(*a*)anthracene DNA adducts in replicating mammalian cells. *Biochem.* **16,** 1499–1503.

Dipple, A., Brookes, P., Mackintosh, D. S., & Rayman, M. P. (1971) Reaction of 7-bromomethylbenz(*a*)anthracene with nucleic acids, polynucleotides, and nucleosides. *Biochem.* **10,** 4323–4330.

Fenwick, R. G., Jr. (1979) A precursor of mammalian hypoxanthine-guanine phosphoribosyltransferase (abstract). *J. Cell. Biol.* **83,** No. 2, Part 2, p. 443a.

Fenwick, R. G. & Caskey, C. T. (1975) Mutant Chinese hamster cells with a thermosensitive hypoxanthine-guanine phosphoribosyltransferase. *Cell* **5,** 115–122.

Fenwick, R. G., Jr., Sawyer, T. H., Kruh, G. D., Astrin, K. H., & Caskey, C. T. (1977a) Forward and reverse mutations affecting the kinetics and apparent molecular weight of mammalian HGPRT. *Cell* **12,** 383–391.

Fenwick, R. G., Jr., Wasmuth, J. J., & Caskey, C. T. (1977b) Mutations affecting the antigenic properties of hypoxanthine-guanine phosphoribosyltransferase in cultured Chinese hamster cells. *Somat. Cell Genet.* **2,** 207–216.

Freed, J. J., Angelosanto, F. A., & Hoess, R. H. (1978) Photoreversal as a tool for study of mutagenesis in haploid frog cell cultures. *J. Cell Biol.* **79,** 390a.

Fujiwara, F., Tatsumi, M., & Sasaki, M. S. (1977) Crosslink repair in human cells and its possible defect in Fanconi's anemia cells. *J. Mol. Biol.* **113,** 635–649.

Gatti, M., Santini, G., Pimpinelli, S., & Olivieri, G. (1979) Lack of spontaneous sister chromatid exchanges in somatic cells of *Drosophila melanogaster*. *Genetics* **91,** 255–274.

Gatti, M., Pimpinelli, S., & Baker, B. S. (1980) Relationship among chromatid interchanges, sister chromatid exchanges, and meiotic recombination in *Drosophila melanogaster*. *Proc. Natl. Acad. Sci. U.S.* **77,** 1575–1579.

Gerchman, L. L. & Ludlum, D. B. (1973) Properties of O^6-methylguanine in templates for RNA polymerase. *Biochim. Biophys. Acta.* **308,** 310–316.

Goth-Goldstein, R. (1977) Repair of DNA damaged by alkylating carcinogens is defective in xeroderma pigmentosum-derived fibroblasts. *Nature (Lond.)* **267,** 81–82.

Grant, E. L., von Borstel, R. C., & Ashwood-Smith, M. J. (1979) Mutagenicity of cross-links and monoadducts of furocoumarins (psoralen and angelicin) induced by 360-nm radiation in excision-repair-defective and radiation-insensitive strains of *Saccharomyces cerevisiae. Environ. Mutat.* **1,** 55–63.

Gupta, R. S. & Siminovitch, L. (1977) The molecular basis of emetine resistance in Chinese hamster ovary cells: alteration in the 40S ribosomal subunit. *Cell* **10,** 61–66.

Hecht, F. & McCaw, B. K. (1977) Chromosome instability syndromes. In *Genetics of Human Cancer* (J. J. Mulvihill, R. W. Miller, & J. F. Fraumeni, Jr., Eds.), Raven Press, New York, pp. 105–123.

Hoo, J. J. & Parslow, M. I. (1979) Relation between the SCE points and the DNA replication bands. *Chromosoma (Berl.)* **73:** 67–74.

Hsie, A. W., Li, A. P., & Machanoff, R. (1977) A fluence response study of lethality and mutagenicity of white, black, and blue fluorescent light, sunlamp, and sunlight irradiation in Chinese hamster cells. *Mutat. Res.* **45,** 333–342.

Hsie, A. W., O'Neill, J. P., Couch, D. B., SanSebastian, J. R., Brimer, P. A., Machanoff, R. Fuscoe, J. C., Riddle, J. C., Li, A. P., Forbes, N. L., & Hsie, M. H. (1978) Quantitative analysis of radiation- and chemical-induced lethality and mutagenesis in Chinese hamster ovary cells. *Radiat. Res.* **76,** 471–492.

Hsie, A. W., O'Neill, J. P., & McElheny, V. K., Eds. (1979) *Banbury Report 2, Mammalian Cell Mutagenesis: The Maturation of Test Systems.* Cold Spring Harbor Laboratory.

Hsu, I. C., Bowden, G. T., & Harris, C. C. (1979) A comparison of cytotoxicity, ouabain-resistant mutation, sister-chromatid exchanges, and nascent DNA synthesis in Chinese hamster cells treated with dihydrodiol epoxide derivatives of benzo(*a*)pyrene. *Mutat. Res.* **63,** 351–359.

Irr, J. D. & Snee, R. D. (1979) Statistical evaluation of mutagenicity in the CHO/HGPRT system. *Banbury Report 2, Mammalian Cell Mutagenesis: The Maturation of Test Systems* (A. W. Hsie, J. P. O'Neill, & V. K. McElheny, Eds.), Cold Spring Harbor Laboratory, pp. 263–273.

Ishizaki, K., Nikaido, O., & Takebe, H. (1980) Photoreactivation of ultraviolet light-induced sister chromatid exchanges in Potorous cells. *Photochem. Photobiol.* **31,** 277–279.

Jones, G. E. & Sargent, P. A. (1974) Mutants of cultured Chinese hamster cells deficient in adenine phosphoribosyl transferase. *Cell* **2,** 43–54.

Jostes, R. F., Dewey, W. C., & Hopwood, L. E. (1977) Mutagenesis by fluorescent light in mammalian cell culture. *Mutat. Res.* **42,** 139–144.

Kato, H. (1974a) Possible role of DNA synthesis in formation of sister chromatid exchanges. *Nature (Lond.)* **252,** 739–741.

Kato, H. (1974b) Spontaneous sister chromatid exchanges detected by a BrdU-labeling method. *Nature (Lond.)* **251,** 70–72.

Kato, H. (1974c) Photoreactivation of sister chromatid exchanges induced by ultraviolet irradiation. *Nature (Lond.)* **249,** 552–553.

Kato, H. (1974d) Induction of sister chromatid exchanges by chemical mutagens and its possible relevance to DNA repair. *Exp. Cell Res.* **85,** 239–247.

Kaufman, E. R. & Davidson, R. L. (1978) Bromodeoxyuridine mutagenesis in mammalian cells: mutagenesis is independent of the amount of bromouracil in DNA. *Proc. Natl. Acad. Sci. U.S.* **75,** 4982–4986.

Kihlman, B. A., Natarajan, A. T., & Anderson, H. C. (1978) Use of 5-bromodeoxyuridine-labelling technique for exploring mechanisms involved in the formation of chromosomal aberrations I. G_2 experiments with root tips of *Vicia faba*. *Mutat. Res.* **52,** 181–198.

Kinsella, A. R. & Radman, M. (1978) Tumor promotor induces sister chromatid exchanges: relevance to mechanisms of carcinogenesis. *Proc. Natl. Acad. Sci. U.S.* **75,** 6149–6153.

Latt, S. A. (1974) Sister chromatid exchanges, indices of human chromosome damage and repair: detection by fluorescence and induction by mitomycin C. *Proc. Natl. Acad. Sci. U.S.* **71,** 3162–3166.

Lawley, P. D. (1974) Some chemical aspects of dose–response relationships in alkylation mutagenesis. *Mutat. Res.* **23,** 283–295.

Lieberman, M. W. & Dipple, A. (1972) Removal of bound carcinogen during DNA repair in nondividing human lymphocytes. *Cancer Res.* **32,** 1855–1860.

Lipsett, M. N. & Weissbach, A. (1965) The site of alkylation of nucleic acids by mitomycin. *Biochem.* **4,** 206–211.

Littlefield, L. G., Colyer, S. P., Joiner, E. E., & DuFrain, R. J. (1979) Sister chromatid exchanges in human lymphocytes exposed to ionizing radiation in G_0. *Radiat. Res.* **78,** 514–521.

Loveday, K. S. & Latt, S. A. (1979) The effect of a tumor promoter, 12-*O*-tetradecanoylphorbol-13-acetate (TPA), on sister chromatid exchange formation in cultured Chinese hamster cells. *Mutat. Res.* **67,** 343–348.

Loveless, A. (1969) Possible relevance of O-6 alkylation of deoxyguanosine to mutagenicity of nitrosamines and nitrosamides. *Nature (Lond.)* **223,** 206–208.

Luria, S. E. & Delbruck, M. (1943) Mutations of bacteria from virus sensitivity to virus resistance. *Genetics* **28,** 491–511.

Mazrimas, J. A. & Stetka, D. G. (1978) Direct evidence for the role of incorporated BrdUrd in the induction of sister chromatid exchanges. *Exp. Cell Res.* **117,** 23–30.

Milman, G., Lee, E., Ghangas, G. S., McLaughlin, J. R., & George, M., Jr. (1976) Analysis of HeLa cell hypoxanthine phosphoribosyltransferase mutants and revertants by two-dimensional polyacrylamide gel electrophoresis: evidence for silent gene activation. *Proc. Natl. Acad. Sci. U.S.* **73,** 4589–4593.

Milman, G., Krauss, S. W., & Oslen, A. S. (1977) Tryptic peptide analysis of normal and mutant forms of hypoxanthine phosphoribosyltransferase from HeLa cells. *Proc. Natl. Acad. Sci. U.S.* **74,** 926–930.

Moore-Brown, M. M. & Clive, D. (1979) The L5178Y/TK$^{+/-}$ mutagen assay system *in situ* results. *Banbury Report 2, Mammalian Cell Mutagenesis: The Maturation of Test Systems* (A. W. Hsie, J. P. O'Neill, & V. K. McElheny, Eds.), Cold Spring Harbor Laboratory, pp. 71–81.

Nagasawa, H. & Little, J. B. (1979) Effect of tumor promoters, protease inhibitors, and repair processes on x-ray-induced sister chromatid exchanges in mouse cells. *Proc. Natl. Acad. Sci. U.S.* **76,** 1943–1947.

Natarajan, A. T., Tates, A. D., van Buul, P. P. W., Meijers, M., & de Vogel, N. (1976) Cytogenetic effects of mutagens/carcinogens after activation in a microsomal system *in vitro* I. Induction of chromosome aberrations and sister chromatid exchanges by diethylnitrosamine (DEN) and dimethylnitrosamine (DMN) in CHO cells in the presence of rat-liver microsomes. *Mutat. Res.* **37,** 83–90.

Natarajan, A. T., van Zeeland, A. A., Verdegaal-Immerzeel, E. A. M., & Filon, A. R. (1980) Studies on the influence of photoreactivation on the frequencies of UV-induced chromosomal aberrations, sister-chromatid exchanges and pyrimidine dimers in chicken embryonic fibroblasts. *Mutat. Res.* **69,** 307–317.

O'Neill, J. P. & Hsie, A. W. (1977) Chemical mutagenesis in mammalian cells can be quantified. *Nature (Lond.)* **269,** 815–817.

Otsuji, N. & Murayama, I. (1972) Deoxyribonucleic acid damage by monofunctional mitomycins and its repair in *Escherichia coli. J. Bacteriol.* **109,** 475–483.

Painter, R. B. (1980) A replication model for sister chromatid exchange. *Mutat. Res.* **70,** 337–341.

Perry, P. & Evans, H. J. (1975) Cytological detection of mutagen–carcinogen exposure by sister chromatid exchange. *Nature (Lond.)* **258,** 121–125.

Reynolds, R. J., Natarajan, A. T., & Lohman, P. H. (1979) *Micrococcus luteus* UV-endonuclease-sensitive sites and sister chromatid exchanges in Chinese hamster ovary cells. *Mutat. Res.* **64,** 353–356.

Schvartzman, J. B., Postigo, R., & Gutierrez, C. (1979a) Analysis of visible light-induced sister chromatid exchanges in 5-bromodeoxyuridine-substituted chromosomes. *Chromosoma (Berl.)* **74,** 317–328.

Schvartzman, J. B., Cortes, F., Gonzalez-Fernandez, A., Gutierrez, C., & Lopez-Saez, J. F. (1979b) On the nature of sister chromatid exchanges in 5-bromodeoxyuridine-substituted chromosomes. *Genetics* **92,** 1251–1264.

Sharp, J. D., Capecchi, N. E., & Capecchi, M. R. (1973) Altered enzymes in drug-resistant variants of mammalian tissue culture cells. *Proc. Natl. Acad. Sci. U.S.* **70,** 3145–3149.

Singer, B. (1976) All oxygens in nucleic acids react with carcinogenic ethylating agents. *Nature (Lond.)* **264,** 333–339.

Sirianni, S. R. & Huang, C. C. (1980) Comparison of induction of sister chromatid exchange, 8-azaguanine- and ouabain-resistant mutants by cyclophosphamide, ifosfamide and 1-(pyridyl-3)-3,3-dimethyltriazene in Chinese hamster cells cultured in diffusion chambers in mice. *Carcinogenesis,* **1,** 353–355.

Skaper, S. D., Spector, E. B., & Seegmiller, J. E. (1977) Reversion in expression of hypoxanthine-guanine phosphoribosyltransferase in 6-thioguanine resistant neuroblastoma: evidence for reduced enzyme levels associated with unaltered catalytic activity. *J. Cell. Physiol.* **92,** 275–284.

Stetka, D. G. & Carrano, A. V. (1977) The interaction of Hoechst 33258 and BrdU substituted DNA in the formation of sister chromatid exchanges. *Chromosoma (Berl.)* **63,** 21–31.

Stetka, D. G. & Wolff, S. (1976) Sister chromatid exchange as an assay for genetic damage induced by mutagen-carcinogens. Part II. *In vitro* test for compounds requiring metabolic activation. *Mutat. Res.* **41,** 343–350.

Taylor, M. W., Hershey, H. V., & Simon, A. E. (1979) An analysis of mutation at the adenine phosphoribosyltransferase locus. *Banbury Report 2 Mammalian Cell Mutagenesis: The Maturation of Test Systems* (A. W. Hsie, J. P. O'Neill, & V. K. McElheny, Eds.), Cold Spring Harbor Laboratory, pp. 211–221.

Thompson, L. H. & Baker, R. M. (1973) Isolation of mutants of cultured mammalian cells. In *Methods in Cell Biology*, Vol. VI (D. M. Prescott, Ed.), Academic Press, New York, pp. 209–281.

Thompson, L. H., Fong, S. & Brookman, K. (1980a) Validation of conditions for efficient detection of HPRT and APRT mutations in suspension-cultured Chinese hamster ovary cells. *Mutat. Res.* **74,** 21–36.

Thompson, L. H., Baker, R. M., Carrano, A. V., & Brookman, K. W. (1980b) Failure of the phorbol ester 12-*O*-tetradecanoylphorbol-13-acetate to enhance sister chromatid exchange, mitotic segregation, or expression of mutations in Chinese hamster cells. *Cancer Res.* **40,** 3245–3251.

Thompson, L. H., Rubin, J. S., Cleaver, J. E., Whitmore, G. F., & Brookman, K. (1980c) A screening method for isolating DNA repair-deficient mutants of CHO cells. *Somat. Cell Genet.* **6,** 391–405.

Tomasz, M., Mercado, C. M., Olson, J., & Chatterjee, N. (1974) The mode of interaction of mitomycin C with deoxyribonucleic acid and other polynucleotides *in vitro. Biochem.* **13,** 4878–4887.

Trosko, J. E., Chang, C. C., Yotti, L. P., & Chu, E. H. Y. (1977) Effect of phorbol myristate acetate on the recovery of spontaneous and ultraviolet light-induced 6-thioguanine and ouabain-resistant Chinese hamster cells, *Cancer Res.* **37,** 188–193.

Turnbull, D., Popescu, N. C., DiPaolo, J. A., & Myhr, B. C. (1979) *cis*-Platinum (II) diamine dichloride causes mutation, transformation and sister chromatid exchanges in cultured mammalian cells. *Mutat. Res.* **66,** 267–275.

van Diggelen, O. P., Donahue, T. F., & Skin, S.-L. (1979) Basis of differential cellular sensitivity to 8-azaguanine and 6-thioguanine. *J. Cell. Physiol.* **98,** 59–72.

Wahl, G. M., Hughes, S. H., & Capecchi, M. R. (1975) Immunological characterization of hypoxanthine-guanine phosphoribosyltransferase mutants of mouse L cells: evidence for mutations at different loci in the HGPRT gene. *J. Cell. Physiol.* **85,** 307–320.

Wolff, S. (1978) Relation between DNA repair, chromosome aberrations, and sister chromatid exchanges. In *DNA Repair Mechanisms* (P. C. Hanawalt, E. C. Friedberg, & C. F. Fox, Eds.), Academic Press, New York, pp. 751–760.

Wolff, S. & Perry, P. (1974) Differential Giemsa staining of sister chromatids and the study of sister chromatid exchanges without autoradiography. *Chromosoma (Berl.)* **48,** 341–353.

Wolff, S., Bodycote, J., & Painter, R. B. (1974) Sister chromatid exchanges induced in Chinese hamster cells by UV irradiation of different stages of the cell cycle: the necessity for cells to pass through S. *Mutat. Res.* **25,** 73–81.

Wolff, S., Rodin, B., & Cleaver, J. E. (1977) Sister chromatid exchanges induced by mutagenic carcinogens in normal and xeroderma pigmentosum cells. *Nature (Lond.)* **265,** 347–349.

Zwelling, L. A., Bradley, M. O., Sharkey, N. A., Anderson, T., & Kohn, K. W. (1979) Mutagenicity, cytotoxicity, and DNA crosslinking in V79 Chinese hamster cells treated with *cis*- and *trans*-Pt(II) diaminedichloride. *Mutat. Res.* **67,** 271–280.

Chapter 5

Induction of Sister Chromatid Exchanges by Chemical Agents

SHIN TAKEHISA
DEPARTMENT OF BIOLOGY
KEIO UNIVERSITY
YOKOHAMA, JAPAN

I. Introduction 88
II. SCEs as a sensitive indicator in detecting direct mutagens 89
    A. Sensitivity of the SCE test and types of direct mutagens 89
    B. Use of the SCE test as a short-term test for direct mutagens 91
III. Introducing metabolic activation into the SCE test 104
    A. Chemical carcinogens and metabolic activation 104
    B. Target cells and metabolic activation: Preexisting problems 106
    C. *In vitro* SCE test coupled with a metabolic activation system 108
        1. Methods 108
            (a) The microsome-mediated method 108
            (b) The cell-mediated method 109
            (c) The host-mediated method 110
        2. Cultured cells 110
        3. Lymphocytes 115
        4. Factors affecting the SCE incidence *in vitro* 117
    D. *In vivo* SCE test 119
        1. The *in vivo* BrdUrd labeling and SCE 120
        2. The *in vitro* BrdUrd labeling and SCE 121
IV. Conclusion 136

I. INTRODUCTION

Sister chromatid exchanges, first described by Taylor et al. (1957), who used autoradiographic techniques, are now easily detectable by recently developed BrdUrd–dye techniques (Latt, 1973; Ikushima & Wolff, 1974; Kato, 1974a; Perry & Wolff, 1974; Dutrillaux et al., 1974; Korenberg & Freedlender, 1974; Goto et al., 1975; Franceschini, 1974; Kim, 1974) that stain sister chromatids differentially and demarcate exchange points between sister chromatids with great accuracy. Using these techniques, several workers have found that the frequency of sister chromatid exchanges (SCEs) can be increased by chemical mutagens at doses far below those that can hardly induce any chromosome aberrations (Latt, 1974; Kato and Shimada, 1975; Solomon and Bobrow, 1975). This behavior was interpreted as suggesting that SCEs can be a sensitive indicator for detecting DNA-damaging chemical agents.

SCE is an exchange of polynucleotide strands of DNA double helices, as is evidenced by a switch of [^{3}H]thymidine label along chromatids. Chemical agents used at the early stage of SCE research were the ones that had been proven in several other systems to interact primarily with DNA. Thus the types of chemical agents capable of inducing SCEs are considered to be DNA-damaging agents. SCE-inducing agents, however, are not necessarily agents that react with DNA directly. We do not know the mechanism of SCE formation but do know that for SCE formation, cells need to pass through the DNA synthesis phase of the cell cycle (Wolff et al., 1974) and/or SCE formation is limited to portions of DNA that are undergoing or have just completed replication (Kato, 1974b). An indirect action of chemicals on DNA, for example, DNA synthesis, may contribute to SCE formation.

There are several excellent reviews (Wolff, 1977a, 1977b; Kato, 1977b; Galloway, 1977a; Latt, 1979; Latt et al., 1979; Latt and Schreck, 1980; Perry, 1980) referring to the SCE test in which the frequency of SCE is used to screen chemical mutagens. However, many data on the induction of SCEs by chemical agents have accumulated since the reviews were written. This chapter tries to bring the reader up to date concerning the validity and limit of the SCE test for detecting the mutagens/carcinogens in our environment by reviewing the reports available so far on SCE induction by chemical agents. The reports discussed here are restricted to those using mammalian cell systems. In this Chapter the chemical mutagens known to cause DNA damage in any way are tentatively divided into direct mutagens and indirect mutagens. The direct mutagens exert their effects without any chemical modification such as metabolic activation. The indirect mutagens need chemical modification to become active.

II. SCEs AS A SENSITIVE INDICATOR IN DETECTING DIRECT MUTAGENS

A. *Sensitivity of the SCE Test and Types of Direct Mutagens*

Latt (1974) first reported that mitomycin C (MMC) induced a significant increase of SCEs with as little as 3 ng/ml in human peripheral lymphocytes. MMC was added for the entire period of culture (73 hours). There were 12 SCEs/cell in controls, 14.3–19.0 at 3 ng/ml, the lowest concentration applied, and 62.0–112.0 at 100 ng/ml, the highest concentration. BrdUrd concentration was 12 μM. At 30 ng/ml MMC there was an average of 40 SCEs/cell, but the number of chromosome or chromatid breaks per cell was fewer than 0.07, that is, same as control level. This did not increase over the range of concentrations applied. The concentrations of MMC required to generate large numbers of breaks were known to range from 0.1 to 1.0 μg/ml in human lymphocytes. Therefore, the results obtained by Latt indicate that the sensitivity of the SCE test is about 100 times that of the standard chromosome aberration test.

Kato and Shimada (1975) treated Chinese hamster cells, line D-6, cultured for 26 hours at 0.5 μg/ml BrdUrd concentration with MMC at concentrations from 0.025×10^{-7} (about 1 ng/ml) to 10.00×10^{-7} M for 8 hours before fixation. The rise in the SCE frequency from the control level (SCEs/cell, 2.42) was detected at 0.05×10^{-7} M MMC (SCEs/cells, 3.28), but the frequency of breaks did not increase until 0.5×10^{-7} M. Therefore, MMC induced SCEs at a dose about one-tenth of that required to produce chromosome aberrations.

Solomon and Bobrow (1975) treated human lymphocytes with chlorambucil, a nitrogen mustard derivative, at concentrations 0.51–3.0 μg/ml, adding it at 72 hours, cultures fixed at 96 hours. BrdUrd (0.09 mM) was added from the initiation of the cultures. There was an average of 30 SCEs/cell at 0.5 μg/ml chlorambucil and 3.3–6.5 SCEs/cell in controls. Chromosome aberrations began to increase at 1.0 μg/ml, but they increased proportionately much less than did SCEs.

Hayashi and Schmid (1975) found in human lymphocytes that at low chromosome aberration rates (10% of the metaphases) induced by the alkylating agent Trenimon, the SCE rate rose quite drastically to an average of 41/metaphase. They also noted that fibroblast chromosomes proved to be far more sensitive to SCE induction at all tested Trenimon concentrations (10^{-6} to 10^{-1} μg/ml).

Beek and Obe (1975) found that SCEs in human lymphocytes are 100–200 times as frequent as the number of breakage events per cell. With 10^{-8} M Trenimon they found 0.14 breaks and 14.65 SCEs/cell. The latter were 100 times more frequent than breakage.

In Chinese hamster ovary (CHO) cells, Perry and Evans (1975) screened 14 proven or suspected mutagenic chemicals. The chemicals and 10 μM BrdUrd were added from the beginning of culture. A dose of mutagen that gave a highly significant doubling of the incidence of SCE produced only a minimal and barely noticeable effect on the incidence of chromosome aberrations. Doses giving tenfold increases in SCE gave increased, but still very low, aberration frequencies.

All these early works demonstrated that analysis of chemically induced SCE was a simple and sensitive method for assessing DNA-damaging agents.

Data have been accumulating that indicate the mechanism of SCE formation is different from that of chromosome aberration formation. If the mechanisms are completely independent of each other, there might be two types of direct mutagens, that is, one that can induce SCEs but not chromosome aberrations and another that cannot induce SCEs but can induce chromosome aberrations. Aniline hydrochloride (Abe and Sasaki, 1977a) may be a chemical of the former type. 8-Ethoxycaffeine (Strobel et al., 1979; Kihlman, 1975) may belong to the latter type. Caffeine has been thought to be of the latter type, but as shown in Table 5.3 it cannot be regarded simply so. *In vivo* treatment with caffeine alone can increase the frequency of SCEs (Basler et al., 1979; Nakanishi and Schneider, 1979). Doses of chemical agents needed for SCE induction are always lower than those needed for chromosome aberrations. In Chinese hamster cells, aniline hydrochloride induces SCEs without producing chromosome aberrations at any dose level up to where it is physiologically toxic, that is, where it brings about mitotic inhibition. At certain dose levels 8-ethoxycaffeine starts to induce chromosome aberrations, but below these levels there is neither SCE nor chromosome aberration induction.

Kato (1977b) grouped the chemical agents into three categories on the basis of their efficiency in inducing SCEs and chromosomal aberrations. The first group, designated X-ray-type agents, exerts its effects in a manner similar to that of ionizing radiations in that the increase in SCE frequency caused by these agents is minimal in spite of their remarkable ability to produce chromosomal aberrations. Kato included caffeine, 8-ethoxycaffeine, and bleomycin in this group. The agents involved in the second group, the UV-type agents, are able to induce SCEs very efficiently. Chromosomal aberrations are also produced to a marked extent. This group includes most alkylating agents. Agents in the third group are metabolic inhibitors such as fluorodeoxyuridine, hydroxyurea, excess thymidine, and cycloheximide. Their DNA-damaging effects are indirect.

The lowest concentration of chemical agent needed to induce SCEs and information on whether chromosomal aberrations can be induced at that concentration or not, are presented in Tables 5.1, 5.2, and 5.3 for MMC, bleomycin, and caffeine, respectively, in various treatments and cell systems. From these tables it is clear that the SCE test is effective in detecting the UV-type and S-dependent mutagens, but ineffective for the G_2-dependent mutagens under certain conditions. The S-dependent mutagens are the chromosome-breaking agents, such as alkylating agents, which require the cell to go through a DNA synthesis phase (S phase) to manifest the aberrations. The G_2-dependent or S-independent mutagens such as bleomycin induce chromatid breaks in G_2 cells or express maximum lethality in G_2 or mitosis (Littlefield et al., 1979; Gebhart and Kappauf, 1978; Natarajan and Obe, 1978; Popescu et al., 1979). As for the G_2-dependent mutagens, an increase of SCEs is not observed if they are applied during the last round of the cell cycle just prior to fixation (cf. data of Gebhart and Kappauf in Table 5.2), because for SCE formation cells need to pass through the S phase.

B. *Use of the SCE Test as a Short-Term Test for Direct Mutagens*

Over 100 assays (Hollstein et al., 1979) have been identified as "short-term" methods to test a chemical's potential DNA-damaging ability as manifested by mutagenicity and carcinogenicity. The results of 24 representative short-term tests on 72 carcinogens and mutagens were compiled for comparison by Hollstein et al. (1979). It is evident from their work that the SCE test picks up the chemicals known to have carcinogenicity or mutagenicity as effectively as Ames's *Salmonella*/microsome test. In some cases the SCE test is more sensitive in detecting carcinogens and mutagens than the Ames test. Urethane has been reported to be mutagenic in other organisms, to cause chromosome damage in rats, and to be converted to metabolites that can react with DNA (McCann and Ames, 1976); but it is negative in the Ames test (McCann and Ames, 1976; McCann et al., 1975). The SCE test can detect it as positive (Abe and Sasaki, 1977a; Csukás et al., 1979). Diethylstilbestrol (DES), a human hormonal carcinogen, is negative in the Ames test (Hollstein et al., 1979) but positive in the SCE test (Rüdiger et al., 1979). Similar differences exist for saccharin (Wolff and Rodin, 1978). Though the Ames test is far more sensitive than any other test in being able to detect the potential mutagenic agents and even trace impurity of the chemicals tested (McCann and Ames, 1976), the SCE test has one advantage, that is, utilization of eukaryote cells. *In vitro* mammalian tests can mimic more accurately what is happening in the mammalian body than can bacterial cells.

TABLE 5.1
In vitro *Treatment with Mitomycin C*

Material	Exposure time[a]	Lowest concentration that induced SCEs	Chromosome aberration induction at given concentration[b]	Ref
Chinese hamster D-6	8 h, b.f.	0.05×10^{-7} M	–	1
Chinese hamster	1 h between first and second cell cycle	3×10^{-6} M	+	2
CHO	39 h	0.009 μg/ml	ND[c]	3
CHO	18 h	10^{-8} M	ND	4
CHO	6–30 h, b.f.	0.03 μg/ml	+	5
CHO	2 h, b.c.	10^{-7} M	+	6
CHO	24–36 h	10^{-8} M	–	17
Mouse (C57Bl/6J) embryo fibroblasts	48 h	2.5 ng/ml	ND	7
Human lymphocytes	73 h	3 ng/ml	–	8
Human lymphocytes	21–24 h, b.f.	4 ng/ml	–	9
Human lymphocytes	2 h, b.s.	0.25 μg/ml	ND	10
Human lymphocytes	76 h	25 ng/ml	+	11
	12 h, b.f.	50 ng/ml	+	

Human lymphocytes	54 h, b.f.	1.2×10^{-8} M	ND	12
Human lymphocytes	24 h, b.f.	1 ng/ml	–	13
Human lymphocytes	24 h, b.f.	0.1 μg/ml	ND	18
Human lymphocytes	24 h, b.f.	0.01 μg/ml	+	19
Human lymphocytes	20 h of first or second cell cycle	0.1 μg/ml	+	14
	48 h, b.f.	0.1 μg/ml	+	
Human lymphocytes	48 and 72 h, b.f.	0.04 μg/ml	–	15
Xeroderma (XP12RO) fibroblasts	48 h	10^{-9} M	ND	16
Bloom's lymphocytes	24 h, b.f.	1 ng/ml	+	13
Rat lymphocytes	24 h, b.f.	0.03 μg/ml	+	20

[a]b.f.: before fixation; b.c.: before BrdUrd culture; b.s.: before stimulation.

[b]+: significant increase; –: the same as control level.

[c]ND: no description.

References. 1. Kato and Shimada, 1975. 2. Palitti and Becchetti, 1977. 3. Majone and Levis, 1979. 4. Carrano et al., 1978. 5. Loveday and Latt, 1978. 6. MacRae et al., 1979b. 7. Kram et al., 1979b. 8. Latt, 1974. 9. Raposa, 1978. 10. Littlefield et al., 1979. 11. Ishii and Bender, 1978a. 12. Faed and Mourelatos, 1978. 13. Shiraishi and Sandberg, 1978. 14. Shiraishi et al., 1979. 15. Novotná et al., 1979. 16. Wolff et al., 1977. 17. Perry and Evans, 1975. 18. Tsuchimoto et al., 1979. 19. Pant et al., 1976. 20. Granberg-Öhman et al., 1980.

TABLE 5.2
In vitro Treatment with Bleomycin

Material	Exposure time[a]	Concentration applied	SCE induction[b]	Chromosome aberration induction[b]	Ref.
Human lymphocytes	2 h, b.s.	0.025 μg/ml	–	ND[c]	1
Human lymphocytes	24 h, b.f.	4 μg/ml	–	+	2
Chinese hamster M-3	1 h between first and second cell cycle	2 μg/ml	+	+	3
Chinese hamster A(T_1)Cl-3	24 h	2.5 μg/ml	+	+	4
Chinese hamster Don D-6	30 h	10^{-7} M	–	+	5
Chinese hamster V79-4	2 h before two cell cycles	1 μg/ml	+	+	6
CHO	30 min in G_2	10 μg/ml	ND	+	7
CHO	30 min in G_2	25 μg/ml	ND	+	8
CHO	24 h	3×10^{-6} M	+	+	9

[a]b.s.: before stimulation; b.f.: before fixation; G_2: post DNA synthetic phase of cell cycle.

[b]+: significant increase, –: the same as control level.

[c]ND: no description.

References. 1. Littlefield et al., 1979. 2. Gebhart and Kappauf, 1978. 3. Vig, 1979. 4. Banerjee and Benedict, 1979. 5. Sono and Sakaguchi, 1978. 6. Popescu et al., 1979. 7. Natarajan and Obe, 1978. 8. Sognier et al., 1979. 9. Perry and Evans, 1975.

Table 5.4 gives the chemical agents that have been applied *in vitro* and without metabolic activation. Results with most of the chemical agents that require metabolic activation are omitted from Table 5.4. This table aims to tell readers at a quick glance whether or not a chemical agent in question is an SCE-inducing agent and whether or not its capability of SCE induction is relatively strong in any experimental condition. The chemical agents are arranged according to their SCE inducibility in each experiment. The SCE inducibility is expressed as the SCE ratio, which is obtained by dividing the mean SCE frequency, that is, the mean number of SCEs per cell or per chromosome, at a given dose of chemical agent by the mean SCE frequency of BrdUrd-labeled control in the same experiment. In cases where various doses of a chemical agent were tested in an experiment, one where the SCE frequency was close to twice that of the control was adopted as a representative dose. Differences in cell type, exposure time of chemical agent, concentration of BrdUrd, and so on, are ignored in Table 5.4.

From Table 5.4 we can draw interesting conclusions on the characteristics inherent in the SCE test. (1) There are no data available at present on aniline hydrochloride that suggest its carcinogenicity and mutagenicity, but it induces SCEs in cultured Chinese hamster cells (Abe and Sasaki, 1977a). This may suggest that the SCE test can detect special kinds of DNA damage other than those that result in mutagenesis or carcinogenesis. (2) 5-Azacytidine (Banerjee and Benedict, 1979; Sono and Sakaguchi, 1978), 5-fluorodeoxyuridine (Galloway, 1977b; Sono and Sakaguchi, 1978; Kato, 1980a), cytosine arabinoside (Raposa, 1978; Sono and Sakaguchi, 1978; Littlefield et al., 1979; Shiraishi and Sandberg, 1979; Ishii and Bender, 1980), 6-mercaptopurine (Sono and Sakaguchi, 1978), methotrexate (Banerjee and Benedict, 1979; Raffetto et al., 1979), and bredinine (Sono and Sakaguchi, 1978; Sono et al., 1978) increased SCE rate. These compounds affect the synthesis and/or functioning of nucleic acids (Hoffman and Post, 1973; Wheeler and Simpson-Herren, 1973; Donehower et al., 1979; Sono and Sakaguchi, 1978) and thus belong to the chemical agents of the third group, that is, metabolic inhibitors, in Kato's classification. BrdUrd induces SCEs with increasing concentration (Wolff and Perry, 1974; Stella et al., 1978). It is, however, incorporated into DNA and known to differ somewhat from those metabolic inhibitors in that it damages DNA (cf Davidson et al., 1980). This behavior implies that the chemical agents positive in the SCE test are not only chemical agents that become covalently bound to DNA, but also chemical agents that affect the machinery of DNA synthesis, thus leading indirectly to DNA damage.

TABLE 5.3
In vitro *Treatment with Caffeine*

Material	Exposure time[a]	Concentration applied	SCE induction[b]	Chromosome aberration induction[b]	Ref.
Human fibroblasts WI-38	24 h	1.3×10^{-3} M	ND[c]	+ (breaks)	1
Human fibroblasts	24 h, b.f.	100 μg/ml	+	ND	2
Human lymphocytes	24 h, b.f.	10^{-3} M	–	+ (gaps)	3
Human lymphocytes	54 h ,b.f.	100 μg/ml	±	+	4
Human lymphocytes	50 h, b.f.	5×10^{-4} M	–	ND	5
Human lymphocytes	48 h, b.f.	10^{-3} M	+	ND	6
Human lymphocytes	54 h, b.f.	5×10^{-4} M	+	ND	14
Human lymphocytes	24 h, b.f.	0.5 mg/ml	ND	+	15
Human lymphocytes	24 h, b.f.	250 μg/ml	+	+	16
Mouse leukemia cell line, L5178Y	30 h	10^{-4} M	–	–	7

Chinese hamster Cl-1	14 h, b.f.	10^{-3} M	−	+	8
Chinese hamster V79-4	24 h	100 μg/ml	−	−	9
Chinese hamster D-6	7 h, b.f.	10^{-3} M	−	±	10
Indian muntjac fibroblasts	22 h, b.f.	100 μg/ml	+	±	11
Vicia faba	21 h, b.f.	5×10^{-4} M	−	+	12
Vicia faba	19 h, b.f.	5×10^{-4} M	−	−	13
Human B-lymphoid cells	46 h	10^{-3} M	±	−	17
Bloom's lymphocytes	48 h, b.f.	10^{-3} M	−	−	18

[a]b.f.: before fixation.

[b]+: significant increase; −: the same as control level; ±: slight increase.

[c]ND: no description.

REFERENCES. 1. Weinstein et al., 1977. 2. Sasaki, 1977. 3. Brøgger and Waksvik, 1978. 4. Mourelatos, 1979. 5. Liebeskind et al., 1979. 6. Brøgger et al., 1979. 7. Sono et al., 1978. 8. Palitti and Becchetti, 1977. 9. Popescu et al., 1979. 10. Kato, 1977a. 11. Ved Brat et al., 1979. 12. Kihlman, 1975. 13. Kihlman and Sturelid, 1978. 14. Faed and Mourelatos, 1978. 15. Tsuchimoto et al., 1979. 16. Pant et al., 1976. 17. Shiraishi and Sandberg, 1980a. 18. Shiraishi and Sandberg, 1980b.

TABLE 5.4
SCE Inducibility of Chemical Agents in* in vitro *Treatment

Level of concentration (M)	SCE ratio[a]
	SCE Ratio = ≤1.0
	0.075% rubber solvent,[1] nitrous oxide and 10 inhalational anesthetics[2] (delivered dose, 0.17–75%, v/v), 1 ml/liter styrene,[97] 0.1%, v/v methanol, ethanol, butanol and propanol[10]
10^{-2} M	Urethane[128]
10^{-3}	Maleic hydrazide,[32] antipain,[99] cysteine, [100,106] glutathione,[106] Pb-acetate,[108] benzanthracene,[111] pyrene,[111] cochineal,[129] dipyrone,[129] furosemide,[129] haloperidol,[129] indomethacin,[129] mefenamic acid,[129] stevioside,[129] caffeine,[3] mercaptoethanol,[3] benzene,[4,85] toluene,[4] xylene,[4] quinoline,[5] *N*-methylurea,[5] ϵ-caprolactone,[5] *N*-*n*-butylurethane,[5] dibutylhydroxytoluene,[5] Na-dehydroacetate,[5] metronidazole,[82] misonidazole[82]
10^{-4}	Aminopyrine,[5] diethylstilbestrol,[5] saccharin,[6] urethane,[7] α-naphthylflavone,[11] caffeine,[54,78] chloramphenicol,[94] antipain,[98] cysteine,[101] glutathione (oxidized),[106] hematoporphyrin,[120] annato,[129] diazepam,[129] dimpylate,[129] ethionamide,[129] *p*-hydroxybenzoic acid isopropyl ester,[129] isoascorbic acid Na salt,[129] rose bengal[129]
10^{-5}	Fluorescent brightener-260,[5] phenanthrene,[7] perylene,[7] hydroxyurea,[7] elastatinal,[8] tilorone,[12] 6-mercaptopurine,[12] Cu^{2+},[13] Mn^{2+},[13] α-naphthoflavone,[14] Pb-acetate,[15] Na-acetate,[15] puromycin,[16] NNN,[80] Na_2SeO_4,[105] imipramine·HCl,[129] nitrofurantoin,[129] progesterone,[129] propyl gallate[129]
10^{-6}	Methylene blue,[7] methotrexate,[9] catalase,[13] vincristine,[17,73] cycloheximide,[18] bleomycin,[68] K-strophantin,[20] $CrCl_3$,[50] NBT,[80] di-*tert*-butyl proflavine[19]
10^{-7}	Cycloheximide,[16,118] Hydergine,[25] catalase,[28] Colcemid,[73] α-naphthoflavone[127]
10^{-8}	Bleomycin[9]

TABLE 5.4
Continued

Level of concentration (M)	SCE ratio^a
	SCE ratio = 1.0–1.5
	1% pine-wood tar resin,[117] 8 mg/ml caramel,[129] 0.05 mg/ml Eulan U-33[129]
10^{-2}	Saccharin,[21,27] ethanol[5]
10^{-3}	Barbital,[5] pyridine,[5] potassium metabisulfite,[5] phenanthrene,[5] di-(2-ethylhexyl)phthalate,[5] sunset yellow FCF,[5] caffeine,[74,79,121] hydralazine,[83] phenol,[85] hydroxylamine,[100] ethylamine,[106] cyclophosphamide,[113] Roundup,[124] acid red,[129] chlorodiazepoxide,[129] glycyrrhizic acid trisodium salt,[129] anthracene,[130] *p*-hydroxybenzoic acid ethyl ester,[129] butylbuthanolamine[130]
10^{-4}	2-Methyl-4-dimethylaminoazobenzene,[5] butylhydroxyanisol,[5] fluorescent brightener-24,[5] -225,[5] EMS,[31] caffeine,[23,71] 4-hydroxybenzo(*a*)pyrene,[92] 2′,3-dideoxythymidine triphosphate,[118] novobiocin,[118] ENPU,[123] 1,1,1-trichloropropene 2,3-oxide,[127] allethrin,[129] curcumin,[129] phenylbutazone,[129] reserpine,[129] tris-(2,3-dichloropyl)-phosphate,[129] MC,[130] *p*-hydroxybenzoic acid methyl ester[129]
10^{-5}	4-*o*-Tolylazo-*o*-toluidine,[5] β-propiolactone,[5] cytosine arabinoside,[9] pyrene,[7] 5-fluorouracil,[16] bredinin,[16] hydroxyurea,[16] aflatoxin B_1,[37] dihydrodiol of PAH (abbr. 4),[44] 5-fluorodeoxyuridine,[61] EMS,[77] chloroquine,[109] ethidium bromide,[109] captan,[124,129] retinoids,[126] buformin HCl,[129] Rhodamine B[129]
10^{-6}	*N*-Hydroxy-2-acetylaminofluorene,[7] quinacrine dihydrochloride,[19] acridine orange,[19] bleomycin,[23] cytosine arabinoside,[39] MMS,[77] Ca-DTPA,[93] lipophilic DTPA derivative,[93] Trp-2,[104] nitrosocarbaryl[123]
10^{-7}	Diethylstilbestrol,[14] bleomycin,[16] cytosine arabinoside,[16] 6-mercaptoguanosine,[16] acriflavine,[19] proflavin,[19] cycloheximide[61]
10^{-8}	Vincristine,[12] neocarzinostatine[118]
10^{-9}	Neocarzinostatine,[16] 4-nitroquinoline 1-oxide[77]
10^{-10}	Azathioprine,[110] cytosine arabinoside,[110] vinblastine,[110] vincristine[110]

TABLE 5.4
Continued

Level of concentration (M)	SCE ratio[a]
	SCE ratio = 1.5–2.0
	Acetone 0.01 mg/ml, 3×10^{-1} M propylen glycol,[129] 5×10^{-1} xylitol[129]
10^{-2}	Potassium sorbate,[5] urethane,[5] sodium benzoate,[5] saccharin sodium[5,27]
10^{-3}	Diphenyl,[5] caffeine,[21,94] vitamin C,[28] EMS,[89,124] chlorpropamide,[88] misonidazole,[103] Dexon,[124] butylbuthanolnitrosamine,[130] *N*-*n*-butylurea,[130] *n*-dibutylamine,[130] dimethylamine HCl[130]
10^{-4}	Dibutyl phthalate,[5] *N*-nitrosodiphenylamine,[5] propane sultone,[5] aniline HCl,[5] MMS,[23] EMS,[29,32–34,122] malathion,[31,129] aflatoxin B_1,[40] 2-aminofluorene,[37] urethane,[36] H_2O_2,[90] 3-hydroxy-BP,[92] misonidazole,[96] ENU,[77,123] hydrazine,[100] 8-methoxypsoralen,[102] 2-mercaptoethanol,[106] propidium iodide,[109] hydroxyurea,[118] levodopa,[129] *n*-dibutylnitrosamine,[130] pyrene,[130] BP[130]
10^{-5}	Methylazoxymethanol acetate,[5] cysteine,[13] dimethylhydrazine,[13] 5-azacytidine,[16] azaserine,[20] *N*-hydroxy-2-acetylaminofluorene,[23,37,62] promazine,[24] MNNG,[30] β-propiolactone,[32] vitamin A palmitate,[35] MNU,[40] aflatoxin B_1,[40] BP,[43] dihydrodiols of PAH (abbr. 1, 5, 9),[43,44] 3-methylcholanthrene,[44] DMBA,[44] $PbCrO_4$,[46] MMS,[74] 4-nitroquinoline-1-oxide,[74,114] DAPA,[80] *N*-nitroso-ETU,[80] hydroquinone,[85] Trp-2,[104] cycloheximide,[119] 5-fluorodeoxyuridine[122]
10^{-6}	Proflavine,[7,23] acridine orange,[7] methylene blue,[24] chloroprothixene,[24] chlorpromazine,[24] 6-mercaptopurine,[16] bleomycin,[32,42] RO 10-9359 (retinoid),[35] carofur,[38] furazolidone,[38] dihydrodiols of PAH (abbr. 3, 8),[43,44] 1,2-dihydrodiol of MBA (abbr. 1),[43] $K_2Cr_2O_7$,[55] $Na_2Cr_2O_7$,[55] thio-TEPA,[74] EMS,[77] captafol,[80] ENNG,[89] BP-4,5-oxide,[92,127,131] *syn*-BP-7,8-diol-9,10-epoxide,[131] Se,[105] Hoechst 33258,[32] 7-bromomethylbenz(*a*)anthracene[131]

TABLE 5.4
Continued

Level of concentration (M)	SCE ratio[a]
10^{-7}	Diepoxy-butane,[32] 4-nitroquinoline-1-oxide,[32,121] TEPA,[49] $K_2Cr_2O_7$,[50] MNNG,[50] K_2CrO_4,[50] BP-diol epoxide II,[51,92] *anti*-BP-7,8-diol-9,10-epoxide,[131] Trp-1,[104] acriflavin,[109] acridine yellow,[109] aphidicolin,[118] 5-fluorodeoxyuridine,[119] EMS[121]
10^{-8}	Hycanthone,[12] diethylstilbestrol-α-β-oxide,[14] adriamycin,[32] 5-fluorodeoxyuridine,[33] 9-aminoacridine derivatives,[48] mitomycin C,[52,95,115] MNNG,[77] *N*-acetoxy-2-acetylaminofluorene,[37] MMS,[84] EMS,[84] cytosine arabinoside,[84] 4-nitroquinoline 1-oxide,[84] *anti*-BP-7,8-diol-9,10-epoxide[92]
10^{-9}	Trenimon,[30] mitomycin C[75,77]
10^{-10}	Trenimon,[60] 4-nitroquinoline 1-oxide[77]
10^{-12}	Trenimon[60]
	SCE ratio = 2.0–3.0
	Cigarette smoke condensate 100 μg/ml,[87] 0.5 ml/liter cyclohexene oxide[97]
10^{-2}	Isoniazid[100,129]
10^{-3}	Nicotinamide,[8] hydrazine,[13] isoniazid,[13] caffeine,[18] *N*-diazoacetylglycine amide,[20] hydroxyurethane,[36] EMS,[74,114,115] vamidothion,[80] vitamin C,[107] lacchaic acid,[129] theophylline,[129] sodium hypochlorite[129]
10^{-4}	*N*-*n*-Butyl-*N*-nitrosourethane,[5] acetaldehyde,[10,56] EMS,[41,61,95] ENU,[41,77,89] methylazoxymethanol acetate,[53] caffeine,[67] EDB,[80] HEH,[80] DDVP,[80] epichlorohydrin,[81] MNU,[89] MMS,[89] cysteamine,[106] 4-nitro-*o*-phenylenediamine,[116] 2-nitro-*p*-phenylenediamine,[116] ethenzamide,[129] fenitrothion,[129] vitamin A acetate,[129] DMBA[130]
10^{-5}	Glutathione,[13] cysteamine,[13] 5-fluorodeoxyuridine,[16] methotrexate,[20] sodium bisulfite,[22] 4-nitroquinoline-1-oxide,[30,114] MMS,[32] dihydrodiols of PAH (abbrs. 2, 6, 7),[43,44] 8,9-dihydrodiol of MBA (abbr. 2),[43] Tris-BP,[63] dimethylsulfate,[74,77] captan,[80] DBCP,[80] catechol,[85] BP,[92] Trp-2,[104] SeO_2,[105] Na_2Se,[105] Na_2SeO_3,[105] carofur,[112] MNU,[123] MNPU,[123] Cl-MNPU,[123] F-MNPU,[123] Br-MNPU,[123] 4-aminoquinoline 1-oxide[130]

TABLE 5.4
Continued

Level of concentration (M)	SCE ratio[a]
10^{-6}	Chlorambucil,[9] 5-azacytidine,[12] bleomycin,[12,32] H_2O_2,[13] proflavin,[24] acridine orange,[24] MBA,[43] non-K-region dihydrodiol of MBA,[43] non-K-region dihydrodiols of MC,[44] busulfan,[47] vitamin A acid,[35] bredinin,[54] MMS,[77] MNNG,[71,91] BP,[35,87,127] ENNG,[91] BP-7,8-oxide,[92] anti-BP-9,10-diol-7,8-epoxide,[92] Trp-2[104]
10^{-7}	Mitomycin C,[9,49,76] thio-TEPA,[12,71] *m*-AMSA,[19] quinacrine mustard,[19,32] MNNG,[20,32,89] nitrogen mustard,[32] adriamycin,[33] busulfan,[39] proflavin,[41] *cis*-DDP,[64] MMS,[84] daunomycin,[84] bleomycin,[84] Trp-1,[104] DMBA,[111] cytosine arabinoside,[118] 4-nitroquinoline 1-oxide[126]
10^{-8}	Melphalan,[12] β-dienoestrol,[14] trenimon,[15] actinomycin D,[16] mitomycin C,[26,30,32,41,48,71,77] lycurim,[39] vincristine,[39] adriamycin,[57] EMS,[84] MNNG,[84] 4-nitroquinoline 1-oxide,[84,130] 7-bromomethylbenzanthracene[111]
10^{-9}	Actinomycin D,[12] alkeran,[39] mitomycin C,[59] trenimon[69]
	SCE ratio = 3.0–5.0
	25 μg/embryo neutral red,[86] 0.1 ml/liter styrene oxide[97]
10^{-3}	Sodium ascorbate,[22] EMS,[36,88] dimethoate,[129] $NaNO_2$[130]
10^{-4}	EMS,[54] MMS,[62] Tris-BP,[129] *N*-*n*-butyl-*N*-nitrosourea[130]
10^{-5}	Thio-TEPA,[9] non-K-region dihydrodiol of MBA,[43] Na_2SeO_3[62]
10^{-6}	4-Nitroquinoline-1-oxide,[7,23] *N*-acetoxy-2-acetylaminofluorene,[7,95] non-K-region dihydrodiols of BP,[43] non-K-region dihydrodiols of DMBA,[44] mitomycin C[74]
10^{-7}	Methotrexate,[12] mitomycin C,[25,45,84,121] chlorambucil,[39] MNNG,[45,66] BP-diol epoxide I,[51] anthramycin,[67] Trp-1,[104] 4-nitroquinoline 1-oxide[127]
10^{-8}	Trenimon,[68] mitomycin C[39,55,48,94]
	SCE ratio = >5.0
	Dimethylnitrosamine (1.4×10^{-2} M)[130]

TABLE 5.4
Continued

Level of concentration (M)	SCE ratio[a]
10^{-3}	MNU[5]
10^{-4}	MMS[29], *trans*-DDP[90]
10^{-5}	*cis*-DDP,[90] captafol,[129] triamterene[129]
10^{-6}	MNNG,[7,23,99] *cis*-DDP,[23] chlorambucil,[70] quinacrine mustard,[70] nitrogen mustard,[114] Trp-1[104]
10^{-7}	MNNG,[28] mitomycin C,[29,33,65,72,84,125] MAC[58]

[a]SCE ratio, mean SCEs/cell or SCEs/chromosome in an agent-treated group is divided by mean SCEs/cell or SCEs/chromosome in control group of the given experiment.

ABBREVIATIONS. *m*-AMSA: 4′-(9-acridinylamino)methanesulphon-*m*-anisidide-HCl. BP-diol epoxide I: (±)-7β,8α-dihydroxy-9α,10α epoxy-7,8,9,10-tetrahydrobenzo(*a*)pyrene. BP-diol epoxide II: (±)-7β,8α-dihydroxy-9β,10β epoxy-7,8,9,10-tetrahydrobenzo(*a*)pyrene. *cis*- or *trans*-DDP:*cis*- or *trans*-platinum(II)diamine dichloride. BP: benzo(*a*)pyrene. DMBA: 7,12-dimethylbenz(*a*)anthracene. MBA: 7-methylbenz(*a*)anthracene. MC: 3-methylcholanthrene. Dihydrodiols of PAH: (1) *trans*-4,5-dihydro-4,5-dihydroxy-BP; (2) *trans*-9,10-dihydro-9,10-dihydroxy-BP; (3) *trans*-5,6-dihydro-5,6-dihydroxy-MBA; (4) *trans*-5,6-dihydro-5,6-dihydroxy-DMBA; (5) *trans*-8,9-dihydro-8,9-dihydroxy-DMBA; (6) *trans*-10,11-dihydro-10,11-dihydroxy-DMBA; (7) *cis*-2α,3-dihydroxy-MC; (8) *trans*-4,5-dihydro-4,5-dihydroxy-MC; (9) *trans*-11,12-dihydro-11,12-dihydroxy-MC. 1,2- and 8,9-Dihydrodiols of MBA: (1) *trans*-1,2-dihydro-1,2-dihydroxy-MBA; (2) *trans*-8,9-dihydro-8,9-dihydroxy-MBA. Non-K-region dihydrodiols of MC: (1) *trans*-7,8-dihydro-7,8-dihydroxy-MC; (2) *trans*-9,10-dihydro-9,10-dihydroxy-MC. Non-K-region dihydrodiol of MBA: *trans*-3,4-dihydro-3,4-dihydroxy-MBA. Non-K-region dihydrodiol of BP: *trans*-7,8-dihydro-7,8-dihydroxy-BP. Non-K-region dihydrodiol of DMBA: *trans*-3,4-dihydro-3,4-dihydroxy-DMBA. EMS: ethyl methanesulfonate. ENU: *N*-ethyl-*N*-nitrosourea. MMS: methyl methanesulfonate. MNU: *N*-methyl-*N*-nitrosourea. MNNG: *N*-methyl-*N*-nitro-*N*-nitrosoguanidine. MAC: *m*-AMSA. Lycurim: 1,4-di-(methylsulfonyl-oxyethylamino)-1,4-didesoxy-*m*-erythrite dimethylsulfonate. TEPA: Triethylene phosphoramide. Thio-TEPA: Triethylene thiophosphoramide. Tris-BP: tris(2,3-dibromopropyl) phosphate. Ca-DTPA: Ca-diethylenetriaminepentaacetic acid. DAPA: *p*-dimethylaminobenzenediazo sodium sulfonate. DBCP: 1,2-dibromo-3-chloropropane. DDVP: dichlorvos. EDB: 1,2-dibromoethane. ENNG: *N*-ethyl-*N*-nitro-*N*-nitrosoguanidine. HEH: 2-hydrazinoethanol. NBT: 2,4-dinitrophenyl thiocyanate. NNN: 5-nitro-1-naphthonitrile. Trp-1: 3-amino-1,4-dimethyl-5*H*-pyrido [4,3-b] indole. Trp-2: 3-amino-1-methyl-5*H*-pyrido [4,3-b] indole. MNPU: 1-methyl-1-nitroso-3-phenylurea. ENPU: 1-ethyl-1-nitroso-3-phenylurea. F-MNPU: 1-methyl-l-nitroso-3-(*p*-fluorophenyl) urea. Cl-MNPU: 1-methyl-1-nitroso-3-(*p*-chlorophenyl) urea. Br-MNPU: 1-methyl-1-nitroso-3-(*p*-bromophenyl) urea. Captan: *cis*-*N*-((trichloromethyl)-thio)-4-cyclohexene-1,2-dicarboximide. Dexon: sodium (4-(dimethylamino) phenyldiazene sulfate. Roundup: isopropylamine salt of *N*-(phosphonomethyl)glycine. Retinoids: 13-*cis*-retinoic acid, retinyl phenyl ether, retinyl methyl ether, retinyl thioether, and axerophthene.

III. INTRODUCING METABOLIC ACTIVATION INTO THE SCE TEST

A. *Chemical Carcinogens and Metabolic Activation*

It has become axiomatic that the induction of cancer by chemical carcinogens results from covalent binding to one or more of the cellular mac-

---

REFERENCES. 1. Altenburg et al., 1979. 2. White et al., 1979. 3. Brøgger and Waksvik, 1978. 4. Gerner-Smidt and Friedrich, 1978. 5. Abe and Sasaki, 1977a. 6. Saxholm et al., 1979. 7. Popescu et al., 1977. 8. Utakoji et al., 1979. 9. Littlefield et al., 1979. 10. Obe and Ristow, 1977. 11. Rüdiger et al., 1976. 12. Banerjee and Benedict, 1979. 13. MacRae and Stich, 1979a. 14. Rüdiger et al., 1979. 15. Beek and Obe, 1975. 16. Sono and Sakaguchi, 1978. 17. Stoll et al., 1976. 18. Sasaki, 1977. 19. Crossen, 1979. 20. Raffetto et al., 1979. 21. Brøgger et al., 1979. 22. MacRae and Stich, 1979b. 23. Popescu et al., 1979. 24. Speit and Vogel, 1979. 25. Tsuchimoto et al., 1979. 26. Latt, 1974. 27. Wolff and Rodin, 1978. 28. Galloway and Painter, 1979. 29. Craig-Holmes and Shaw, 1977. 30. Faed and Mourelatos, 1978. 31. Nicholas et al., 1979. 32. Perry and Evans, 1975. 33. Galloway, 1977b. 34. Stetka and Wolff, 1976b. 35. Juhl et al., 1978. 36. Csukás et al., 1979. 37. Takehisa and Wolff, 1977. 38. Cohen and Sagi, 1979. 39. Raposa, 1978. 40. Vijayalaxmi, 1979. 41. Carrano et al., 1978. 42. Vig, 1979. 43. Pal et al., 1978. 44. Pal et al., 1979. 45. MacRae et al., 1979b. 46. Douglas et al., 1980. 47. Honeycombe, 1978. 48. Gibas and Limon, 1979. 49. Novotná et al., 1979. 50. MacRae et al., 1979d. 51. Hsu et al., 1979a. 52. Kato and Shimada, 1975. 53. Evans et al., 1977. 54. Sono et al., 1978. 55. Majone and Levis, 1979. 56. Ristow and Obe, 1978. 57. Nevstad, 1978. 58. Deaven et al., 1978. 59. Shiraishi and Sandberg, 1978. 60. Hayashi and Schmid, 1975. 61. Sono and Sakaguchi, 1979. 62. Ray et al., 1978. 63. Furukawa et al., 1978. 64. Wiencke et al., 1979. 65. Shiraishi et al., 1979. 66. MacRae et al., 1979c. 67. Ved Brat et al., 1979. 68. Gebhart and Kappauf, 1978. 69. De Raat, 1979. 70. Solomon and Bobrow, 1975. 71. Mourelatos, 1979. 72. Loveday and Latt, 1978. 73. Morgan and Crossen, 1980. 74. Palitti and Becchetti, 1977. 75. Kram et al., 1979b. 76. Ishii and Bender, 1978a. 77. Wolff et al., 1977. 78. Liebeskind et al., 1979. 79. Kato, 1977a. 80. Tezuka et al., 1980. 81. White, 1980. 82. Prosser and Hesketh, 1980. 83. Cavaglia, 1980. 84. Shiraishi and Sandberg, 1979. 85. Morimoto and Wolff, 1980. 86. Byrnes and Bloom, 1980. 87. Hopkin and Perry, 1980. 88. Brown and Wu, 1977. 89. Raff and Sugahara, 1980. 90. Bradley et al., 1979. 91. Umezawa et al., 1979. 92. Pal et al., 1980. 93. Prosser, 1978. 94. Pant et al., 1976. 95. Schneider and Monticone, 1978. 96. Clare and Joslin, 1978. 97. De Raat, 1978. 98. Fujiwara et al., 1980. 99. Kinsella and Radman, 1980. 100. Speit et al., 1980a. 101. Abramovsky et al., 1978. 102. Faed and Peterson, 1980. 103. Geard and Rutledge-Freeman, 1980. 104. Sasaki et al., 1980b. 105. Ray and Altenburg, 1980. 106. Speit et al., 1980b. 107. Speit et al., 1980c. 108. Stella et al., 1978. 109. Raj and Heddle, 1980. 110. Kučerová and Polívková, 1977. 111. Perry, 1980. 112. Vig et al., 1977. 113. Allen and Latt, 1976b. 114. Perry et al., 1978. 115. Latt et al., 1975. 116. Perry and Searle, 1977. 117. Athanasiou and Bartsocas, 1980. 118. Ishii and Bender, 1980. 119. Kato, 1980a. 120. Moan et al., 1980. 121. Shiraishi and Sandberg, 1980a. 122. Sono and Sakaguchi, 1980. 123. Thust et al., 1980. 124. Vigfusson and Vyse, 1980. 125. Vijayalaxmi, 1980. 126. Tetzner et al., 1980. 127. Schürer et al., 1980. 128. Roberts and Allen, 1980. 129. Sasaki et al., 1980a. 130. Abe and Sasaki, 1977b. 131. Connell, 1979.

romolecules such as DNA, RNA, and proteins of the cells in target tissues (Miller, 1970; Heidelberger, 1975). Ames and his collaborators (Ames et al., 1973; McCann et al., 1975; McCann and Ames, 1976) have presented the data that carcinogens are mutagens. Their data support strongly the theory that DNA is the target macromolecule of carcinogens. Though there is evidence for multifactoriality in carcinogenesis (Frei, 1976) and it is too early to adopt any unitary theory of chemical carcinogenesis (Miller, 1979), the damage of DNA by chemical carcinogens seems most likely to be the initiation of chemical carcinogenesis. If this is so, the SCE test for detecting chemical agents that damage DNA is effective in screening not only potential mutagens, but also potential carcinogens.

Many known carcinogens are inert themselves and cannot interact with DNA. Most of these classes of chemical carcinogens are metabolized in the host to a chemically reactive form (Miller, 1970; Heidelberger, 1975; Weisburger and Williams, 1975), which then reacts to form covalent binding to DNA (Lutz, 1979). We know now that most xenobiotics (foreign organic compounds) are metabolized in the human body to products that are readily excreted in urine, air, or bile. Mixed-function oxidases (MFOs), which are found in the liver and several other organs (Conney, 1967; Bend and Hook, 1977), are located in the endoplasmic reticulum and, with much less activity, in the nuclear envelope (Bend and Hook, 1977; Zbarsky, 1978), play an important role in the overall biotransformation of xenobiotics. The MFO system not only detoxifies chemicals, but also activates those that have carcinogenic potential.

If the SCE test is to be a useful tool in screening potential mutagens and carcinogens in our environment, enzyme systems related to biotransformation of xenobiotics should be incorporated into it. Miller and others have discovered that metabolic activation of chemical carcinogens is carried out primarily by the microsomal MFOs (Cramer et al., 1960; Malling, 1971). Ames et al. (1973) have coupled their *Salmonella* back mutation test with rat liver microsomes containing MFOs to obtain metabolic activation and showed that this becomes a general screening test for mutagenic and carcinogenic activities of chemical agents. Natarajan et al. (1976) and Stetka and Wolff (1976b) found that coupling of the *in vitro* SCE test system with rat liver extract containing microsomes provides a sensitive *in vitro* mammalian test system for the detection of carcinogens that require metabolic activation as well as those that do not.

To determine whether or not the administered chemical agents are metabolized in the human body to ultimate mutagens, the SCE analysis of cultured cells following an incubation in 5–20% human urine may be helpful. Guerrero et al. (1979) found that the SCE frequency in WI-38 human fibroblasts was increased after exposure to urine specimens collected from subjects 2 and 24 hours after cyclophosphamide administration.

The relationship of metabolic activation with the SCE test *in vitro* and that *in vivo* is discussed in the following sections.

B. Target Cells and Metabolic Activation: Preexisting Problems

The SCE test system consists primarily of the chemical agent being tested and the SCE indicator cells, that is, target cells. For the screening of indirect mutagens, metabolic activation lies between the chemical agent and the target cells (see below). The target cells are various types of cultured cells, blood peripheral lymphocytes, bone-marrow cells, regenerating liver cells, and so on.

$$\text{direct mutagens} \longrightarrow \text{target cells}$$

$$\text{indirect mutagens} \xrightarrow[\uparrow]{} \text{target cells}$$

$$\text{metabolic activation}$$

For the *in vivo* SCE test system, there is no need to incorporate the metabolic activation system. However, it is generally believed that the *in vitro* SCE test needs an exogenous supplement of the metabolic activation system because the test system by itself does not provide it. This implies that the target cells in the *in vitro* system, cultured cells and lymphocytes, lack the ability for metabolic activation. In CHO cells, cyclophosphamide, a chemical agent known to require metabolic activation, cannot induce SCEs when applied alone, but can induce SCEs when coupled with the *in vitro* metabolic activation system added exogenously (Stetka and Wolff, 1976b). Some types of cultured cells, however, are capable of metabolic activation. Appearance of the activity is influenced by the passage of culture, the species difference of cultured cells, and other factors. BHK21 cells, isolated from Syrian hamster kidney tissue, can be subcultured indefinitely with no attenuation of mixed-function oxidase activity (Newbold et al., 1977). Unlike most early passage fibroblasts, Chinese hamster cell line V79 has no mixed-function oxidase activity (Huberman & Sachs, 1974). In hamster fetus cultures the fibroblasts progressively lose their mixed-function oxidase activity and after a few passages become unable to metabolize carcinogenic polycyclic hydrocarbons (Nebert & Gelboin, 1968). The capacity of fibroblasts to activate carcinogenic polycyclic hydrocarbons such as 7,12-dimethylbenz(*a*)anthracene or benzo(*a*)pyrene to DNA-damaging forms varies among species, suggesting that mixed-function oxidase activity differs among cultured fibroblasts of different species (Schwartz and Moore, 1979). The SCE-inducing effect of the several metabolic activation-dependent mutagens varies among different types of cells. Some cell types seemed to activate some of the indirect mutagens more efficiently than others (Perry, 1980). Human fibroblasts from healthy individuals, which were between the 5th and the 10th passages, showed the capacity to activate benzo(*a*)pyrene, as it increased SCEs (Juhl et al., 1978). As to whether or not the cell line used in the test contains a low level of mixed-function oxidase activity, there are conflicting results. In inducing SCEs in CHO cells, benzo(*a*)pyrene has been reported to be inactive in the absence of a microsomal activation system (Takehisa and Wolff, 1977; Hopkin and Perry, 1980; De Raat, 1979), but it turned out in another study to be active (Pal et al., 1978).

In the *in vitro* SCE test that uses cultured lymphocytes as target cells, the use of an exogenous metabolic activation system becomes complicated because peripheral blood lymphocytes are stimulated to divide. Following mitogenic stimulation, the blastogenic response, or transformation, of the lymphocyte takes place. It is accompanied by vast changes in biochemical events (Cooper, 1973). There may be a new production of mixed-function oxidases. Freshly isolated human lymphocytes do not contain measurable amounts of aryl hydrocarbon hydroxylase (AHH) activity, a component of the microsomal mixed-function oxidases. However, after culture in the presence of mitogens, the resulting blast cells contain appreciable amounts of AHH, which is further inducible by polycyclic aromatic hydrocarbons such as benz(*a*)anthracene (Whitlock et al., 1972). It has been shown that AHH activity exists in unstimulated peripheral blood monocyte cultures (Bast et al., 1974; Ptashne et al., 1974; Lake et al., 1977). Mixed-function oxidases are inducible in a variety of cultured cells, as well as in lymphocytes, by the chemical agent that is under test (Nebert and Gelboin, 1968). This may be worth recalling when the treatment time with a chemical agent is long. Galloway et al. (1980) exposed cultured mouse embryos to benzo(*a*)pyrene in the absence of an exogenous *in vitro* metabolic activation system. The mouse embryos, which were explanted at 7 or 8 days gestation, were genetically different in their degree of the inducibility of AHH. A high frequency of SCEs was induced in embryos from those strains with a high degree of AHH inducibility.

As to the metabolic activation system, we now know the kinetics of the microsomal mixed-function oxidases are rather variable, depending on species, age, sex, climate, food, genetic history, circadian rhythms, and so on, and that numerous foreign compounds can stimulate the induction of the microsomal enzymes (Conney, 1967; Estabrook et al., 1978; Uehleke, 1974; Nebert et al., 1978; Thorgeirsson and Nebert, 1977). It has been known that the aromatic amines and their *N*-acetyl derivatives are activated *in vivo* by multiple pathways (Irving, 1979). It is likely that multiple pathways for activation of other classes of carcinogens, for example, the polycyclic aromatic hydrocarbons and the dialkylnitrosamines, also operate. Since so many factors can readily influence the enzyme composition of the metabolic activation system, it should be kept in mind that in *in vitro* metabolic activation the metabolic profiles may differ somewhat from those that existed previously.

C. **In vitro** *SCE Test Coupled with a Metabolic Activation System*

1. *Methods*

Several different methods of metabolic activation have been employed for the *in vitro* SCE test. They are divided into (a) the microsome-mediated method, (b) the cell-mediated method, and (c) the host-mediated method.

(a) THE MICROSOME-MEDIATED METHOD. The microsome-mediated method utilizes microsomal or postmitochondrial fractions isolated from liver or other tissue homogenates, being supplemented with an NADPH-generating system. Generally the procedure of Ames (Ames et al., 1975) has been followed in preparing the microsome fraction (S-9, abbreviation of the 9000 g supernatant fraction) and the NADPH-generating system. The preparation thus made is called S-9 Mix. Microsomal fraction S-9, as the source of enzyme activity, is heterogenous and subject to wide variability in enzyme composition. S-9 Mix may not retain the integrated interplay of the multiple *in vivo* enzymes and cofactors involved in the biotransformation of xenobiotics. The following factors must be kept in mind. The metabolic activation of xenobiotics not only is carried out by microsomes, but also occurs by means of cytosol enzymes (Thorgeirsson and Nebert, 1977). The steady-state level of reactive metabolites formed in S-9 Mix is determined by the interaction of a set of "activating" and "detoxifying" reactions. Variation in its balance could alter the mutagenic potency inherent in a chemical agent (Ashby and Styles, 1978a,b). Matters that demand special attention in using S-9 Mix can be found elsewhere (Mattern and Greim, 1978). The S-9 microsomal preparations are usually stored at or below −70°C. However, storage at −196°C is now preferable for retaining full activities of the enzyme in the S-9 (Tredger and Chhabra, 1976; Ashwood-Smith, 1980).

The mammalian liver is usually the source of the S-9 microsome preparations because it possesses the highest tissue concentration of mixed-function oxidases. However, the source of the metabolic activation system is not necessarily confined to only the mammalian liver. It may be worthwhile to utilize, for instance, cell-free extracts of the intestinal microbial flora instead of the mammalian liver S-9 (Callen, 1978). The intestinal microflora has a role in the etiology of colon and rectal cancer and metabolizes a large variety of substrates before they are absorbed through the digestive tract (Callen, 1978).

Hydroxylation, one of the metabolic activation reactions performed by the mixed-function oxidases system, can be mimicked by a nonenzymatic hydroxylation system (Udenfriend et al., 1954). This system, which consists of ascorbic acid, Fe^{2+}, and oxygen, can bring about the hydroxylation of aromatic compounds. There is no report on the SCE test coupled with the system alone, but it was shown that the nonenzymatic hydroxylation system, in which 2×10^{-2} M ascorbic acid and 10^{-3} M $FeSO_4$ activated cyclophosphamide, could generate an increase of chromatid aberrations in human leucocytes (Madle and Obe, 1978). It should also be remembered that 10^{-3} and 10^{-2} M ascorbate can induce chromosome aberrations in CHO cells and 10^{-4} M Fe^{2+} strongly enhances the chromosome-damaging capacity of ascorbate (Stich et al., 1979).

(b) THE CELL-MEDIATED METHOD. This method principally consists of cocultivation of target cells lacking the capacity for metabolic activation with lethally irradiated cells that can metabolize chemically nonreactive carcinogens (Huberman and Sachs, 1974). Popescu et al. (1977) tested three carcinogenic [benzo(*a*)pyrene, 7,12-dimethylbenz(*a*)anthracene, and 3-methylcholanthrene at 1.0 μg/ml] and two noncarcinogenic (pyrene and phenanthrene at 10 μg/ml) aromatic polycyclic hydrocarbons for their ability to induce SCEs in V79-4 cells when the latter were grown on an irradiated feeder layer of secondary Syrian hamster embryonic cells. In the presence of the feeder layer, benzo(*a*)pyrene, 7,12-dimethylbenz(*a*)anthracene, and 3-methylcholanthrene-treated cells exhibited a significant increase in SCE frequency, but no increase was observed with pyrene and phenanthrene. Hsu et al. (1979b) cocultivated Chinese hamster V79 cells with human pulmonary macrophages (PAMs) obtained from lung tissues. PAMs contain aryl hydrocarbon hydroxylase activity. The cells were exposed to the chemical procarcinogens benzo(*a*)pyrene (BP) and (±) *trans*-7,8-dihydroxy-7,8-dihydrobenzo(*a*)pyrene (7,8-diol). Seventy hours after the treatment was initiated, the cocultivated V79 cells were reseeded in medium containing

BrdUrd. An increase in the number of SCE by either BP or 7,8-diol after their activation by PAM was observed in V79 cells. However, it should be noted that in the experiments there was a considerable time lag between metabolic activation and cell fixation. Kligerman et al. (1980) cocultured freshly isolated rat hepatocytes with confluent human fibroblasts and treated them with cyclophosphamide at doses of 0.125–2 μg/ml. This cocultivation system activated cyclophosphamide, resulting in an increase of SCEs in the human fibroblasts.

(c) THE HOST-MEDIATED METHOD. The host-mediated method comprises the inoculation of foreign cells into laboratory mammals and the subsequent recovery of the cells after treatment of the host with exogenous chemicals (see a review by Mohn, 1977). Furukawa and Huang (1978), Sirianni and Huang (1978), and Furukawa et al. (1978) cultured Chinese hamster V79 cells in diffusion chambers (DC) and implanted them into the peritoneal cavity of mice. After implantation of the chambers with V79 cells, mice received injections of chemical agents. Then, after a certain period, the diffusion chambers were removed and the V79 cells were processed for SCE analysis. Injection of cyclophosphamide at 2.5, 5, 10, and 15 μg/g of body weight into the host caused a linear increase in SCE frequency (Furukawa and Huang, 1978; Sirianni and Huang, 1978). In addition, Sirianni and Huang (1978) tested cyclophosphamide, 1-(pyridyl-3)-3,3-dimethyltriazene, dimethylnitrosamine, diethylnitrosamine, benzo(*a*)pyrene, 7,12-dimethylbenz(*a*)anthracene, 3-methylcholanthrene, pyrene, anthracene, and perylene for SCE induction in this system. All the substances except diethylnitrosamine, pyrene, perylene, and anthracene caused an increase in the frequency of SCE. Diethylnitrosamine was effective only at the highest dose (6×100 μg/g of body weight) and only after pretreatment with Aroclor 1254. A significant dose-dependent increase in SCE was observed in V79 cells in DC in mice given injections of the flame retardant tris(2,3-dibromopropyl)-phosphate at doses of 60–500 μg/g body weight (Furukawa et al., 1978). However, this chemical agent can induce SCEs in V79 cells cultured *in vitro* without metabolic activation.

2. *Cultured Cells*

In the experiments shown in Table 5.5, S-9 Mix was added with a chemical during short "pulse" treatments. After treatment, cells were cultured in medium containing only BrdUrd for one (Tohda et al., 1980) or two cell cycles. The source of the S-9 was liver homogenates from polychlorinated biphenyl (Aroclor)-treated male rats except in the case of the work of

Strobel et al. (1979), who used mouse liver S-9 Mix, and the work of Thust et al. (1978), who used human liver extracts prepared from two patients and rat liver extracts from noninduced animals. There is a report that *Drosophila melanogaster* microsomes (S-20) can activate dimethylnitrosamine and increase the SCE rate in CHO cells (Baars et al., 1980). Thust et al. (1978) reported that human S-9 fractions, standardized for protein content, vary in their ability to carry out metabolic activation. The liver microsomes of one patient, who had received medazepam therapy, were highly active for dimethylnitrosamine (DMN) and cyclophosphamide, while those from the second patient failed to activate cyclophosphamide but did activate DMN. Takehisa and Wolff (1978a) reported that both the $9000 \times g$ and the $800 \times g$ supernatant prepared from rat liver pretreated with Aroclor were active, but $800 \times g$ supernatants derived from uninduced and phenobarbital-induced rat livers were very poor in metabolic activation of 2-acetylaminofluorene (2-AAF). De Raat (1977) showed in CHO cells that the formation of SCE by cyclophosphamide depends strongly on the induction of the microsomal enzyme systems used. Aroclor 1254 and phenobarbital are both good enzyme inducers. On the other hand, 3-methylcholanthrene hardly induces any extra SCEs. De Raat (1978) also showed in CHO cells that the chromosomal toxicity of styrene depends above all on the ratio of the activities of the monooxygenases (styrene → styrene oxide; activation) and the epoxide hydrases (decomposition of styrene oxide; detoxification). The amount of the S-9 (N.B., not the S-9 Mix) contained in the "pulse" treatment medium ranges from 0.17 to 21% in Table 5.5. Differences in amount of the S-9 microsome fractions in the culture medium do not influence the rate of SCE induction, as is evident in case of the treatments with cyclophosphamide when the results of Thust et al. (1978) are compared with those of Stetka and Wolff (1976b). Takehisa and Wolff (1977) examined optimum S-9 concentrations for metabolic activation between 0.2 and 10%, and found that the 0.5% level has the peak activation in inducing SCEs (see also Wolff and Takehisa, 1977). Since it was demonstrated that the electrophilic reactive metabolites of 2-AAF, such as *N*-hydroxy-2-acetylaminofluorene and *N*-acetoxy-2-acetylaminofluorene, which can react rapidly with proteins, quickly lose their SCE-inducing ability as the concentration of S-9 Mix increases (Takehisa & Wolff, 1977), it does not seem desirable to use high concentrations of the S-9, which is rich in proteins. However, it was found that the level of S-9 present in the metabolic system greatly affected the quantitative mutagenic response and this varied considerably from chemical to chemical in the bacteria mutagenicity test (Gatehouse and Delow, 1979). The time of "pulse" treatment is usually 1 hour, as shown in Table 5.5.

TABLE 5.5
SCE Induction in the Culture Cell System Coupled with S-9 Mix

Chemical agent	Indicator cell	Time of treatment	S-9 (%)	Dose of chemicals applied +S−9	Dose of chemicals applied −S−9	SCEs/cell (except where indicated) +S−9	SCEs/cell (except where indicated) −S−9	Ref.
DMN	CHO	1 h	21	2.7 mM	135 mM	85	10	3
DEN	CHO	1 h	21	4 mM	200 mM	23	11	
BP	CHO	1 h	3	25 μg/ml	25 μg/ml	37	11	1
Cigarette smoke condensate	CHO	1 h	3	300 μg/ml	—	23	11	
CP	CHO	20 min	0.17–0.5	10^{-3} M	10^{-3} M	1.2–2.3[a]	0.7[a]	5
MH	CHO	20 min	0.17–0.5	10^{-3} M	10^{-3} M	0.4[a]	0.4[a]	
AFB_1	CHO	1 h	1	10^{-5} M	10^{-5} M	44	13	9
CP	CHO	1 h	1	10^{-4} M	—	70	9	
MNU	CHO	1 h	1	10^{-4} M	10^{-4} M	44	44	
9,10-DMBA	CHO	35–45 min	1.5	10^{-3} M	10^{-3} M	17	14	2
CP	CHO	40 min	1.5	10^{-3} M	—	90	10	
AFB_1	CHO	30 min	0.2–10	10^{-6} M	10^{-6} M	1.8[a]	0.6[a]	7
BP	CHO	30 min	0.2–10	10^{-5} M	10^{-5} M	1.0[a]	0.5[a]	
AF	CHO	30 min	0.2–10	10^{-4} M	10^{-4} M	0.6[a]	0.6[a]	
AAF	CHO	30 min	0.2–10	10^{-4} M	10^{-4} M	0.5[a]	0.5[a]	

AAF	CHO	5 h	1	10^{-4} M	10^{-4} M	1.6[a]	0.6[a]	8
Divinyl ether	CHO	1 h	1	2% (v/v)	2% (v/v)	0.7[a]	0.5[a]	12
Fluoxene	CHO	1 h	1	2.5% (v/v)	2.5% (v/v)	1.1[a]	0.5[a]	
Ethyl vinyl ether	CHO	1 h	1	2% (v/v)	2% (v/v)	1.3[a]	0.5[a]	
CP	Chinese hamster fibroblasts PIE 71-2	30 min	21	1 mM	1 mM	2.0–3.8[a]	0.6[a]	11
DMN	Chinese hamster fibroblasts PIE 71-2	30 min	21	40 mM	120 mM	1.5–2.2[a]	0.6[a]	
Citrinin	Chinese hamster V79-E	1 h	21	5×10^{-4} M	5×10^{-4} M	0.4–0.6[a]	0.4[a]	10
BP	Mouse embryo blastcysts	30 min	1	1 μg/ml	—	0.77[a]	0.22[a]	4
EOC	CHO	1 h	6	1.7×10^{-2} M	1.7×10^{-2} M	7.15	7.50	6

TABLE 5.5
Continued

Chemical agent	Indicator cell	Time of treatment	S-9(%)	Dose of chemicals applied +S−9	Dose of chemicals applied −S−9	SCEs/cell (except where indicated) +S−9	SCEs/cell (except where indicated) −S−9	Ref.
CP	Syrian hamster cells A(T_1)Cl-3	1 h	21	10 μg/ml	—	35.6	6.6	13
	Mouse cells C3H/10T1/2CL8	1 h	21	10 μg/ml	—	63.0	12.6	
Trp-P-1	Human lymphoblastoid	2 h	0.25	10^{-5} M	10^{-5} M	20.1	12.0	14
Trp-P-2	Ibid.	2 h	0.25	10^{-5} M	10^{-5} M	22.1	10.3	
Glu-P-1	Ibid.	2 h	0.25	5×10^{-5} M	5×10^{-5} M	19.3	10.2	
2 Amino-α-carboline	Ibid.	2 h	0.25	10^{-4} M	10^{-4} M	17.3	7.2	

[a]SCEs/chromosome.

ABBREVIATIONS. AAF: 2-acetylaminofluorene. AFB_1: aflatoxin B_1. AF: 2-aminofluorene. BP: benzo(*a*)pyrene. CP: cyclophosphamide. DEN: *N*-nitrosodiethylamine. 9,10-DMBA: 9,10-dimethyl-1,2-benzanthracene. DMN: *N*-nitrosodimethylamine. EOC: 8-ethoxycaffeine. MH: maleic hydrazide. MNU: *N*-methyl-*N*-nitrosourea. Trp-P-1: 3-amino-1,4-dimethyl-5*H*-pyrido[4,3-*b*] indole. Trp-P-2: 3-amino-1-methyl-5*H*-pyrido[4,3-*b*] indole. Glu-P-1: 2-amino-6-methyldipyrido[1,2-*a*:3′,2′-*d*] imidazole. 2-Amino-α-carboline: 2-amino-9*H*-pyrido[2,3-*b*] indole.

REFERENCES. 1. De Raat, 1979. 2. Galloway and Wolff, 1979. 3. Natarajan et al., 1976. 4. Pedersen et al., 1978. 5. Stetka & Wolff, 1976b. 6. Strobel et al., 1979. 7. Takehisa and Wolff, 1977. 8. Takehisa and Wolff, 1978a. 9. Vijayalaxmi et al., 1979. 10. Thust and Kraist, 1979. 11. Thust et al., 1978. 12. White et al., 1979. 13. Benedict et al., 1978. 14. Tohda et al., 1980.

Takehisa and Wolff (1978a) showed that a 30 min short incubation time of 2-AAF with S-9 Mix did not lead to an increase in SCEs, but that longer incubation times up to 5 hours could bring about an increase. There is a report that in the bacterial mutagenicity test, the metabolic activity of the S-9 Mix lasts for the initial 5 hour period of incubation until the NADPH-generating system is exhausted (Gatehouse and Delow, 1979). However, other evidence suggests that while the microsomes remain viable for prolonged periods, metabolites are produced almost immediately or not at all in similar bacterial tests (Rickard et al., 1978). The SCE-inducing activity of *N-methyl-N*-nitrosourea (MNU) is not influenced by S-9 Mix (Vijayalaxmi et al., 1979). MNU is a direct mutagen and a monofunctional alkylating agent. This contrasts with the fact that *N*-acetoxy-2-acetylaminofluorene and *N*-hydroxy-2-acetylaminofluorene lose their SCE-inducing ability when incubated with S-9 Mix (Takehisa and Wolff, 1977), although they are highly reactive compounds and are effective in inducing SCEs. MNU, when incubated in culture medium with 15% fetal bovine serum, became inactive with a half-life of 8 min (Jensen et al., 1977). Whether the alkylating agents in general behave like MNU or *N*-acetoxy-2-acetylaminofluorene remains to be seen. The addition of S-9 Mix does not necessarily activate chemical agents, for we know that it deactivates some clastogenic chemicals (Hsu and Au, 1979) and reduces the SCE-inducing activity of some chemicals (Csukás et al., 1979; White and Hesketh, 1980).

3. *Lymphocytes*

Experiments that use lymphocytes as target cells and incorporate the metabolic activation system are not so numerous. Dimethylnitrosamine (DMN) was activated and produced a significant increase of chromosomal aberrations in human lymphocytes cultured for 24 hours after a 45 min exposure to microsome fractions prepared from mouse liver (Bimboes and Greim, 1976). Cyclophosphamide (CP) has been activated by S-9 Mix to induce chromatid translocations in human lymphocytes (Madle and Obe, 1977). In this experiment, CP was incubated with rat liver S-9 Mix for 1 hour. This mixture was put into a dialysis bag that was submerged in human whole blood cultures for 3 hours at 48 hours of culture time. The activated products diffused out of the bag into the cultures.

As to the SCE test, a 1 hour exposure of human lymphocytes to aflatoxin B_1 (AFB_1) in the presence of the rat liver S-9 Mix resulted in a marked enhancement of the action of AFB_1 in inducing SCEs (Vijayalaxmi et al., 1979). The treatment was applied at 48 hour culture time and the cells were cultured for an additional 24 hours. Saxholm et al. (1979) treated human lymphocytes for 2 hours with saccharin and S-9 Mix in a dialysis bag at 24 hours before harvest and obtained no elevated SCE frequency. Csukás et al. (1979) treated human lymphocytes at 20 hours before harvest with rat liver S-9 Mix and the chemical agent CP, urethane, or hydroxyurethane for 1 hour. CP was activated and increased the SCE frequency. However, S-9 Mix deactivated urethane and hydroxyurethane and thus reduced the induced SCE frequency. The same detoxifying effect of S-9 Mix on the SCE frequency was observed in human lymphocytes exposed to 1-chloro-2,3-epoxypropane (epichlorohydrin) for 2 hours at 48–50 hours of culture time (White, 1980). Human peripheral lymphocytes were exposed to CP (10^{-4} to 10^{-6} M) and rat liver S-9 Mix for 60–120 min at 0 or 48 hours of culture. CP was activated and increased SCEs. Exposure after 48 hours from the commencement of culture was more effective for SCE induction (White and Hesketh, 1980). Takehisa and Wolff (1978b) found that in rabbit lymphocytes the metabolic changes brought about by PHA stimulation do affect the induction of enzymes involved in the activation of the procarcinogens 2-aminofluorene (2-AF) and 2-acetylaminofluorene (2-AAF). In 30 min treatments with 2-AF after PHA stimulation an increase of SCEs occurred only with the S-9 Mix. In 30 min treatments before stimulation no increase in SCEs was observed either with or without the S-9 Mix. 2-AAF did not show any increase in SCEs in the same treatments. Rabbit blood cultures were treated with 2-AF and 2-AAF either during the entire period of PHA stimulation or during the period of culture with BrdUrd. With both 2-AF and 2-AAF there was no increase in SCEs when treatment occurred during the PHA-stimulation period, but the SCEs increased when the treatment took place during the period of culture with BrdUrd. Takehisa and Wolff observed that benzo(*a*)pyrene (BP) can be activated by rabbit lymphocytes independently of PHA stimulation. The addition of S-9 Mix reduced the BP-induced increase in SCEs somewhat. The results described above suggest that the stage of the treatment with chemical agents in lymphocyte culture is a crucial factor for activation of chemical agents leading to SCE induction, and that the whole blood culture provides the enzymes of metabolic activation for some types of the chemical agents.

As for the treatment time in human lymphocyte cultures, a "pulse" treatment with chemical agents, but without *in vitro* metabolic activation, prior to PHA stimulation showed that mitomycin C (MMC) persisted in cultures and continued to increase SCE frequencies for about 30 hours (Ishii and Bender, 1978a,b) and that the alkylating chemicals MMC, chlorambucil, and thiotepa increased SCEs, while such increases were not observed in cultures exposed to methotrexate, cytarabin (cytosine arabinoside), and bleomycin (Littlefield et al., 1979). The latter results were interpreted as indicating that not all chemical agents that induce

SCEs in proliferative cells are capable of inducing long-lived lesions in the DNA of G_0 lymphocytes. When human G_0 lymphocytes exposed to mitomycin C for 1 hour were stored for various periods, up to 9 days, before stimulation with PHA, induced SCEs increased with increasing storage time (Evans and Vijayalaxmi, 1980).

Several workers have made reports that are suggestive of what types of activation systems lymphocyte or whole blood cultures provide in themselves. Rüdiger et al. (1976) and Schönwald et al. (1977) treated human lymphocytes with BP and observed increases in SCEs. The rate of SCE did not correlate with different rates of BP metabolism in lymphocytes between individuals (Rüdiger et al., 1976). Craig-Holmes and Shaw (1977) observed that DMN and diethylnitrosamine (DEN) did not produce an increase of SCEs, but depending on time of exposure, BP did. This behavior implies that the human whole blood culture system lacks the metabolizing system for DMN and DEN but possess the metabolizing system for BP that is inducible by BP. The fact that BP failed to induce SCEs when it was present for 30 hour culture, instead of 70 hour culture (Rüdiger et al., 1976), is also indicative of the induction of BP-metabolizing enzymes by BP.

There are probably more complicated metabolic activation systems intrinsic to whole blood cultures. Ray and Altenburg (1978) showed that the presence of red blood cells, especially red blood cell lysate, can activate Na_2SeO_3 in producing an elevation of SCE frequencies.

4. *Factors Affecting the SCE Incidence* in vitro

Several factors that affect the incidence of SCE *in vitro* have been documented. Among these factors, the importance of the difference in BrdUrd concentration, cell lines, the age of the culture, and the culture conditions, especially the serum component, were summarized by Kato (1977b).

The data indicate that composition of the culture medium is an important factor influencing the frequency of SCEs. Several experiments recognized amino acids in the culture medium as one of the contributing factors. The incidence of UV-induced SCEs was reduced in CHO cells that had been arrested in medium deficient in arginine (MacRae et al., 1979a), while in cells treated with *N*-methyl-*N′*-nitro-*N*-nitrosoguanidine (MNNG) similar SCE levels were observed in arginine-deprived and proliferating cells (MacRae et al., 1979b). An increase in the rate of SCEs was found when V79 Chinese hamster cells were exposed to increasingly severe degrees of arginine and lysine deficiency (Schempp and Krone, 1979). The importance of serum factors had been pointed out earlier by

Kato (1977b). Kato and Sandberg (1977) reported that serum contained some heat-labile and dialyzable SCE-inducing factors. The serum or plasma used in the culture medium influences cell cycle time (Auf der Maur and Berlincourt-Böhni, 1979). Human lymphocytes grown in medium supplemented with fetal calf and human AB serum showed higher yields of SCE than those cultured with autologous serum (Ghosh and Nand, 1979). Different types of culture medium could be the cause of variability in the rate of SCE incidence (Bianchi et al., 1979). Differences in types of culture medium are also known to influence the incidence of chromosomal breakage. Human lymphocyte cultures incubated with TC medium 199 showed higher chromosomal breakage than those incubated with Eagle's MEM or RPMI 1629 (Keck and Emerit, 1979). The elevated incidence of chromosome breakage in patients with Crohn's disease was much more striking in cultures set up with TCM 199 than in cultures set up with RPMI 1629 rich in L-cysteine (Emerit et al., 1979). In cocultivation experiments with different cell lines, an unknown agent(s) produced by one of the cocultivated cell lines could be important. Tice et al. (1978) concluded that Bloom's syndrome cells produced an agent(s) capable of causing an increased SCE response in normal cells. This has not been accepted fully (Bryant et al., 1979). Bartram et al. (1979) and Rüdiger et al. (1980) obtained results indicating that normal cell cultures were able to supply Bloom's syndrome cells with an unknown factor that reduced their high rate of SCEs. Alhadeff et al. (1980) also obtained data that the high rate of SCE in Bloom's syndrome cells had to be considered the consequence of a lost normal function.

Culturing cells in media that had been exposed to fluorescent light resulted in a significant increase in SCE levels in human fetal lung fibroblasts (Monticone and Schneider, 1979). Medium irradiated with either 300 rad (Pant et al., 1976) or up to 2 Mrad (Vijayalaxmi, 1980) of cobalt gamma rays did not induce SCEs in human lymphocytes. Hyperthermia (44°C for 20–60 min) elevated the frequency of SCE in CHO cells (Livingston and Dethlefsen, 1979). Culturing Chinese hamster cells at temperatures above 39°C enhanced SCE formation (Kato, 1980b).

In human peripheral lymphocyte cultures, considerable variation in the frequency of SCEs seems to exist not only among cells, but also among individuals (Lezana et al., 1977), and even among repeated cultures at similar BrdUrd concentrations. This may be attributable in part to the existence of several, or at least two, subpopulations of cells in culture (Beek and Obe, 1979). Cells differing in their cycle time may differ in their sensitivity to an SCE-inducing agent. Santesson et al. (1979) reported that (1) rate of SCEs in T lymphocytes was higher than that in B lymphocytes, (2) T lymphocytes were less tolerant to BrdUrd treatment than B lymphocytes, (3) the ratio of T lymphocytes to B lymphocytes was different for first, second, and third generation mitoses, and (4) this ratio changed depending on differences in BrdUrd concentration. Trenimon-treated human lymphocyte cultures, however, indicated a homogenous sensitivity of early- and late-proliferating cells for the induction of SCEs (Beek and Obe, 1979), but in another report, there was a clear drop of the Trenimon-induced SCE frequencies at fixation times 91 hours and later (Riedel and Obe, 1980). The incidence of SCE appeared to be independent of the proliferative properties of cultured lymphocytes, such as length of cell cycle, fast or delayed reponse to PHA and the number of divisions performed *in vitro* (Giulotto et al., 1980).

Differences in cell types are related to the SCE response to chemical agents. Embryonal carcinoma cells did not show the marked increase in MMC-induced SCEs that differentiated cells did (Fabricant et al., 1979). For SCE induction by Trenimon, human fibroblasts are far more sensitive than human lymphocytes (Hayashi and Schmid, 1975). Human fibroblasts are more sensitive than CHO cells in inducing SCEs by chromate (MacRae et al., 1979d). There is also a tissue-specific difference in SCE frequency. The incidence of SCE was significantly lower in human bone marrow cells labeled with BrdUrd *in vitro* than in human lymphocytes (Knuutila et al., 1978; Becher et al., 1979). The sequence of carcinogen and BrdUrd treatment is responsible for differential SCE frequency in MNNG-treated V79 cells (Popescu et al., 1980). The addition of SV40 to four human diploid fibroblast cultures increased the SCE frequency from the time when cells exhibited tumor antigen (Nichols et al., 1978). Mouse embryo fibroblasts infected by Rauscher leukemia virus showed an increase of the SCE frequency (Brown and Crossen, 1976). Cellular aging resulted in an altered SCE response. The SCE frequencies in human fetal lung fibroblasts (IMR-90, WI38) induced by mitomycin C, ethyl methanesulfonate, and *N*-acetoxy-2-acetylaminofluorene declined as a function of serial passage (*in vitro* aging) (Schneider and Monticone, 1978).

D. In vivo *SCE Test*

The significance of the *in vivo* SCE test resides in utilizing the host's metabolism in activating and inactivating the chemical agents. In the *in vivo* SCE test systems there are two procedures of BrdUrd labeling that differ in time of their application. One is the *in vivo* BrdUrd labeling. In it the test drugs are administered by intraperitoneal injection or other ways either before or after the BrdUrd treatment. The mitoses of indicator cells are examined directly without any cell culture. The other procedure is the *in vitro* BrdUrd labeling. The indicator cells are isolated from the host animal, which has previously received the test drugs. Then the cells are labeled with BrdUrd in cell culture. The former method is suitable to see the *in vivo* induction of SCEs by chemical agents. The latter is suitable for examining *in vivo* persistence of chemical-agent-induced DNA damage represented as SCEs.

1. *The* in vivo *BrdUrd Labeling and SCE*

Four different methods for applying *in vivo* BrdUrd labeling have been used: (1) mice received a subcutaneous implantation of a 50–55 mg tablet of BrdUrd and were sacrificed 26 hours later to obtain bone marrow cells (Allen et al., 1977); (2) mice or rats were continuously infused through a tail vein for 26 hours with 50 mg BrdUrd/kg body weight/hour (in the first report, 100–120 mg/kg/hour) (Schneider et al., 1976); (3) mice were injected intraperitoneally with 0.4 mg FUdR/g body weight and 3 hours later they received six or nine times 0.04 mg BrdUrd/g body weight during 6 hours (Vogel and Bauknecht, 1976); and (4) mice received 14 hourly intraperitoneal (0.2 ml) injections of BrdUrd (10^{-2} M) together with deoxycytidine (5×10^{-3} M) (Allen and Latt, 1976a). Similarly, mice were given 9 hourly intraperitoneal injections (0.2 ml each) of 2×10^{-2} M BrdUrd (Conner et al., 1978). These procedures overcome the difficulty in *in vivo* BrdUrd labeling. The *in vivo* BrdUrd level decreases immediately after BrdUrd injection because of the high metabolic activity of liver enzymes. After 1 hour, only 20% of the original BrdUrd concentration is left in the serum (see Bauknecht et al., 1977). The data relevant to the *in vivo* BrdUrd labeling are shown in Table 5.6.

Indirect mutagens such as cyclophosphamide, benzo(*a*)pyrene, 2-acetylaminofluorene, and 7,12-dimethylbenzanthracene were activated and an increase of SCEs *in vivo* resulted. Without losing their SCE-inducing activity, direct mutagens such as mitomycin C reached the tissue where the indicator cells were isolated. Mitomycin C, methylmethanesulfonate, and cyclophosphamide showed a transplacental activity and mitomycin C affected spermatogonia in mice, but it was suggested that mutagenic drugs differ in their transplacental activities (Kram et al., 1980). In regard to the influence of the mode of exposure to a chemical, Kram et al. (1979b) reported that administration of mitomycin C intravenously was as consistent as intraperitoneal administration but yielded significantly higher frequencies of SCEs. It is interesting that *N*-acetoxy-2-acetylaminofluorene, which is highly electrophilic and reacts rapidly with proteins (Miller, 1970), produced an increase of SCEs in bone marrow cells in spite of its intravenous administration and that caffeine brought about a slight increase in SCEs. No significant difference in baseline level of SCE frequencies was observed among the cell types or between hepatectomized and nonhepatectomized animals (Conner et al., 1979a). However, 2-acetylaminofluorene was more activated in hepatectomized mice than in nonhepatectomized ones (Schreck et al., 1979).

There were a few reports on how much host conditions contribute to the agent-induced SCE frequencies. Dietary fluoride did not affect the SCE frequencies induced by mitomycin C (Kram et al., 1978a). Mitomycin C-induced SCEs decline with age (Kram et al., 1978b; Nakanishi et al., 1979). An increase in adriamycin-, cyclophosphamide-, and mitomycin C-induced SCE levels was seen with post embryonic development (Nakanishi et al., 1980); however, it was shown in mice that cyclophosphamide-induced SCE frequencies remained constant in maternal cells but declined in the fetus throughout development (Kram et al., 1980). Studies of female C57B1/6J mice and male Wistar rats indicated that male/female differences in SCE frequencies were not present in the case of the treatment with mitomycin C (Kram et al., 1978b).

2. *The* in vitro *BrdUrd Labeling and SCE*

In the *in vivo* SCE test with *in vitro* BrdUrd labeling, the indicator cells taken from the animal exposed to the test drugs are mostly peripheral lymphocytes. This method has the advantages of enabling the examination of the *in vivo* effect of a chemical agent on DNA without sacrificing the test animal and the tracking of the persistence of an SCE effect in the test animal. The analysis of SCE in peripheral lymphocytes, however, is complicated by the time lag between the induction of DNA damage and its demonstration. It should be remembered that at the time of *in vivo* exposure the peripheral lymphocytes are in the G_0 phase and that to demonstrate SCE the cells must be stimulated and then pass through two cell cycles *in vitro*.

Stetka and Wolff (1976a) first reported that the frequency of SCEs in peripheral lymphocytes one day after intraperitoneal injection of rabbits with methylmethanesulfonate (MMS), ethylmethanesulfonate (EMS), and cyclophosphamide (CP) increased and these increases were short-lived. Within 2 weeks the SCE frequency declined to the preexposure level. The same pattern of SCE response was found in rabbits following the intraperitoneal injection of the carcinogens 2-aminofluorene and 2-acetylaminofluorene: an increase 1 day after injection and a subsequent gradual decrease to the control level (Takehisa and Wolff, 1978b). In rabbits receiving a high dose of 2-aminofluorene (200 mg/kg body weight), the SCE frequency was slower in returning than it was after lower doses.

TABLE 5.6

In vivo *Induction of SCEs by Chemical Agents in the* in vivo *BrdUrd Labeling*

Chemical agent				Mean SCEs/cell		
Name and dose (body weight)	Mode of administration[a]	Animal[b]	Indicator cells	Control	Treated	Ref.
Cyclophosphamide						
5 mg/kg	i.p.	Ch	Cheek pouch	4.8	9.5	17
			Bone marrow	5.4	17.4	
2.5 mg/kg	i.p.	m	Regenerating liver	5.3	14.8	16
			Bone marrow			
			Hepatectomized	4.0	13.9	
			Nonhepatectomized	2.8	10.1	
12.5 mg/kg	i.p.	m	Bone marrow	4	15	18
5 mg/kg	i.p.	Ch	Bone marrow	3.3	8.8	2
			Embryonic liver	4.1	8.8	
1 mg/kg	i.p.	m	Bone marrow	3.7	7.2	4
2 × 5 mg/kg[c]	i.p.	Ch	Bone marrow	3.8	12.4	3
45 mg/kg	i.v.	m	Bone marrow	4.8	123.0	12
35 mg/kg	i.p.	m	Bone marrow	1.5	23.3	22
			Salivary gland	2.3	24.2	
20 mg/kg	i.p.	Ch	Bone marrow	4	20	23
20 mg/kg	i.v.	m	Bone marrow	5.0	51.5	13
10 mg/kg	i.v.	m	Bone marrow	4.8	22.0	10
			Embryo	3.6	25.0	

5 mg/kg	i.p.	m	Bone marrow	7.7	23.4	27
			Spleen	6.7	19.3	
5 mg/kg	i.p.	m	Spermatogonia	1.7	4.9	28
10 mg/kg	i.v.	m	Fetal cells	4.0	31.1	32
			Fetal liver	4.0	35.8	
			Fetal lung	4.0	21.0	
			Fetal gut	4.0	24.9	
			Maternal bone marrow	5.0	21.4	
Mitomycin C						
0.5 mg/kg	i.p.	m	Spermatogonia	1.8	7.2	1
2.5 mg/kg	i.v.	m	Bone marrow	4.3	39.2	8
3 mg/kg	i.v.	m and rats	Bone marrow	4.5	40.0	9
5 mg/kg	i.v.	m	Bone marrow	4.8	99.5	12
1 mg/kg	i.v.	m	Bone marrow	5.0	21.2	13
1 mg/kg	i.v.	m	Bone marrow	4.8	21.2	10
			Embryo	3.6	15.8	
1 mg/kg	i.v.	m	Fetal cells	4.0	12.2	32
			Maternal bone marrow	5.0	15.4	
0.75 mg/kg	i.p.	Ch	Bone marrow	4	10	23
AAF						
25 mg/kg	i.p.	m	Regenerating liver	6.7	11.4	16

TABLE 5.6
Continued

Chemical agent				Mean SCEs/cell		
Name and dose (body weight)	Mode of administration[a]	Animal[b]	Indicator cells	Control	Treated	Ref.
			Bone marrow			
			Hepatectomized	5.1	11.6	
			Nonhepatectomized	3.3	6.7	
Styrene						
565 ppm	Inhaled (6 h/day for 4 days)	m	Bone marrow			
			Hepatectomized	3.0	11.9	7
			Nonhepatectomized	3.3	11.0	
			Liver	3.8	12.2	
Triaziquon						
0.125 mg/kg	i.p.	m	Bone marrow	4	30	18
0.125 mg/kg	i.p.	m	Bone marrow	3.7	17.2	4
0.025 mg/kg	i.p.	Ch	Bone marrow	4	13	23
MMS						
20 mg/kg	i.p.	Ch	Bone marrow	3.3	10.4	2
			Embryonic liver	4.1	11.8	
10 mg/kg	i.p.	Ch	Bone marrow	3.3	9.0	11
20 mg/kg	i.p.	Ch	Bone marrow	4	20	23
80 mg/kg	i.p.	Ch	Bone marrow	4.8	12.6	30
100 mg/kg	i.p.	Ch	Bone marrow	4.8	34.6	

cis-DDP						
13.85 mg/kg	i.p.	m	Bone marrow	6.5	17.7	19
DMN						
2 × 1 mg/kg[c]	i.p.	m	Bone marrow	3.7	10.6	4
DEN						
200 mg/kg	i.p.	Ch	Bone marrow	3.3	4.0	5
2 × 100 mg/kg[c]	i.p.	m	Bone marrow	3.7	4.2	4
MNNG						
2 × 0.15 mg/kg[c]	i.p.	m	Bone marrow	3.7	7.2	4
50 mg/kg	i.v.	m	Bone marrow	5.0	35.7	13
EMS						
200 mg/kg	i.v.	m	Bone marrow	5.0	19.0	13
DḾBA						
0.5%	0.1 ml painted	Ch	Cheek pouch	5.6	11.2	17
			Bone marrow	5.2	8.5	
7.5 mg/kg	i.p.	Ch	Bone marrow	3.4	7.5	6
Benzo(*a*)pyrene						
2 × 450 mg/kg[c]	i.p.	Ch	Bone marrow	3.9	10.6	3, 15
50 mg/kg	i.p.	Ch	Bone marrow	3.4	6.0	6

TABLE 5.6
Continued

Chemical agent				Mean SCEs/cell		
Name and dose (body weight)	Mode of administration[a]	Animal[b]	Indicator cells	Control	Treated	Ref.
Anthracene						
2 × 450 mg/kg[c]	i.p.	Ch	Bone marrow	3.9	3.7	15
Dibenzanthracene						
2 × 450 mg/kg[c]	i.p.	Ch	Bone marrow	3.9	4.9	15
Chrysene						
2 × 450 mg/kg[c]	i.p.	Ch	Bone marrow	3.9	5.1	15
Phenanthrene						
2 × 450 mg/kg[c]	i.p.	Ch	Bone marrow	3.9	5.5	15
50 mg/kg	i.p.	Ch	Bone marrow	3.4	4.0	6
Benzo(*e*)pyrene						
2 × 450 mg/kg[c]	i.p.	Ch	Bone marrow	3.9	5.5	15
Benzo(*b*)fluoranthene						
2 × 450 mg/kg[c]	i.p.	Ch	Bone marrow	3.9	5.6	15
Benzanthracene						
2 × 450 mg/kg[c]	i.p.	Ch	Bone marrow	3.9	6.1	15
Natulan						
200 mg/kg	i.p.	Ch	Bone marrow	3.3	12.0	5

PDMT						
2 × 30 mg/kg[c]	i.p.	m	Bone marrow	3.7	6.9	4
Caffeine						
2 × 100 mg/kg[c]	Stomach tube	Ch	Bone marrow	3.8	5.4	3
50 mg/kg	i.v.	m	Bone marrow	5.0	8.8	13
Daunomycin						
6 mg/kg	i.v.	m	Bone marrow	5.0	13.0	13
5 mg/kg	i.v.	m	Fetal cells	4.0	7.1–8.7	32
	i.v.	m	Maternal bone marrow	5.0	19.9	
Aflatoxin B_1						
20 mg/kg	i.v.	m	Bone marrow	5.0	10.4	13
4-Nitroquinoline 1-oxide						
60 mg/kg	i.v.	m	Bone marrow	5.0	62.1	13
Streptonigrin						
5 mg/kg	i.v.	m	Bone marrow	5.0	30.0	13
Dibromomannitol						
90 mg/kg	i.v.	m	Bone marrow	5.0	16.7	13
AAAF						
4 mg/kg	i.v.	m	Bone marrow	5.0	8.9	13

TABLE 5.6
Continued

Chemical agent				Mean SCEs/cell		
Name and dose (body weight)	Mode of administration[a]	Animal[b]	Indicator cells	Control	Treated	Ref.
Tris-BP						
1500 mg/kg	i.v.	m	Bone marrow	5.0	10.1	13
Hoechst 33258						
90 mg/kg	i.v.	m	Bone marrow	5.0	4.9	13
Cytosine arabinoside						
600 mg/kg	i.v.	m	Bone marrow	5.0	6.6	13
Proflavine						
6 mg/kg	i.v.	m	Bone marrow	5.0	4.1	13
Ethanol						
10% (v/v)	3–16 weeks, in drinking water	m	Bone marrow	4.7	6.0–7.9	14
20% (v/v)	12–16 weeks, in drinking water	m	Bone marrow	4.7	6.6–7.6	
10% (v/v)	3–6 weeks, in drinking water	Rat	Bone marrow	5.8	5.9	29
20% (v/v)	3–6 weeks, in drinking water	Rat	Bone marrow	5.8	5.9	

Acetaldehyde						
1 ml (10^{-4}% v/v)	i.p.	m	Bone marrow	4.7	7.9	14
Chlorpropamide						
7.1 mg/kg	Stomach tube	m	Bone marrow	4.5	7.0	20
		Ch	Bone marrow	3.8	5.6	
Tolbutamide						
28.6 mg/kg	Stomach tube	m	Bone marrow	4.5	7.9	20
		Ch	Bone marrow	3.8	5.7	
β-Cytotropic sulphonylureas						
0.2–28.6 mg/kg	Stomach tube	m	Bone marrow	4.5	4.9	20
		Ch	Bone marrow	3.8	3.8–4.3	
Benzene						
3100 ppm	Inhaled for 4 h	m	Bone marrow	4.6	8.5	21
Phenobarbital						
6 × 50 mg/kg (twice daily for 3 days)	i.p.	m	Bone marrow	4.6	4.2	
Phenobarbital pretreatment						
6 × 50 mg/kg			m			
plus benzene (3100 ppm)		m	Bone marrow	4.6	8.3 ♂ 12.7 ♀	
Procarbazine						
100 mg/kg	i.p.	Ch	Bone marrow	4	14	23
Bleomycin						
20 mg/kg	i.p.	Ch	Bone marrow	4	9	23

TABLE 5.6
Continued

Chemical agent				Mean SCEs/cell		
Name and dose (body weight)	Mode of administration[a]	Animal[b]	Indicator cells	Control	Treated	Ref.
Adriamycin						
12 mg/kg	i.v.	m	Bone marrow	4.8	32.9	13
5 mg/kg	i.v.	m	Bone marrow	5.0	15.5	13
5 mg/kg	i.v.	m	Bone marrow	4.8	17.6	10
			Embryo	3.6	6.5	
Na_2SeO_3						
3–6 mg/kg	i.p.	Ch	Bone marrow	4.8	4.6–11.4	30
BCNU						
8.8 mg/kg	i.v.	m (AKR/J)	Bone marrow	5.9	23.2	31
		m (BDF_1)	Bone marrow	5.4	28.0	
Urethane						
50 mg/kg	i.p.	m	Liver, hepatectomized	9.9	22.9	33
			Bone marrow			
			Hepatectomized	7.9	12.7	
			Nonhepatectomized	7.7	12.7	
50 mg/kg	i.p.	m	Spermatogonia	1.7	1.7	

400 mg/kg	i.p.	m	Spermatogonia	1.7	3.3	
Vinyl chloride						
1.25%	Inhaled for 24 h	Ch	Bone marrow	4.4	22.3	24
2.5%	Inhaled for 24 h	Ch	Bone marrow	4.5	33.2	
Isoniazid						
100 mg/kg	i.p.	Ch	Bone marrow	3.0	3.2	25
Ascorbic acid						
10 g/kg	i.p.	Ch	Bone marrow	3.0	2.8	26
	Stomach tube	Ch	Bone marrow	3.0	3.2	

[a]i.p., intraperitoneal injection; i.v., intravenous injection.

[b]Ch, Chinese Hamster; m, mice.

[c]Injection, twice.

ABBREVIATIONS. AAF: 2-acetylaminofluorene. AAAF: *N*-acetoxy-2-acetylaminofluorene. DMBA: 7,12-dimethylbenzanthracene. DMN: dimethylnitrosamine. DEN: Diethylnitrosamine. *cis*-DDP: *cis*-dichlorodiamminoplatinum-II. EMS: ethyl methanesulfonate. MMS: methyl methanesulfonate. MNNG: *N*-methyl-*N'*-nitro-*N*-nitrosoguanidine. Natulan: *N*-isopropyl-α-(2-methylhydrazino)-*p*-toluamide. PDMT: dimethylphenyltriazene. Tris-BP: tris-(2,3-dibromopropyl)phosphate. β-Cytotropic sulfonylureas: carbutamide, tolazamide, acetohexamide, glymidine, gliquidone, glipizide, glibornuride, glibenclamide, glisoxepide. BCNU: 1,3-bis(2-chloroethyl)-1-nitrosourea.

REFERENCES. 1. Allen and Latt, 1976a. 2. Basler, 1979. 3. Basler et al., 1979. 4. Bauknecht et al., 1977. 5. Bayer, 1978. 6. Bayer and Bauknecht, 1977. 7. Conner et al., 1979b. 8. Kram et al., 1978a. 9. Kram et al., 1978b. 10. Kram et al., 1979a. 11. Marquardt and Bayer, 1977. 12. Nakanishi et al., 1979. 13. Nakanishi and Schneider, 1979. 14. Obe et al., 1979. 15. Roszinsky-Köcher et al., 1979. 16. Schreck et al., 1979. 17. Shuler and Latt, 1979. 18. Vogel and Bauknecht, 1976. 19. Wiencke et al., 1979. 20. Renner and Münzer, 1980. 21. Tice et al., 1980. 22. Ramirez, 1980. 23. Renner, 1979. 24. Basler and Röhrborn, 1980. 25. Speit et al., 1980a. 26. Speit et al., 1980c. 27. Allen et al., 1977. 28. Allen and Latt, 1976b. 29. Tates et al., 1980. 30. Norppa et al., 1980b. 31. Biegel et al., 1980. 32. Kram et al., 1980. 33. Roberts and Allen, 1980.

If the injection of 2-aminofluorene was repeated twice after an interval of 4 weeks, when the frequency of SCEs returned to its control level, the same transient increase in SCEs was produced. The same was true after injection of 2-acetylaminofluorene. Stetka et al. (1978) studied extensively the effect of repeated exposure to the direct mutagen MMC. They injected MMC, 0.5 mg/kg, intraperitoneally into rabbits once a week for 8 weeks. This treatment initially induced small increases in SCE frequency within 1 day of injection, which were followed by a return to control levels within 1 week. After the fourth injection, however, the frequency failed to return to normal. After the fifth injection it showed a four-fold increase over the control, which was sustained for the remaining 3 weeks of treatment and for an additional 2 weeks thereafter. The frequency then dropped to twice the control value and remained at this level for more than 4 months. Ohtsuru et al. (1980) observed the same pattern in the SCE responses in cancer patients receiving i.v. injections of MMC at a dose of 4 mg. The frequencies of SCEs in lymphocytes of cancer patients were similar to those of healthy controls. After the first and second i.v. injections of MMC, however, the frequencies of SCEs increased with time, reached a peak in about 24 hours, and returned to the preinjection level in about 48 hours. After the third and fourth injections, the frequency increased further and remained high throughout the 3 day period. The SCE values in lymphocytes of the two patients were at the pretreatment level when they were examined about 3 months later. Littlefield et al. (1980) did not observe any reduction in SCE frequency in cultured lymphocytes from a patient within 8 days after MMC treatment. As to the possibility that an active mutagen is transferred with the blood inoculum into the *in vitro* culture, DuFrain et al. (1979) showed in cyclophosphamide-treated rabbits that this hypothesis is not true. The work of Stetka et al. (1978) suggests that the analysis of lymphocyte SCEs may be an excellent assay for repeated low-level exposure of man to mutagenic/carcinogenic chemicals, even when the individual is not available for testing until long after the last exposure, while the transient SCE increase following acute exposure precludes the use of this assay as an indicator of human-chromosome damage unless the individual is available for testing soon after the suspected exposure. Raposa (1978) examined SCEs in lymphocytes from patients receiving cytostatic drugs according to some therapeutic protocol. In one patient, who received CP as monotherapy, 100 mg daily for 22 days, the increase in SCEs was evident 13 days after the start of therapy, whereas no increase was found 3 weeks after therapy ceased. The other patient, who received one large dose of CP and vinblastine, showed an increase in SCEs from the fifth day after the drug injection. The level remained elevated for 2 days, after which the same cytostatics were given at the same dose. There then was an extreme increase in SCEs, which then returned to the preexposure level 10 days later. Seventeen days after the second injection, a third injection of the same drugs was given. The frequency of SCEs increased again. Cytosine arabinoside in combination with perorally administered thioguanine had no effect on SCE frequency in four patients. Nevstad (1978) observed the transient nature of the increase in SCEs in a patient who received 125 mg of adriamycin intravenously as a single dose. An increase in the SCE frequency (15.5 SCEs/cell with a control level of 5.0) was observed 10 min after the injection. Six hours after the injection the number of SCEs decreased to 10.0/cell and after 1 month it was down to the control level. Musilová et al. (1979) reported that in patients who received 2.2 and 3.0 g of CP perorally for 3 weeks there was an increase in SCEs, and the elevated SCE rate was maintained for 4 months after the therapy. In a patient who was treated with a total peroral dose of 150 mg busulfan in the course of 5 weeks, the increase of SCE frequency reached its peak 5 weeks after the termination of the treatment. In a patient treated with a single intravenous 800 mg dose of adriamycin the SCE rate increased and remained elevated for 3 months. No rise in the SCE frequency was found in two patients treated with a single intravenous dose of 5000 mg of fluorouracil. Lambert et al. (1979b) have studied the SCE frequencies in patients receiving various types of chemotherapy against melanoma and other malignant disorders. Lymphocytes for SCE analysis were sampled during drugfree intervals of the treatment schedule. In the majority of the patients the frequency of SCE 2–8 weeks after drug administration was similar to that before treatment or in control subjects. However, all patients treated with 1-(2-chloroethyl)-3-cyclohexyl-1-nitrosourea (CCNU) had increased SCE frequencies, which in some cases remained high up to 8 weeks after single administration of the drug. The patients who were treated with 5-fluorouracil, vincristine, adriamycin, methotrexate, CP, thiotepa, bleomycin, and prednimustine as a single drug or in combination therapy, and the patients subjected to treatment with 5-(3,3-dimethyl-1-triazeno)imidazole-4-carboxamide and actinomycin D did not show any significant alterations in SCE frequency. Lambert et al. (1978b) considered that CCNU and melphalan, in contrast to several other chemotherapeutic agents, induce very long-lived DNA lesions. Patients suffering for more than 1 year with chronic lymphatic leukemia and receiving mono- or polychemotherapy with one or several drugs, such as chlorambucil, melphalan, vincristine, CP, oxymetholone, and prednisone, had elevated SCE frequencies compared to untreated patients or normal donors (McDonald & Fitzgerald, 1979). Kakati et al. (1978) obtained bone marrow samples from 11 patients with chronic myelocytic leukemia and cultured the bone marrow cells *in vitro* with BrdUrd. All patients had received polychemotherapy in the past and also close to the time of sampling. Except for two patients, the SCE values were within the normal range. In a patient with acute myeloblastic leukemia, the SCE incidence in peripheral lymphocytes increased when therapy was changed to daunorubicin and vincristine. Similar studies on the SCE incidence in bone marrow cells taken from acute nonlymphocytic leukemia patients suggested different susceptibilities of leukemic and normal cells to antileukemic treatments (Abe and Sandberg, 1980). Generally, the SCE frequency in the aneuploid leukemic cells of the patients was lower than the control values even after chemotherapeutic treatment, whereas it was higher in the normal diploid cells. There are indications that lymphocytes in patients with acute lymphoblastic leukemia (ALL) have elevated SCE levels that remain high during periods of remission maintained with chemotherapy. However, long-term survivors of ALL who are in remission and off chemotherapy did not demonstrate significantly increased SCE frequencies (Otter et al., 1979). Burgdorf et al. (1977) found a significant increase in SCE rate in lymphocytes of six individuals who had taken arsenic (1% $KAsO_2$ in H_2O) 1–56 years ago for 4 months to 27 years and later developed stigmata, including multiple cutaneous malignancies, of arsenic use. In lymphocytes taken at least 2 months after the last treatment from patients who had received single or combination treatment with CP, thiotepa, vincristine, and CCNU, the frequencies of SCEs was significantly higher in two patients, but the same as control in another two (Natarajan et al., 1978). A CP-treated (100 mg/day) patient showed higher frequencies of SCEs during therapy, and the frequency came back to control level 25 days after the termination of therapy. In rats administered dimethylnitrosamine (DMN), diethylnitrosamine (DEN), and methylnitrosourea (MNU) the SCE rate in lymphocytes taken after the treatment was high for MNU and DMN, but not for DEN (Natarajan et al., 1978). The frequency of SCE was increased in lymphocytes of nephrotic children and some other juvenile patients treated with CP or azathioprine, and 3–4 months after cessation of CP medication it returned to the normal level (Torigoe, 1980). In children with urinary tract infections treated with nalidixic acid, there was a more than twofold increase of the SCE frequency in lymphocytes 10 days after treatment (Kowalczyk, 1980). Na_2SeO_3 was given to neuronal ceroid lipofuscinosis patients (0.004–0.015 mg/kg body weight daily, for 13.5 months) and healthy individuals (0.025 mg/kg body weight daily, for 2 weeks) (Norppa et al., 1980a). There was no rise in SCE frequency in lymphocytes in either group. However, Na_2SeO_3 increased the SCE frequency after *in vivo* BrdUrd labeling of Chinese hamster bone marrow cells (Norppa et al., 1980b). Male Wistar rats exposed to ethanol in the drinking water showed elevated SCE frequencies in the blood lymphocytes labeled *in vitro* with BrdUrd, but not in the bone marrow cells labeled *in vivo* (Tates et al., 1980).

The reports cited above deal mostly with cancer patients. SCE response to chemical agents might be different in cancer patients and normal subjects. Microsomal drug metabolism is known to be depressed in animals bearing ascites tumors (Litterst et al., 1977). However, the elevated SCE rate appears to be a function of the therapy and not the leukemia (McDonald and Fitzgerald, 1979). Following i.p. injection of 2-acetylaminofluorene, a tumor-bearing rabbit and normal rabbits responded with a similar increase of SCE (Takehisa and Wolff, 1978b).

Studies on the frequencies of SCEs in peripheral lymphocytes might be useful for monitoring human exposure to environmental mutagens and carcinogens. The frequency of SCEs was significantly increased in a group of 12 technicians working in laboratories performing hormone analysis (Funes-Cravioto et al., 1978). Four children of two female technicians working during pregnancy also had a significantly increased frequency of SCEs. However, there was a considerable overlapping between the exposed and control groups in SCE frequency (Funes-Cravioto et al., 1978). Husum and Wulf (1980) examined the SCE frequency in lymphocytes of hospital personnel exposed to anesthetics. There was no indication of an SCE increase due to long-term exposure to waste anesthetic gases such as halothane and nitrous oxide. Examination of the SCE frequencies in lymphocytes from 18 workers currently exposed to epoxy resins of the type diglycidyl ether bisphenol-A showed that there was no difference in SCE frequency between the controls and the group exposed to epoxy resin (Mitelman et al., 1980). Crossen et al. (1978) studied lymphocytes of individuals with heavy occupational exposures to herbicides and pesticides. There were no differences in the SCE rate between the control groups and the sprayers. Kučerová et al. (1979) analyzed blood samples from nine workers occupationally exposed for 10–27 years to relatively high mean annual doses of vinyl chloride monomer (about 20–150 ppm of air). The frequency of SCEs was significantly higher in lymphocytes of the five vinyl chloride workers than in the controls. Murthy and Prema (1979) reported that some oral contraceptive users, who had been taking the oral contraceptive that contained 150 μg *d*-norgestrel and 50 or 30 μg ethenylestradiol/day for periods ranging from 6 to 24 months, showed significantly higher SCE frequencies when compared with both normal and pregnant women. Stella et al. (1978) found a normal SCE frequency in lymphocytes of a lead-poisoning patient who had a blood lead level of 800 μg/liter (4×10^{-5} M).

Since there is evidently a considerable interindividual variability in the extent and kinetics of SCE response to drugs, we should be careful in interpreting the data that lack the preexposure level of SCE frequency.

Takehisa and Wolff (1978b) observed that there were great differences among rabbits in their SCE response of lymphocytes to *in vivo* administered 2-acetylaminofluorene.

Norppa et al. (1979) exposed Chinese hamsters to styrene oxide. The animals were sacrificed after the last exposure and bone marrow cells were isolated. The bone marrow cells were cultured with BrdUrd for two cell cycles. No difference in the mean number of SCEs was found between controls and groups exposed by inhalation to 25–100 ppm styrene oxide for either 9 or 21 hours. After i.p. injection of styrene oxide at 500 mg/kg body weight, one animal out of six showed slightly elevated SCE values. MMS injection, 80 mg/kg body weight, increased SCE frequencies in bone marrow cells that were isolated and cultured 7 hours after the injection. Rats treated with phenacetin, phenazone, and caffeine, alone and in different combinations, in the diet showed an increased SCE frequency in their blood cultures (Granberg-Öhman et al., 1980).

The physical condition of the test animals may have some influence on generating SCEs in cultured lymphocytes. Knuutila et al. (1979) reported a decreased level of SCE in patients with Down's syndrome after measles vaccination. Lambert et al. (1979a) found that measles vaccination induced an initial decrease in the SCE frequency. When the immunological reactions subsided the SCE frequency rose to a higher level than before vaccination. The SCE frequencies counted in 12-year-old children 1.5 and 8 months after smallpox revaccination, however, did not differ from controls (Kučerovà et al., 1980). Cigarette smokers were found in several studies to have significantly higher SCE frequencies than nonsmokers (Lambert et al., 1978a; Husgafvel-Pursiainen et al., 1980). However, Crossen and Morgan (1980) could not find any differences in the SCE rate between smokers and nonsmokers. Ardito et al. (1980) studied the effect of smoking habits on SCE frequency in pregnant women and their newborns. There was no significant difference in SCE frequency between smoking and nonsmoking females. However, in both smoking and nonsmoking subjects, a significant difference was found between mothers and newborns. Young children showed a significantly lower frequency of SCEs than adults (Husgafvel-Pursiainen et al., 1980). As to severe protein calorie malnutrition in children, there was no significant difference between the number of SCEs in the malnourished children than in an adequate control group (Mutchinick et al., 1979).

IV. CONCLUSION

As has been claimed by many workers, SCEs are excellent indicators for detecting chemical agents that are suspected mutagens or carcinogens.

Now we know that carcinogenesis is a two-stage process, initiation and promotion (Berenblum, 1979). According to this theory, the carcinogens discussed in this chapter belong to the initiator category. Promoters are recognized to be neither carcinogenic nor mutagenic when tested alone, but greatly increase tumor frequency and shorten the lag time for tumor appearance when applied after an initiator. Is the SCE test also applicable for screening promoters?

Simultaneously, Kinsella and Radman (1978) exposed Chinese hamster V79-4 fibroblasts to 30 μM BrdUrd and 12-O-tetradecanoylphorbol 13-acetate (TPA), a powerful tumor promoter. TPA induced an increase of SCEs, whereas the nonpromoting derivative 4-O-methyl-TPA did not. Inhibitors of tumor promotion—antipain, leupeptin, and fluocinolone acetonide—inhibited formation of such TPA-induced SCEs. Similar results were obtained by Nagasawa and Little (1979) and Little et al. (1979) with X-irradiated C3H 10T$\frac{1}{2}$ mouse embryo-derived fibroblasts. However, Loveday and Latt (1979) obtained results indicating that TPA did not increase the SCE frequency in either CHO or V79 cells. In their experiments TPA did not alter the induction of SCEs in CHO cells by mitomycin C during the first three cycles following addition of the alkylating agent. Failure of TPA to enhance SCE formation was also reported by Fujiwara et al. (1980) and by Thompson et al. (1980). Thus the applicability and response of the SCE test to promoters remains to be seen in the future.

Inhibitors of tumor promotion, antipain and leupeptin, have no effect on the induction of SCEs (Nagasawa and Little, 1979; Little et al., 1979). In conjunction with inhibitors of carcinogenesis, Wattenberg (1978, 1979) has examined strategies for preventing cancer and possible inhibitors of chemical carcinogenesis. He cited several chemical agents as having the capacity to inhibit the occurrence of neoplasia when administered prior to or simultaneously with exposure to cancer-producing agents. Among them, butylated hydroxyanisol, ascorbic acid, and cysteamine have been tested for the induction of SCE (cf. Table 5.4). They were more or less positive for SCE inducibility. Thus whether the SCE test is applicable for detection of the inhibitors of carcinogenesis is an open question just as in case of the detection of promoters.

References

Abe, S. & Sandberg, A. A. (1980) *Cancer Res.* **40,** 1292–1299.

Abe, S. & Sasaki, M. (1977a) *J. Natl. Cancer Inst.* **58,** 1635–1641.

Abe, S. & Sasaki, M. (1977b) *Proc. Jap. Acad.* **53,** 46–49.

Abramovsky, I., Vorsanger, G., & Hirschhorn, K. (1978) *Mutat. Res.* **50,** 93–100.

Alhadeff, B., Velivasakis, M., Pagan-Charry, I., Wright, W. C., & Siniscalco, M. (1980) *Cytogenet. Cell Genet.* **27,** 8–23.

Allen, J. W. & Latt, S. A. (1976a) *Nature (Lond.)* **260,** 449–451.

Allen, J. W. & Latt, S. A. (1976b) *Chromosoma (Berl.)* **58,** 325–340.

Allen, J. W., Shuler, C. F., Mendes, R. W., & Latt, S. A. (1977) *Cytogenet. Cell Genet.* **18,** 231–237.

Altenburg, L. C., Ray, J. H., Smart, C. E., & Moore, F. B. (1979) *Mutat. Res.* **67,** 331–341.

Ames, B. N., Durston, W. E., Yamasaki, E., & Lee, F. D. (1973) *Proc. Natl. Acad. Sci. U.S.* **70,** 2281–2285.

Ames, B. N., McCann, J., & Yamasaki, E. (1975) *Mutat. Res.* **31,** 347–364.

Ardito, G., Lamberti, L., Ansaldi, E., & Ponzetto, P. (1980) *Mutat. Res.* **78,** 209–212.

Ashby, J. & Styles, J. A. (1978a) *Nature (Lond.)* **271,** 452–455.

Ashby, J. & Styles, J. A. (1978b) *Mutat. Res.* **54,** 105–112.

Ashwood-Smith, M. J. (1980) *Mutat. Res.* **69,** 199–200.

Athanasiou, K. & Bartsocas, C. S. (1980) *Mutat. Res.* **79,** 79–80.

Auf der Maur, P. & Berlincourt-Böhni, K. (1979) *Hum. Genet.* **49,** 209–215.

Baars, A. J., Blijleven, W. G. H., Mohn, G. R., Natarajan, A. T., & Breimer, D. D. (1980) *Mutat. Res.* **72,** 257–264.

Banerjee, A. & Benedict, W. F. (1979) *Cancer Res.* **39,** 797–799.

Bartram, C. R., Rüdiger, H. W., & Passarge, E. (1979) *Hum. Genet.* **46,** 331–334.

Basler, A. (1979) *Cytogenet. Cell Genet.* **24,** 193–196.

Basler, A. & Röhrborn, G. (1980) *Arch. Toxicol.* **45,** 1–7.

Basler, A., Bachmann, U., Roszinsky-Köcher, G., & Röhrborn, G. (1979) *Mutat. Res.* **59,** 209–214.

Bast, R. C., Whitlock, J. P., Miller, H., Rapp, H. J., & Gelboin, H. V. (1974) *Nature (Lond.)* **250,** 664–665.

Bauknecht, Th., Vogel, W., Bayer, U., & Wild, D. (1977) *Hum. Genet.* **35,** 299–307.

Bayer, U. (1978) *Mutat. Res.* **56,** 305–309.

Bayer, U. & Bauknecht, T. (1977) *Experientia* **33,** 25.

Becher, R., Schmidt, C. G., Theis, G., & Hossfeld, D. K. (1979) *Hum. Genet.* **50,** 213–216.

Beek, B. & Obe, G. (1975) *Humangenetik* **29,** 127–134.

Beek, B. & Obe, G. (1979) *Hum. Genet.* **49,** 51–61.

Bend, J. R. & Hook, G. E. R. (1977) In *Handbook of Physiology* (D. H. K. Lee, H. L. Falk, S. D. Murphy, & S. R. Geiger, Eds.), American Physiological Society, Bethesda, Md., pp. 419–440.

Benedict, W. F., Banerjee, A., & Venkatesan, N. (1978) *Cancer Res.* **38,** 2922–2924.

Berenblum, I. (1979) In *Carcinogens: Identification and Mechanisms of Action* (A. C. Griffin & C. R. Shaw, Eds.), Raven Press, New York, pp. 25–36.

Bianchi, N. O., Bianchi, M. S., & Larramendy, M. (1979) *Mutat. Res.* **63,** 317–324.

Biegel, J. A., Boggs, S. S., & Conner, M. K. (1980) *Mutat. Res.* **79,** 87–90.

Bimboes, D. & Greim, H. (1976) *Mutat. Res.* **35,** 155–160.

Bradley, M. O., Hsu, I. H., & Harris, C. C. (1979) *Nature (Lond.)* **282,** 318–320.

Brøgger, A. & Waksvik, H. (1978) *Hereditas* **89,** 131–132.

Brøgger, A., Ardito, G., & Waksvik, H. (1979) *Hereditas* **91,** 135–138.

Brown, R. F. & Wu, Y. (1977) *Mutat. Res.* **56,** 215–217.

Brown, R. L. & Crossen, P. E. (1976) *Exp. Cell Res.* **103,** 418–420.

Bryant, E. M., Hoehn, H., & Martin, G. M. (1979) *Nature (Lond.)* **279,** 795–796.

Burgdorf, W., Kurvink, K., & Cervenka, J. (1977) *Hum. Genet.* **36,** 69–72.

Byrnes, J. & Bloom, S. E. (1980) *Mutat. Res.* **70,** 203–210.

Callen, D. F. (1978) *Mutat. Res.* **55,** 153–163.

Carrano, A. V., Thompson, L. H., Lindl, P. A., & Minkler, J. L. (1978) *Nature (Lond.)* **271,** 551–553.

Cavaglià, A. M. V. (1980) *Mutat. Res.* **77,** 383–385.

Clare, M. G. & Joslin, C. A. (1978) *Br. J. Radiol.* **51,** 926–927.

Cohen, M. M. & Sagi, M. (1979), *Mutat. Res.* **59,** 139–142.

Connell, J. R. (1979) *Int. J. Cancer* **24,** 485–489.

Conner, M. K., Boggs, S. S., & Turner, J. H. (1978) *Chromosoma (Berl.)* **68,** 303–311.

Conner, M. K., Alarie, Y., & Dombroske, R. L. (1979a) *Chromosoma (Berl.)* **74,** 51–55.

Conner, M. K., Alarie, Y., & Dombroske, R. L. (1979b) *Toxicol. Appl. Pharmacol.* **50,** 365–367.

Conney, A. H. (1967) *Pharmacol. Rev.* **19,** 317–366.

Cooper, H. L. (1973) In *Drugs and the Cell Cycle* (A. M. Zimmerman, G. M. Padilla, & I. L. Cameron, Eds.), Academic Press, New York and London, pp. 137–194.

Craig-Holmes, A. P. & Shaw, M. W. (1977) *Mutat. Res.* **46,** 375–384.

Cramer, J. W., Miller, J. A., & Miller, E. C. (1960) *J. Biol. Chem.* **235,** 885–888.

Crossen, P. E. (1979) *Mutat. Res.* **68,** 295–299.

Crossen, P. E. & Morgan, W. F. (1980) *Hum. Genet.* **53,** 425–426.

Crossen, P. E., Morgan, W. F., Horan, J. J., & Stewart, J. (1978) *N. Z. Med. J.* **88,** 192–195.

Csukás, I., Gungl, E., Fedorcsák, I., Vida, G., Antoni, F., Turtóczky, I., & Solymosy, F. (1979) *Mutat. Res.* **67,** 315–319.

Davidson, R. L., Kaufman, E. R., Dougherty, C. P., Ouellette, A. M., DiFolco, C. M., & Latt, S. A. (1980) *Nature (Lond.)* **284,** 74–76.

Deaven, L. L., Oka, M. S., & Tobey, R. A. (1978) *J. Natl. Cancer Inst.* **60,** 1155–1161.

De Raat, W. K. (1977) *Chem.-Biol. Interact.* **19,** 125–131.

De Raat, W. K. (1978) *Chem.-Biol. Interact.* **20,** 163–170.

De Raat, W. K. (1979) *Mutat. Res.* **66,** 253–259.

Donehower, R. C., Myers, C. E., & Chabner, B. A. (1979) *Life Sci.* **25,** 1–14.

Douglas, G. R., Bell, R. D. L., Grant, C. E., Wytsma, J. M., & Bora, K. C. (1980) *Mutat. Res.* **77,** 157–163.

DuFrain, R. J., Littlefield, L. G., & Wilmer, J. L. (1979) *Environ. Mutagenesis* **1,** 283–286.

Dutrillaux, B., Fosse, A. M., Prieur, M., & Lejeune, J. (1974) *Chromosoma (Berl.)* **48,** 327–340.

Emerit, I., Emerit, J., Levy, A., & Keck, M. (1979) *Hum. Genet.* **50,** 51–57.

Estabrook, R. W., Werringloer, J., Capdevila, J., & Prough, R. A. (1978) In *Polycyclic Hydrocarbons and Cancer* Vol. 1, (H. V. Gelboin & P.O.P. Ts'o, Eds.), Academic Press, New York, San Francisco, and London, pp. 285–319.

Evans, H. J. & Vijayalaxmi (1980) *Nature (Lond.)* **284,** 370–372.

Evans, L. A., Kevin, M. J., & Jenkins, E. C. (1977) *Mutat. Res.* **56,** 51–58.

Fabricant, J. D., Hofnung, M. J., & Kelly, F. (1979) *Mutat. Res.* **63,** 215–219.

Faed, M. J. W. & Mourelatos, D. (1978) *Mutat. Res.* **49,** 437–440.

Faed, M. J. W. & Peterson, S. (1980) *Mutat. Res.* **78,** 389–391.

Franceschini, P. (1974) *Exp. Cell Res.* **89,** 420–421.

Frei, J. V. (1976) *Chem.-Biol. Interact.* **13,** 1–25.

Fujiwara, Y., Kano, Y., Tatsumi, M. & Paul, P. (1980) *Mutat. Res.* **71,** 243–251.

Funes-Cravioto, F., Zapata-Gayon, C., Kolmodin-Hedman, B., Lambert, B., Lindsten, J., Norberg, E., Nordenskjöld, M., Olin, R., & Swensson, A. (1978) In *Mutagen-induced Chromosome Damage in Man* (H. J. Evans & D. C. Lloyd, Eds.), Yale University Press, New Haven, pp. 275–289.

Furukawa, M., & Huang, C. C. (1978) *Mutat. Res.* **57,** 233–239.

Furukawa, M., Sirianni, S. R., Tan, J. C., & Huang, C. C. (1978) *J. Natl. Cancer Inst.* **60,** 1179–1181.

Galloway, S. M. (1977a) In *DNA Repair Processes* (W. W. Nichols & D. G. Murphy, Eds.), Symposia Specialists Inc., Miami, Florida, pp. 191–201.

Galloway, S. M. (1977b) *Mutat. Res.* **45,** 343–349.

Galloway, S. M. & Painter, R. B. (1979) *Mutat. Res.* **60,** 321–327.

Galloway, S. M. & Wolff, S. (1979) *Mutat. Res.* **61,** 297–307.

Galloway, S. M., Perry, P. E., Meneses, J., Nebert, D. W., & Pedersen, R. A. (1980) *Proc. Natl. Acad. Sci. U.S.* **77,** 3524–3528.

Gatehouse, D. G. & Delow, G. F. (1979) *Mutat. Res.* **60,** 239–252.

Geard, C. R. & Rutledge-Freeman, M. H. (1980) *Mutat. Res.* **78,** 289–294.

Gebhart, E. & Kappauf, H. (1978) *Mutat. Res.* **58,** 121–124.

Gerner-Smidt, P. & Friedrich, U. (1978) *Mutat. Res.* **58,** 313–316.

Ghosh, P. K. & Nand, R. (1979) *Hum. Genet.* **51,** 167–170.

Gibas, Z. & Limon, J. (1979) *Mutat. Res.* **67,** 93–96.

Giulotto, E., Mottura, A., Giorgi, R., de Carli, L., & Nuzzo, F. (1980) *Mutat. Res.* **70,** 343–350.

Goto, K., Akematsu, T., Shimizu, H., & Sugiyama, T. (1975) *Chromosoma (Berl.)* **53,** 223–230.

Granberg-Öhman, I., Johansson, S., & Hjerpe, A. (1980) *Mutat. Res.* **79,** 13–18.

Guerrero, R. R., Rounds, D. E., & Hall, T. C. (1979) *J. Natl. Cancer Inst.* **62,** 805–809.

Hayashi, K. & Schmid, W. (1975) *Humangenetik* **29,** 201–206.

Heidelberger, C. (1975) *Annu. Rev. Biochem.* **44,** 79–121.

Hoffman, J. & Post, J. (1973) In *Drugs and the Cell Cycle* (A. M. Zimmerman, G. M. Padilla, & I. L. Cameron, Eds.), Academic Press, New York and London, pp. 219–247.

Hollstein, M., McCann, J., Angelosanto, F. A., & Nichols, W. W. (1979) *Mutat. Res.* **65,** 133–226.

Honeycombe, J. R. (1978) *Mutat. Res.* **57,** 35–49.

Hopkin, J. M. & Perry, P. E. (1980) *Mutat. Res.* **77,** 377–381.

Hsu, I. C., Bowden, G. T., & Harris, C. C. (1979a) *Mutat. Res.* **63,** 351–359.

Hsu, I. C., Harris, C. C., Yamaguchi, M., Trump, B. F., & Schafer, P. W. (1979b) *J. Clin. Invest.* **64,** 1245–1252.

Hsu, T. C. & Au, W. (1979) In *Effects of Drugs on the Cell Nucleus* (H. Busch, S. T. Crooke, & Y. Daskal, Eds.), Academic Press, New York, pp. 315–332.

Huberman, E. & Sachs, L. (1974) *Int. J. Cancer* **13,** 326–333.

Husgafvel-Pursiainen, K., Mäki-Paakkanen, J., Norppa, H., & Sorsa, M. (1980) *Hereditas* **92,** 247–250.

Husum, B. & Wulf, H. C. (1980) *Acta Anaesth. Scand.* **24,** 22–24.

Ikushima, T. & Wolff, S. (1974) *Exp. Cell Res.* **87,** 15–19.

Irving, C. C. (1979) In *Carcinogens: Identification and Mechanisms of Action* (A. C. Griffin, & C. R. Shaw, Eds.), Raven Press, New York, pp. 211–227.

Ishii, Y. & Bender, M. A. (1978a) *Mutat. Res.* **51,** 411–418.

Ishii, Y. & Bender, M. A. (1978b) *Mutat. Res.* **51,** 419–425.

Ishii, Y. & Bender, M. A. (1980) *Mutat. Res.* **79,** 19–32.

Jensen, E. M., LaPolla, R. J., Kirby, P. E., & Haworth, S. R. (1977) *J. Natl. Cancer Inst.* **59,** 941–944.

Juhl, H. J., Schürer, C. C., Bartram, C. R., Kohl, F.-V., Melderis, H., Wichert, P.v., & Rüdiger, H. W. (1978) *Mutat. Res.* **58,** 317–320.

Kakati, S., Abe, S., & Sandberg, A. A. (1978) *Cancer Res.* **38,** 2918–2921.

Kato, H. (1974a) *Nature (Lond.)* **251,** 70–72.

Kato, H. (1974b) *Nature (Lond.)* **252,** 739–741.

Kato, H. (1977a) *Chromosoma (Berl.)* **59,** 179–191.

Kato, H. (1977b) *Int. Rev. Cytol.* **49,** 55–97.

Kato, H. (1980a) *Cancer Genet. Cytogenet.* **2,** 69–77.

Kato, H. (1980b) *Cancer Genet. Cytogenet.* **2,** 61–67.

Kato, H. & Sandberg, A. A. (1977) *Exp. Cell Res.* **109,** 445–448.

Kato, H. & Shimada, H. (1975) *Mutat. Res.* **28,** 459–464.

Keck, M. & Emerit, I. (1979) *Hum. Genet.* **50,** 277–283.

Kihlman, B. A. (1975) *Chromosoma (Berl.)* **51,** 11–18.

Kihlman, B. A. & Sturelid, S. (1978) *Hereditas* **88,** 35–41.

Kim, M. A. (1974) *Humangenetik* **25,** 179–188.

Kinsella, A. R. & Radman, M. (1978) *Proc. Natl. Acad. Sci. U.S.* **75,** 6149–6153.

Kinsella, A. R. & Radman, M. (1980) *Proc. Natl. Acad. Sci. U.S.* **77,** 3544–3547.

Kligerman, A. D., Strom, S. C., & Michalopoulos, G. (1980) *Environ. Mutagenesis* **2,** 157–165.

Knuutila, S., Helminen, E., Vuopio, P., & De La Chapelle, A. (1978) *Hereditas* **88,** 189–196.

Knuutila, S., Harkki, A., Ellimäki, K., & Salunen, R. (1979) *Hereditas* **90,** 147–149.

Korenberg, J. R. & Freedlender, E. F. (1974) *Chromosoma (Berl.)* **48,** 355–360.

Kowalczyk, J. (1980) *Mutat. Res.* **77,** 371–375.

Kram, D., Schneider, E. L., Singer, L., & Martin, G. R. (1978a) *Mutat. Res.* **57,** 51–55.

Kram, D., Schneider, E. L., Tice, R. R., & Gianas, P. (1978b) *Exp. Cell Res.* **114,** 471–475.

Kram, D., Bynum, G. D., Senula, G. C., & Schneider, E. L. (1979a) *Nature (Lond.)* **279,** 531.

Kram, D., Schneider, E. L., Senula, G. C., & Nakanishi, Y. (1979b) *Mutat. Res.* **60,** 339–347.

Kram, D., Bynum, G. D., Senula, G. C., Bickings, C. K., & Schneider, E. L. (1980) *Proc. Natl. Acad. Sci.* U.S. **77,** 4784–4787.

Kučerová, M. & Polívková, Z. (1977) In *Progress in Genetic Toxicology* (D. Scott, B. A. Bridges, & F. H. Sobels, Eds.), Elsevier/North-Holland, Amsterdam, pp. 319–325.

Kučerová, M., Polívková, Z., & Bátora, J. (1979) *Mutat. Res.* **67,** 97–100.

Kučerová, M., Polívková, Z., & Matoušek, V. (1980) *Mutat. Res.* **71,** 263–267.

Lake, R. S., Pezzutti, M. R., Kropko, M. L., Freeman, A. E., & Igel, H. J. (1977) *Cancer Res.* **37,** 2530–2537.

Lambert, B., Lindblad, A., Nordenskjöld, M., & Werelius, B. (1978a) *Hereditas* **88,** 147–149.

Lambert, B., Ringborg, U., Harper, E., & Lindblad, A. (1978b) *Cancer Treat. Rep.* **62,** 1413–1419.

Lambert, B., Ehrnst, A., Hansson, K., Lindblad, A., Morad, M., & Werelius, B. (1979a) *Hum. Genet.* **50,** 291–296.

Lambert, B., Ringborg, U., & Lindblad, A. (1979b) *Mutat. Res.* **59,** 295–300.

Latt, S. A. (1973) *Proc. Natl. Acad. Sci. U.S.* **70,** 3395–3399.

Latt, S. A. (1974) *Proc. Natl. Acad. Sci. U.S.* **71,** 3162–3166.

Latt, S. A. (1979) *Genetics* **92,** (Suppl.) s83–s95.

Latt, S. A. & Schreck, R. R. (1980) *Am. J. Hum. Genet.* **32,** 297–313.

Latt, S. A., Stetten, G., Juergens, L. A., Buchanan, G. R., & Gerald, P. S. (1975) *Proc. Natl. Acad. Sci. U.S.* **72,** 4066–4070.

Latt, S. A., Schreck, R. R., Loveday, K. S., & Shuler, C. F. (1979) *Pharmacol. Rev.* **30,** 501–535.

Lezana, E. A., Bianchi, N. O., Bianchi, M. S., & Zabala-Suarez, J. E. (1977) *Mutat. Res.* **45,** 85–90.

Liebeskind, D., Bases, R., Mendez, F., Elequin, F., & Koenigsberg, M. (1979) *Science* **205,** 1273–1275.

Litterst, C. L., Mimnaugh, E. G., & Gram, T. E. (1977) *J. Natl. Cancer Inst.* **59,** 1737–1739.

Little, J. B., Nagasawa, H., & Kennedy, A. R. (1979) *Radiat. Res.* **79,** 241–255.

Littlefield, L. G., Colyer, S. P., Sayer, A. M., & DuFrain, R. J. (1979) *Mutat. Res.* **67,** 259–269.

Littlefield, L. G., Colyer, S. P., & DuFrain, R. J. (1980) *Mutat. Res.* **69,** 191–197.

Livingston, G. K. & Dethlefsen, L. A. (1979) *Radiat. Res.* **77,** 512–520.

Loveday, K. S. & Latt, S. A. (1978) *Nucleic Acids Res.* **5,** 4087–4104.

Loveday, K. S. & Latt, S. A. (1979) *Mutat. Res.* **67,** 343–348.

Lutz, W. K. (1979) *Mutat. Res.* **65,** 289–356.

McCann, J. & Ames, B. N. (1976) *Proc. Natl. Acad. Sci.* U.S. **73,** 950–954.

McCann, J., Choi, E., Yamasaki, E., & Ames, B. N. (1975) *Proc. Natl. Acad. Sci. U.S.* **72,** 5135–5139.

McDonald, M. A. & Fitzgerald, P. H. (1979) *J. Natl. Cancer Inst.* **62,** 1169–1171.

MacRae, W. D. & Stich, H. F. (1979a) *Mutat. Res.* **68,** 351–365.

MacRae, W. D. & Stich, H. F. (1979b) *Toxicology* **13,** 167–174.

MacRae, W. D., MacKinnon, E. A., & Stich, H. F. (1979a) *Chromosoma (Berl.)* **72,** 15–22.

MacRae, W. D., MacKinnon, E. A., & Stich, H. F. (1979b) *In Vitro* **15,** 555–564.

MacRae, W. D., MacKinnon, E. A., & Stich, H. F. (1979c) *Mutat. Res.* **62,** 495–504.

MacRae, W. D., Whiting, R. F., & Stich, H. F. (1979d) *Chem.-Biol. Interact.* **26,** 281–286.

Madle, S. & Obe, G. (1977) *Mutat. Res.* **56,** 101–104.

Madle, S. & Obe, G. (1978) *Mutat. Res.* **49,** 149–151.

Majone, F. & Levis, A. G. (1979) *Mutat. Res.* **67,** 231–238.

Malling, H. V. (1971) *Mutat. Res.* **13,** 425–429.

Marquardt, H. & Bayer, U. (1977) *Mutat. Res.* **56,** 169–176.

Mattern, I. E. & Greim, H. (1978) *Mutat. Res.* **53,** 369–378.

Miller, J. A. (1970) *Cancer Res.* **30,** 559–576.

Miller, J. A. (1979) In *Carcinogens: Identification and Mechanisms of Action* (A. C. Griffin & C. R. Shaw, Eds.), Raven Press, New York, pp. 455–469.

Mitelman, F., Fregert, S., Hedner, K., & Hillbertz-Nilsson, K. (1980) *Mutat. Res.* **77,** 345–348.

Moan, J., Waksvik, H., & Christensen, T. (1980) *Cancer Res.* **40,** 2915–2918.

Mohn, G. R. (1977) *Arch. Toxicol.* **38,** 109–133.

Monticone, R. E. & Schneider, E. L. (1979) *Mutat. Res.* **59,** 215–221.

Morgan, W. F. & Crossen, P. E. (1980) *Mutat. Res.* **77,** 283–286.

Morimoto, K. & Wolff, S. (1980) *Cancer Res.* **40,** 1189–1193.

Mourelatos, D. C. (1979) *Experientia* **35,** 822–824.

Murthy, P. B. K. & Prema, K. (1979) *Mutat. Res.* **68,** 149–152.

Musilová, J., Michalová, K., & Urban, J. (1979) *Mutat. Res.* **67,** 289–294.

Mutchinick, O., Lisker, R., Ruz, L., Salamanca, F., & Armendares, S. (1979) *Ann. Génét.* **22,** 129–132.

Nagasawa, H. & Little, J. B. (1979) *Proc. Natl. Acad. Sci. U.S.* **76,** 1943–1947.

Nakanishi, Y. & Schneider, E. L. (1979) *Mutat. Res.* **60,** 329–337.

Nakanishi, Y., Kram, D., & Schneider, E. L. (1979) *Cytogenet. Cell Genet.* **24,** 61–67.

Nakanishi, Y., Dein, R. A., & Schneider, E. L. (1980) *Cytogenet. Cell Genet.* **27,** 82–87.

Natarajan, A. T. & Obe, G. (1978) *Mutat. Res.* **52,** 137–149.

Natarajan, A. T., Tates, A. D., Van Buul, P. P. W., Meijers, M., & De Vogel, N. (1976) *Mutat. Res.* **37,** 83–90.

Natarajan, A. T., Van Buul, P. P. W., & Raposa, T. (1978) In *Mutagen-induced Chromosome Damage in Man* (H. J. Evans & D. C. Lloyd, Eds.), Yale University Press, New Haven, pp. 268–274.

Nebert, D. W. & Gelboin, H. V. (1968) *J. Biol. Chem.* **243,** 6250–6261.

Nebert, D. W., Atlas, S. A., Guenthner, T. M., & Kouri, R. E. (1978) in *Polycyclic Hydrocarbons and Cancer* Vol. 2 (H. V. Gelboin & P. O. P. Ts'o, Eds.), Academic Press, New York, San Francisco, and London, pp. 345–390.

Nevstad, N. P. (1978) *Mutat. Res.* **57,** 253–258.

Newbold, R. F., Wigley, C. B., Thompson, M. H., & Brookes, P. (1977) *Mutat. Res.* **43,** 101–116.

Nicholas, A. H., Vienne, M., & Van Den Berghe, H. (1979) *Mutat. Res.* **67,** 167–172.

Nichols, W. W., Bradt, C. I., Toji, L. H., Godley, M., & Segawa, M. (1978) *Cancer Res.* **38,** 960–964.

Norppa, H., Elovaara, E., Husgafvel-Pursiainen, K., Sorsa, M., & Vainio, H. (1979) *Chem.-Biol. Interact.* **26,** 305–315.

Norppa, H., Westermarck, T., Laasonen, M., Knuutila, L., & Knuutila, S. (1980a) *Hereditas* **93,** 93–96.

Norppa, H., Westermarck, T., & Knuutila, S. (1980b) *Hereditas* **93,** 101–105.

Novotná, B., Goetz, P., & Surkova, N. I. (1979) *Hum. Genet.* **49,** 41–50.

Obe, G. & Ristow, H. (1977) *Mutat. Res.* **56,** 211–213.

Obe, G., Natarajan, A. T., Meyers, M., & Den Hertog, A. (1979) *Mutat. Res.* **68,** 291–294.

Ohtsuru, M., Ishii, Y., Takai, S., Higashi, H., & Kosaki, G. (1980) *Cancer Res.* **40,** 477–480.

Otter, M., Palmer, C. G., & Baehner, R. L. (1979) *Hum. Genet.* **52,** 185–192.

Pal, K., Tierney, B., Grover, P. L., & Sims, P. (1978) *Mutat. Res.* **50,** 367–375.

Pal, K., Grover, P. L., & Sims, P. (1979) *Cancer Lett.* **7,** 45–49.

Pal, K., Grover, P. L., & Sims, P. (1980) *Mutat. Res.* **78,** 193–199.

Palitti, F. & Becchetti, A. (1977) *Mutat. Res.* **45,** 157–159.

Pant, G. S., Kamada, N., & Tanaka, R. (1976) *Hiroshima J. Med. Sci.* **25,** 99–105.

Pedersen, R. A., Spindle, A. I., & Takehisa, S. (1978) In *Developmental Toxicology of Energy-related Pollutants* (D. D. Mahlum, M. R. Sikov, P. L. Hackett, & F. D. Andrew, Eds.), Technical Information Center, U.S. Department of Energy, pp. 152–161.

Perry, P. E. (1980) In *Chemical Mutagens,* Vol. 6 (F. J. DeSerres & A. Hollaender, Eds.), Plenum Press, New York, pp. 1–39.

Perry, P. & Evans, H. J. (1975) *Nature (Lond.)* **258,** 121–125.

Perry, P. E. & Searle, C. E. (1977) *Mutat. Res.* **56,** 207–210.

Perry, P. & Wolff, S. (1974) *Nature (Lond.)* **251,** 156–158.

Perry, P. E., Jager, M., & Evans, H. J. (1978) In *Mutagen-Induced Chromosome Damage in Man* (H. J. Evans & D. C. Lloyd, Eds.), Yale University Press, New Haven, pp. 201–207.

Popescu, N. C., Turnbull, D., & DiPaolo, J. A. (1977) *J. Natl. Cancer Inst.* **59,** 289–293.

Popescu, N. C., Amsbaugh, S. C., & DiPaolo, J. A. (1979) *Mutat. Res.* **60,** 313–320.

Popescu, N. C., Amsbaugh, S. A., & DiPaolo, J. A. (1980) *Chromosoma (Berl.)* **76,** 329–338.

Prosser, J. S. (1978) *Br. J. Ind. Med.* **35,** 174–176.

Prosser, J. S. & Hesketh, L. C. (1980) *Br. J. Radiol.* **53,** 376–377.

Ptashne, K., Brothers, L., Axline, S. G., & Cohen, S. N. (1974) *Proc. Soc. Exp. Biol. Med.* **146,** 585–589.

Raff, P. & Sugahara, T. (1980) *Mutat. Res.* **78,** 51–58.

Raffetto, G., Parodi, S., Faggin, P., & Maconi, A. (1979) *Mutat. Res.* **63,** 335–343.

Raj, S. A. & Heddle, J. A. (1980) *Mutat. Res.* **78,** 253–260.

Ramírez, P. M. (1980) *Mutat. Res.* **74,** 61–69.

Raposa, T. (1978) *Mutat. Res.* **57,** 241–251.

Ray, J. H. & Altenburg, L. C. (1978) *Mutat. Res.* **54,** 343–354.

Ray, J. H. & Altenburg, L. C. (1980) *Mutat. Res.* **78,** 261–266.

Ray, J. H., Altenburg, L. C., & Jacobs, M. M. (1978) *Mutat. Res.* **57,** 359–368.

Renner, H. W. (1979) *Arzneim.-Forsch.* **29(II),** 1871–1875.

Renner, H. W. & Münzner, R. (1980) *Mutat. Res.* **77,** 349–355.

Rickard, R. W., Dorough, H. W., & Pan, S. H. (1978) *J. Agric. Food Chem.* **26,** 141–146.

Riedel, L. & Obe, G. (1980) *Mutat. Res.* **73,** 125–131.

Ristow, H. & Obe, G. (1978) *Mutat. Res.* **58,** 115–119.

Roberts, G. T. & Allen, J. W. (1980) *Environ. Mutagenesis* **2,** 17–26.

Roszinsky-Köcher, G., Basler, A., & Röhrborn, G. (1979) *Mutat. Res.* **66,** 65–67.

Rüdiger, H. W., Kohl, F., Mangels, W., Wichert, P. von, Bartram, C. R., Wöhler, W., & Passarge, E. (1976) *Nature (Lond.)* **262,** 290–292.

Rüdiger, H. W., Haenisch, F., Metzler, M., Oesch, F., & Glatt, H. R. (1979) *Nature (Lond.)* **281,** 392–394.

Rüdiger, H. W., Bartram, C. R., Harder, W., & Passarge, E. (1980) *Am. J. Hum. Genet.* **32,** 150–157.

Santesson, B., Lindahl-Kiessling, K., & Mattsson, A. (1979) *Clin. Genet.* **16,** 133–135.

Sasaki, M., Sugimura, K., Yoshida, M. A., & Abe, S. (1980a) *La Kromosomo* (Tokyo) **II-20,** 574–584.

Sasaki, M., Sugimura, K., Yoshida, M. A., & Kawachi, T. (1980b) *Proc. Jap. Acad. Ser. B* **56,** 332–337.

Sasaki, M. S. (1977) *Nature (Lond.)* **269,** 623–625.

Saxholm, H. J. K., Iversen, O. H., Reith, A., & Brøgger, A. (1979) *Eur. J. Cancer* **15,** 509–513.

Schempp, W. & Krone, W. (1979) *Hum. Genet.* **51,** 315–318.

Schneider, E. L. & Monticone, R. E. (1978) *Exp. Cell Res.* **115,** 269–276.

Schneider, E. L., Chaillet, J. R., & Tice, R. R. (1976) *Exp. Cell Res.* **100,** 396–399.

Schönwald, A. D., Bartram, C. R., & Rüdiger, H. W. (1977) *Hum. Genet.* **36,** 261–264.

Schreck, R. R., Paika, I. J., & Latt, S. A. (1979) *Mutat. Res.* **64,** 315–328.

Schürer, C. C., Bartram, C. R., Glatt, H. R., Kohl, F. V., Mangels, W., Oesch, F., & Rüdiger, H. W. (1980) *Biochim. Biophys. Acta* **609,** 272–277.

Schwartz, A. G. & Moore, C. J. (1979) *Fed. Proc.* **38,** 1989–1992.

Shiraishi, Y. & Sandberg, A. A. (1978) *Mutat. Res.* **49,** 233–238.

Shiraishi, Y. & Sandberg, A. A. (1979) *J. Natl. Cancer Inst.* **62,** 27–33.

Shiraishi, Y. & Sandberg, A. A. (1980a) *Mutat. Res.* **72,** 251–256.

Shiraishi, Y. & Sandberg, A. A. (1980b) *Mutat. Res.* **72,** 265–272.

Shiraishi, Y., Yamamoto, K., & Sandberg, A. A. (1979) *Mutat. Res.* **62,** 139–149.

Shuler, C. F. & Latt, S. A. (1979) *Cancer Res.* **39,** 2510–2514.

Sirianni, S. R. & Huang, C. C. (1978) *Proc. Soc. Exp. Biol. Med.* **158,** 269–274.

Sognier, M. A., Hittelman, W. N., & Rao, P. N. (1979) *Mutat. Res.* **60,** 61–72.

Solomon, E. & Bobrow, M. (1975) *Mutat. Res.* **30,** 273–278.

Sono, A. & Sakaguchi, K. (1978) *Cell Struct. Funct.* **3,** 341–347.

Sono, A. & Sakaguchi, K. (1979) *Cell Struct. Funct.* **4,** 75–80.

Sono, A. & Sakaguchi, K. (1980) *Cell Struct. Funct.* **5,** 175–182.

Sono, A., Matsumoto, K., Hayano, K., & Sakaguchi, K. (1978) *Cell Struct. Funct.* **3,** 249–254.

Speit, G. & Vogel, W. (1979) *Mutat. Res.* **59,** 223–229.

Speit, G., Wick, C., & Wolf, M. (1980a) *Hum. Genet.* **54,** 155–158.

Speit, G., Wolf, M., & Vogel, W. (1980b) *Mutat. Res.* **78,** 267–272.

Speit, G., Wolf, M., & Vogel, W. (1980c) *Mutat. Res.* **78,** 273–278.

Stella, M., Rossi, R., Martinucci, G. B., Rossi, G., & Bonfante, A. (1978) *Biochem. Exp. Biol.* **14,** 221–231.

Stetka, D. G. & Wolff, S. (1976a) *Mutat. Res.* **41,** 333–342.

Stetka, D. G. & Wolff, S. (1976b) *Mutat. Res.* **41,** 343–350.

Stetka, D. G., Minkler, J., & Carrano, A. V. (1978) *Mutat. Res.* **51,** 383–396.

Stich, H. F., Wei, L., & Whiting, R. F. (1979) *Cancer Res.* **39,** 4145–4151.

Stoll, C., Borgaonkar, D. S., & Levy, J.-M. (1976) *Cancer Res.* **36,** 2710–2713.

Strobel, R., Roszinsky-Köcher, G., & Röhrborn, G. (1979) *Mutat. Res.* **60,** 349–355.

Takehisa, S. & Wolff, S. (1977) *Mutat. Res.* **45,** 263–270.

Takehisa, S. & Wolff, S. (1978a) *Mutat. Res.* **58,** 103–106.

Takehisa, S. & Wolff, S. (1978b) *Mutat. Res.* **58,** 321–329.

Tates, A. D., de Vogel, N., & Neuteboom, I. (1980) *Mutat. Res.* **79,** 285–288.

Taylor, J. H., Woods, P. S., & Hughes, W. L. (1957) *Proc. Natl. Acad. Sci. U.S.* **43,** 122–138.

Tetzner, C., Juhl, H. J., & Rüdiger, H. W. (1980) *Mutat. Res.* **79,** 163–167.

Tezuka, H., Ando, N., Suzuki, R., Terahata, M., Moriya, M., & Shirasu, Y. (1980) *Mutat. Res.* **78,** 177–191.

Thompson, L. H., Baker, R. M., Carrano, A. V., & Brookman, K. W. (1980) *Cancer Res.* **40,** 3245–3251.

Thorgeirsson, S. S. & Nebert, D. W. (1977) *Adv. Cancer Res.* **25,** 149–193.

Thust, R. & Kneist, S. (1979) *Mutat. Res.* **67,** 321–330.

Thust, R., Warzok, R., Grund, E., & Mendel, J. (1978) *Mutat. Res.* **51,** 397–402.

Thust, R., Mendel, J., Schwarz, H., & Warzok, R. (1980) *Mutat. Res.* **79,** 239–248.

Tice, R., Windler, G., & Rary, J. M. (1978) *Nature (Lond.)* **273,** 538–540.

Tice, R. R., Costa, D. L., & Drew, R. T. (1980) *Proc. Natl. Acad. Sci. U.S.* **77,** 2148–2152.

Tohda, H., Oikawa, A., Kawachi, T., & Sugimura, T. (1980) *Mutat. Res.* **77,** 65–69.

Torigoe, K. (1980) *Jap. J. Hum. Genet.* **25,** 15–21.

Tredger, J. M. & Chhabra, R. S. (1976) *Drug Metabolism Disposition* **4,** 451–459.

Tsuchimoto, T., Matter, B. E., & Deyssenroth, H. (1979) *Mutat. Res.* **67,** 39–45.

Udenfriend, S., Clark, C. T., Axelrod, J., & Brodie, B. B. (1954) *J. Biol. Chem.* **208,** 731–739.

Uehleke, H. (1974) *Mutat. Res.* **25,** 159–167.

Umezawa, K., Sawamura, M., Matsushima, T., & Sugimura, T. (1979) *Chem.-Biol. Interact.* **24,** 107–110.

Utakoji, T., Hosoda, K., Umezawa, K., Sawamura, M., Matsushima, T., Miwa, M., & Sugimura, T. (1979) *Biochem. Biophys. Res. Commun.* **90,** 1147–1152.

Ved Brat, S., Verma, R. S., & Dosik, H. (1979) *Mutat. Res.* **63,** 325–334.

Vig, B. K. (1979) *Mutat. Res.* **61,** 309–317.

Vig, B. K., Natarajan, A. T., & Zimmermann, F. K. (1977) *Mutat. Res.* **42,** 109–116.

Vigfusson, N. V. & Vyse, E. R. (1980) *Mutat. Res.* **79,** 53–57.

Vijayalaxmi (1980) *Int. J. Rad. Biol.* **37,** 581–583.

Vijayalaxmi, Thomson, E., & Evans, H. J. (1979) *Mutat. Res.* **67,** 47–53.

Vogel, W. & Bauknecht, T. (1976) *Nature (Lond.)* **260,** 448–449.

Wattenberg, L. W. (1978) *J. Natl. Cancer Inst.* **60,** 11–18.

Wattenberg, L. W. (1979) In *Carcinogens: Identification and Mechanisms of Action* (A. C. Griffin & C. R. Shaw, Eds.), Raven Press, New York, pp. 299–316.

Weinstein, D., Katz, M. L., & Kazmer, S. (1977) *Mutat. Res.* **46,** 297–304.

Weisburger, J. H. & G. M. Williams (1975) In *Cancer, a Comprehensive Treatise* Vol. 1. (F. F. Becker, Ed.), Plenum, New York, pp. 185–234.

Wheeler, G. P. & Simpson-Herren, L. (1973) In *Drugs and the Cell Cycle* (A. M. Zimmerman, G. M. Padilla, & I. L. Cameron, Eds.), Academic Press, New York and London, pp. 249–306.

White, A. D. (1980) *Mutat. Res.* **78,** 171–176.

White, A. D. & Hesketh, L. C. (1980) *Mutat. Res.* **69,** 283–291.

White, A. E., Takehisa, S., Eger, E. I., Jr., Wolff, S., & Stevens, W. C. (1979) *Anesthesiology* **50,** 426–430.

Whitlock, J. P., Jr., Cooper, H. L., & Gelboin, H. V. (1972) *Science* **177,** 618–619.

Wiencke, J. K., Cervenka, J., & Paulus, H. (1979) *Mutat. Res.* **68,** 69–77.

Wolff, S. (1977a) *Annu. Rev. Genet.* **11,** 183–201.

Wolff, S. (1977b) In *Research in Photobiology* (A. Castellani, ed.), Plenum, New York, pp. 721–732.

Wolff, S. & Perry, P. (1974) *Chromosoma (Berl.)* **48,** 341–353.

Wolff, S. & Rodin, B. (1978) *Science* **200,** 543–545.

Wolff, S. & Takehisa, S. (1977) In *Progress in Genetic Toxicology* (D. Scott, B. A. Bridges, & F. H. Sobels, Eds.), Elsevier/North-Holland Biomedical Press, Amsterdam, pp. 193–200.

Wolff, S., Bodycote, J., & Painter, R. B. (1974) *Mutat. Res.* **25,** 73–81.

Wolff, S., Rodin, B., & Cleaver, J. E. (1977) *Nature (Lond.)* **265,** 347–349.

Zbarsky, I. B. (1978) *Int. Rev. Cytol.* **54,** 295–360.

Chapter 6

The Use of Sister Chromatid Exchange to Monitor Human Populations for Exposure to Toxicologically Harmful Agents

B. LAMBERT, A. LINDBLAD, K. HOLMBERG, AND D. FRANCESCONI
DEPARTMENT OF CLINICAL GENETICS
KAROLINSKA HOSPITAL
STOCKHOLM, SWEDEN

I. Introduction 150
II. Methodological considerations 151
  A. Standard technique 151
  B. Variations in the culture conditions 152
  C. Effect of BrdUrd concentration 155
  D. Staining procedures 156
III. Factors influencing the SCE frequency in healthy control subjects 157
  A. The individual SCE frequency 157
  B. The proportion of B and T lymphocytes 158
  C. Smoking 159
  D. Other factors influencing the individual SCE frequency 165
IV. Patients under medical treatment 167
  A. PUVA treatment 168
  B. Metronidazole and nalidixic acid 169
  C. Cancer chemotherapy 170

The authors' work was supported by grants from the Swedish Work Environmental Fund, the Swedish Medical Research Council, the Swedish Council for Research Coordination, and the Swedish Environment Protection Board. One of the authors (D.F.) was a scholar of the Rusconi Foundation.

V.	Occupational exposure	173
	A. Laboratory workers	173
	B. Operating room personnel	175
	C. Ethylene oxide	175
	D. Various occupations and types of exposure	176
VI.	Conclusions	177

I. INTRODUCTION

Since the discovery and development of the new staining techniques for the differentiation between sister chromatids in metaphase chromosomes (Latt, 1973; Perry and Wolff, 1974), these methods have been extensively utilized for the analysis of SCE formation in cells exposed to genotoxic agents. Many chemicals that give rise to mutations and/or chromosome aberrations in eukaryotic cells have been shown to induce SCE as well (e.g., Perry and Evans, 1975). The dose of chemicals needed for the detection of an increase in the SCE frequency is generally very low. The analysis of SCE in mammalian cells is therefore a useful contribution to the battery of test systems for the evaluation of mutagenic properties of chemicals and drugs.

SCE is easy to study in human lymphocytes and there are many reports dealing with the formation of SCE in short-term human lymphocyte cultures as a response to chemical exposure *in vivo* or *in vitro*. In comparison with the conventional chromosome aberration analysis the SCE technique appears to be simpler, faster, and less expensive. The SCE technique therefore stands out as one of the few practically useful means by which human subjects exposed to genotoxic agents in the environment can be monitored. However, the use of the SCE technique for this purpose demands a better knowledge about the mechanism(s) that give rise to SCE. The variation in SCE frequency that can be observed with standardized methods in reference populations calls for an explanation. Further questions concern which types of exposure or agents are capable of inducing SCE *in vivo* and for how long an increase of SCE is persistent under *in vivo* conditions.

This chapter deals with studies of SCE in lymphocyte cultures from *in vivo* exposed subjects. Experiments with mammalian cells *in vitro* are discussed elsewhere (Chapter 8) and are not considered unless they have obvious relevance for the understanding of SCE formation *in vivo*. The first section presents methodological problems, mainly with regard to SCE counts in human lymphocytes cultured under various conditions. The second section deals with the individual variation in the SCE frequency in healthy subjects. The third and fourth sections contain data on the SCE frequency in patients under chemotherapy for various disorders and in occupationally exposed subjects.

II. METHODOLOGICAL CONSIDERATIONS

Different base levels of the SCE frequency in lymphocyte cultures of healthy subjects have been presented by various investigators. This difference partly depends on the methodological procedure used. In this section we present data explaining some of the variation due to methodological differences.

A. *Standard Technique*

Our routine procedure for lymphocyte cultivation has been the so-called macro technique as described by Buckton and Evans (1973) with some modifications (Lambert et al., 1976). For staining we have used the principle of the fluorescence plus Giemsa (FPG) technique (Perry and Wolff, 1974). Unless otherwise stated the outline of the procedure is as follows. Venous blood is collected in 10 ml vacutainer tubes (Becton and Dickinson) containing 143 USP units of sodium heparin. Phytohemagglutinin (0.25 ml PHA, HA 15, Wellcome) is added to each tube and the blood is kept for 15 min in an ice bath. After gentle centrifugation the plasma and leucocytes are removed and transferred to 20 ml sterile penicillin bottles with rubber caps. In each bottle 2 ml of the cell-plasma suspension is mixed with 6 ml of medium 199 (Flow Laboratories) containing 150 μg/ml of streptomycin (Novo, Copenhagen) and 150 IU/ml of benzyl penicillin (KABI, Stockholm). Bromodeoxyuridine (BrdUrd, B 5002, Sigma) is added to each bottle to make the final concentration 100 μM. After the mixture is incubated in the dark for 73 hours at 37°C, 0.125 μg/ml of colchicine (Apoteksbolaget, Stockholm) is added and the incubation is allowed to continue for 2 hours longer. Harvesting is then carried out by gentle centrifugation, after which the medium is discarded. The cells are resuspended in 0.075 M KCl, centrifuged, and finally fixed in methanol/acetic acid (3 : 1) overnight in the refrigerator. Fixed cells are dropped on clean, wet slides and allowed to dry in the air.

The slides are stained in 1 μg/ml of 33258 Hoechst in distilled water for 10–30 min, rinsed, and covered with cover slips. Irradiation is carried out immediately under a sunlamp (Ultra-Vitalux, 300 W, Osram, Germany), which is allowed to burn for 5 min before use. The irradiation time is 4 min and the distance from the lamp is 40 cm. The cover slips are then carefully removed and the slides are treated for 1.5 hours in 2 × SSC at 60°C and then stained with Giemsa (5% Giemsa, Searle, England, in phosphate buffer, pH 6.8) for 3 min or longer if necessary. After they are washed in buffer and water, the slides are mounted, coded, and analyzed blindly.

Analysis is carried out at the highest magnification in the light microscope. The slide positions of the cells are recorded to permit later checkup if considered necessary. Only cells containing 46 chromosomes are used. Terminal SCEs are recorded as a single SCE and interstitial SCEs as two exchanges. Centromeric exchanges are included in the analysis unless there is uncertainty whether the chromatids might be twisted. As a rule the first 20 complete metaphases suitable for SCE counting are analyzed in each set of three to six slides from a given culture. The average number of SCEs based on these 20 cells, called the individual SCE frequency, is used for further calculations of group mean values. It has been noted that the SCE numbers in a given culture are not normally distributed (Pedersen et al., 1979) whereas the individual means are (Morgan and Crossen, 1977). Student's t-test was used for the statistical analysis unless otherwise stated.

B. Variations in the Culture Conditions

It is not always possible to start the lymphocyte culture immediately after blood sampling. Therefore it is of interest to know if storage of the blood under various conditions has an influence on the basal SCE frequency. As is shown in Table 6.1 there was no dramatic effect due to storage of the blood for 8–24 hours at room temperature or in the refrigerator. The effect, if any, was a small decrease in the mean SCE frequency and a reduction of cells with higher SCE counts.

TABLE 6.1
The Frequency of SCE in Lymphocytes from Healthy Donors: Effect of Storing the Blood Sample of a Single Donor

Storage conditions	SCE/cell Mean ± SD	SCE/cell Range	No. of cells
Immediate culture preparation	13.2 ± 4.2	6–22	20
Room temperature, 8 h	10.4 ± 2.9	6–16	20
Room temperature, 24 h	11.0 ± 3.5	5–16	20
Refrigerator, 24 h	10.3 ± 2.6	5–17	20

Another factor of importance for the standardization of the culture technique concerns the addition of PHA. When PHA was added immediately to the heparinized tubes containing 10 ml of the blood sample (thereby facilitating the separation of white blood cells from erythrocytes) the SCE frequency was significantly higher than if PHA was added directly to the culture (Table 6.2). The cause of this difference is not clear to us, but one possibility could be that PHA in the blood before centrifugation yields a different population of lymphocytes.

Most cytogenetic laboratories use the whole blood micro culture technique (Buckton and Evans, 1973), which is considered to be more convenient than the macro technique. We have compared the SCE frequencies in parallel micro and macro cultures from the same donors. In all 19 micro cultures studied the SCE frequency was lower than in the corresponding macro cultures. The average SCE frequency was about 40% lower in the micro cultures (Table 6.3). This difference between micro and macro cultures could be due to the type of serum used. Macro cultures usually make use of autologous human serum, whereas micro cultures are run in the presence of fetal or newborn calf serum. As shown in Table 6.4 there is a significantly lower SCE frequency in macro cultures in the presence of fetal calf serum compared to the parallel macro cultures from the same donors run in the presence of the autologous human serum. On the other hand, we could not find any difference between micro cultures with fetal calf serum and new born calf serum (Table 6.4). Other work on SCE in Chinese hamster cells (Kato and Sandberg, 1977) has shown that untreated fetal calf serum contains some SCE-inducing factors that are heat labile and to some extent dialyzable. It is not unlikely that human serum also contains factors that may influence the SCE frequency. This could constitute a source of variation in the SCE frequency among individuals.

TABLE 6.2
The Frequency of SCE in Lymphocytes of Healthy Donors: Effect of PHA

Treatment conditions	No. of donors	SCE/cell Mean ± SD	SCE/cell Range	No. of cells
0.15 ml of PHA in culture	4	10.1 ± 2.5[a]	7.4–13.0	80
0.25 ml of PHA in tube	4	15.5 ± 1.8[a]	13.6–18.0	80

[a]Significantly different according to t-test ($P < 0.02$).

TABLE 6.3
The Frequency of SCE in Lymphocytes From Healthy Donors: Difference Between Micro and Macro Cultures

Culture conditions[a]	No. of donors	SCE/cell Mean ± SD	SCE/cell Range	No. of cells
Macro	19	14.4 ± 3.5[b]	10.2–25.7	380
Micro	19	10.5 ± 2.2[b]	8.0–15.8	379

[a]Micro cultures contain 0.5 ml of whole blood in 4 ml of medium 199, to which 0.1 ml of PHA, 0.6 ml of fetal calf serum, and BrdUrd (100 μM) is added. Otherwise the micro procedure is identical to the macro procedure as described in the text.

[b]Significantly different according to t-test ($P < 0.001$).

Genotoxic damage usually inhibits DNA replication and cell propagation in culture. It is therefore often necessary to prolong the incubation time of cultures from exposed subjects to obtain a large enough number of cells for SCE counting. The effect of culture time on the SCE frequency was studied in parallel cultures from 16 healthy donors. The cultures were harvested after 75 and 96 hours. As shown in Table 6.5 there was no difference in the SCE frequency among cultures incubated for different periods of time.

TABLE 6.4
The Frequency of SCE in Lymphocytes from Healthy Donors: Cultivation in the Presence of Different Types of Serum

Culture conditions[a]	Type of serum[b]	No. of donors	SCE/cell Mean ± SD	SCE/cell Range	No. of cells
Macro	AHS	4	18.9 ± 0.7[c]	18.2–19.8	80
	FCS	4	12.6 ± 2.0[c]	9.8–14.7	80
Micro	FCS	4	10.2 ± 0.9	9.6–11.5	80
	NBC	4	10.4 ± 1.0	9.5–11.6	80

[a]The macro and micro experiments are not comparable since they were carried out with different donors.

[b]AHS = autologous human serum, FCS = fetal calf serum, NBC = newborn calf serum.

[c]Significantly different according to t-test ($P < 0.001$).

TABLE 6.5
The Frequency of SCE in Lymphocytes from Healthy Donors

		SCE/cell/individual		
Culture time (h)	No. of donors	Group mean ± SD	Range	No. of cells
75	16	16.6 ± 3.6	10.2–23.6	325
96	16	16.5 ± 4.1	10.3–23.8	321

C. *Effect of BrdUrd Concentration*

The frequency of SCE increases with increasing concentration of BrdUrd during cultivation (e.g., Wolff and Perry, 1974; Kato, 1974; Latt, 1974; Lambert et al., 1976; Mazrimas and Stetka, 1978). Figure 6.1 shows the SCE frequency in the lymphocytes from five healthy donors cultivated in various concentrations of BrdUrd. The relative increase, approximately 75%, is of about the same magnitude in all five subjects when the BrdUrd concentration is increased from 10 to 100 μM. Moreover, the relation between the SCE frequency and the BrdUrd concentration appears to be almost linear in the range tested and there is no indication of a decreased BrdUrd dependence at lower concentrations (Figure 6.1). However, in micro cultures there was no change in the SCE frequency when the concentration of BrdUrd was increased from 20 to 100 μM, although the highest concentration of BrdUrd obviously has an inhibitory effect on cell propagation in the culture (Table 6.6).

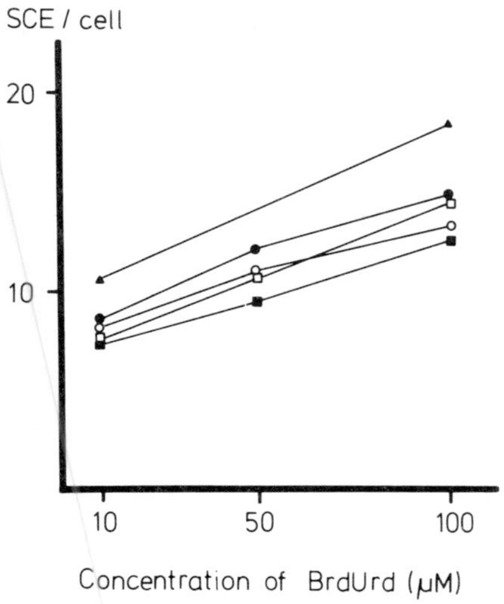

Fig. 6.1. The dependence of the SCE frequency on the concentration of BrdUrd in lymphocyte cultures from five donors.

TABLE 6.6
The Frequency of SCE in Lymphocytes from Healthy Donors: the Effect of Increasing BrdUrd Concentration in Micro Cultures

Concentration of BrdUrd (μM)	No. of donors	SCE/cell Mean ± SD	SCE/cell Range	No. of cells	Percentage of cells in div. no.[a] I	II	III
20	3	9.3 ± 1.0	8.2–10.1	60	24	46	30
50	3	9.1 ± 1.0	8.0–9.8	60	24	48	28
100	3	10.1 ± 1.3	9.2–11.6	60	37	55	8

[a]Data based on the analysis of 200 cells per given concentration of BrdUrd.

Human B and T lymphocytes differ in their sensitivity to SCE induction by BrdUrd (Santesson et al., 1979), and it is possible that other subpopulations of human lymphocytes also may show such differences. Recent information suggests that the induction of SCE by BrdUrd in Chinese hamster ovary cells is largely independent of the BrdUrd content of DNA, whereas the concentration of BrdUrd in the medium appears to be of great importance (Davidson et al., 1980). It seems possible that the effect of BrdUrd on the SCE frequency to some extent reflects BrdUrd induced changes in the precursor supply for DNA synthesis. Our observation of a difference in the sensitivity to BrdUrd between macro and micro cultures suggests that factors such as the cell density in the culture, variations in subpopulations of lymphocytes, and the composition of the culture medium (especially with regard to DNA precursors) could influence the BrdUrd-dependent induction of SCE. In any case the BrdUrd concentration in macro cultures appears to be one factor of great importance for the SCE frequency, and it should therefore be carefully controlled.

D. *Staining Procedures*

A number of different modifications of the original staining procedures (Latt, 1973; Perry and Wolff, 1974) have been presented. The method described by Alves and Jonasson (1978) is as reliable and convenient as the original procedures. This technique makes use of alkaline treatment of fixed chromosomes and does not need fluorescent stain or irradiation. It is rapid and yields good results. We have scored SCEs on slides from the same cultures of five donors. The results obtained with the FPG-staining and the technique of Alves and Jonasson (1978) correspond very well (Table 6.7).

TABLE 6.7
The Frequency of SCE in Lymphocytes from Healthy Donors: Effects of Two Different Staining Procedures

Staining technique[a]	No. of donors	SCE/cell Mean ± SD	SCE/cell Range	No. of cells
FPG	5	16.2 ± 8.2	8.6–29.2	100
APG	5	16.6 ± 7.5	10.1–29.0	100

[a]FPG: fluorescence-plus-Giemsa technique described by Perry and Wolff (1974) and modified as described in this Chapter; APG: alkaline treatment plus Giemsa technique by Alves and Jonasson (1978).

III. FACTORS INFLUENCING THE SCE FREQUENCY IN HEALTHY CONTROL SUBJECTS

The use of the SCE technique in monitoring groups of human subjects for exposure to genotoxic agents requires, among other things, knowledge about the basal SCE frequency in a reference population. A few reports deal specifically with the distribution of SCE in the lymphocytes of healthy, human subjects (Daoud et al., 1976; Lambert et al., 1976, 1978a; Morgan and Crossen, 1977; Crossen et al., 1977; Pedersen et al., 1979). The general conclusion from these reports is that there is a considerable variation in the SCE frequency in the human population.

A. *The Individual SCE Frequency*

The individual SCE frequency (based on the number of SCE in 20 cells) was found to differ by a factor of 2 in a group of nine healthy donors. In contrast the SCE frequency in two or three parallel cultures from the same individual was relatively constant (Figure 6.2). The SCE frequency was followed in a group of donors by repeat analyses at various time intervals. Again there was a clear difference among subjects, but the SCE frequency was less variable in one and the same donor (Table 6.8). These results suggest that constitutional or relatively constant environmental factors play a role in the determination of the individual SCE frequency. It was therefore of interest to study the SCE frequency in twins. Figure 6.3 shows the results in a group of nine monozygotic, female twin pairs. The average SCE frequency in this group was 16.0 ± 2.5 (mean ± S.D., $n = 18$), range 12.0–20.6. The variance within pairs was not significantly smaller than among pairs ($0.25 > p > 0.1$). These results suggest that

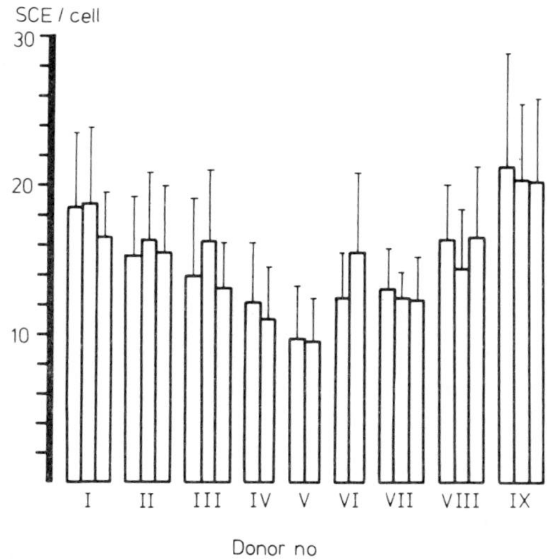

Fig. 6.2. SCE in parallel cultures from nine donors. Mean of 20 cells (columns) and 1 SD (bars).

genetic factors do not play a major role for the frequency of SCE in normal subjects. Pedersen et al. (1979) made the same observation in a more thorough study of monozygotic and dizygotic twin pairs.

B. The Proportion of B and T Lymphocytes

Santesson et al. (1979) showed that human B lymphocytes have a faster rate of cell proliferation and an approximately twofold lower SCE frequency than T lymphocytes. Individuals may differ with regard to the proportions of B and T cells in the peripheral blood. Thus the difference between B and T cells could theoretically cause a variation in the individual SCE frequency if it is expressed in the conventional heterogeneous lymphocyte culture. We have therefore determined the proportion of B and T cells in the peripheral blood of 24 healthy subjects in parallel to the conventional SCE analysis. As is shown in Figure 6.4 there was no correlation between the proportion of T or B cells and the SCE frequency. This observation suggests that different rates of B and T lymphocytes in the peripheral blood have a minor, if any, influence on the individual SCE frequency. The SCE analysis was made in conventional lymphocyte cultures, which probably mainly reflects the T cell SCE frequency. On the other hand, there may exist distinct subpopulations of human lymphocytes other than B and T cells that have different rates of cell proliferation and different basal SCE frequencies (Snope and Rary, 1979; Lindblad

TABLE 6.8
The SCE Frequency in Lymphocytes from Healthy Donors: Repeat Analyses at Various Time Intervals

Donor no.	Culture no.	Months after first culture	SCE/cell Mean ± SD	SCE/cell Range	No. of cells
I	1	—	14.4 ± 3.4	9–21	20
	2	8	14.4 ± 3.8	9–23	20
II	1	—	18.6 ± 4.7	9–28	20
	2	6	23.2 ± 4.5	12–30	20
III	1	—	16.9 ± 4.7	6–28	25
	2	18	14.0 ± 4.9	7–25	20
	3	19	12.8 ± 4.0	4–20	20
	4	53	12.2 ± 3.8	8–18	20
IV	1	—	14.9 ± 5.1	8–23	18
	2	1	14.8 ± 6.1	6–27	20
	3	12	14.8 ± 5.8	6–22	20
	4	13	14.0 ± 3.9	7–20	20
	5	16	18.9 ± 4.9	10–27	20
V	1	—	20.1 ± 4.8	12–28	14
	2	15	17.5 ± 7.3	8–40	20

and Lambert, 1981). This possibility is supported by the observation of a considerable change in the SCE frequency in human lymphocytes after vaccination (Knuutila et al., 1978; Lambert et al., 1979a).

C. *Smoking*

Cigarette smokers have a significantly higher SCE frequency than nonsmokers. This original observation (Lambert et al., 1978a) has now been confirmed by others (Bala Krishna Murthy, 1979; Hopkin and Evans, 1980), although there remains some uncertainty as to the significance and reproducibility of the increased SCE frequency among smokers (Schönwald et al., 1977; Hollander et al., 1978; Bridges et al., 1979).

Table 6.9 shows the difference in SCE frequency between smokers and

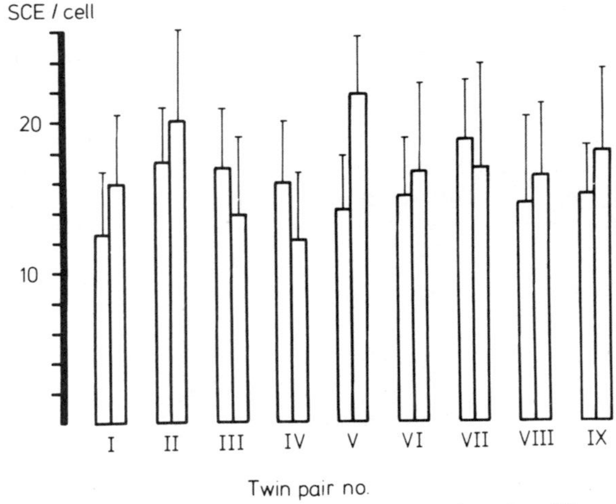

Fig. 6.3. SCE in lymphocytes from 9 pairs of monozygotic twins. Mean of 20 cells (columns) and 1 SD (bars).

nonsmokers in five reference groups. Some of these groups were studied for purposes other than to investigate the effect of tobacco smoking (see below). The SCE frequency was higher in the smokers of all groups than in the corresponding nonsmokers. The SCE frequency among the nonsmokers of the various groups did not differ statistically (Table 6.9). The average increase of SCE was about 20% in smokers. This may seem small, but the difference is highly significant in the total population of 219 subjects investigated ($P < 0.001$). The distribution of the individual SCE frequencies in smokers shows a considerable shift towards higher SCE values compared to that of nonsmokers. About 10% of the smokers had an SCE frequency of 22 SCEs/cell or more, that is, values not seen among nonsmokers (Figure 6.5). Cells with $\geqslant$ 30 SCEs were also encountered in 28 (31%) of the 91 smokers compared to only 9 (7%) among the 128 nonsmokers. Thus the effect of smoking on the SCE frequency is apparent not only as an increase of the individual means, but also as a shift of the SCE in single cells toward higher numbers (Figure 6.6).

Detailed information about the smoking habits was not available for all the smokers, but in 64 subjects it was possible to differentiate between smokers of 10 or less cigarettes daily (moderate smokers) and those smoking more than 10 cigarettes daily (heavy smokers). As shown in Table 6.10 there was a significantly higher SCE frequency in the heavy smokers than in the moderate smokers.

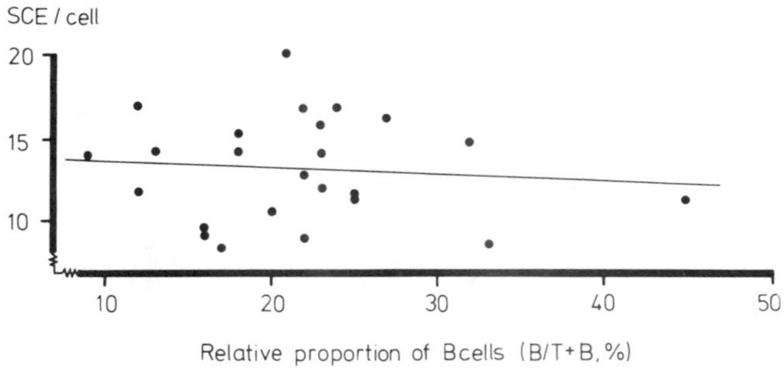

(a)

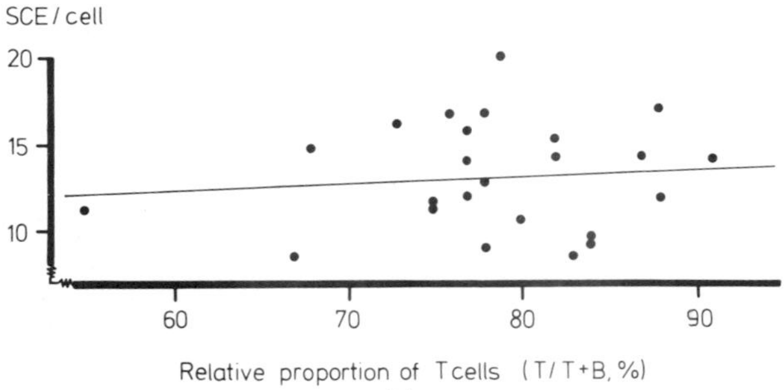

(b)

Fig. 6.4. Absence of a correlation between the proportions of (*a*) B lymphocytes or (*b*) T lymphocytes and the SCE frequency. SCE was determined in 20 cells in conventional lymphocyte cultures from 24 subjects. The proportions of B and T cells in the peripheral blood of the same donors were estimated by standard EAC- and E-rosetting techniques, respectively.

The effect of smoking on the SCE frequency could be due to *in vivo* induced alterations in the genetic material, or to tentative serum factors inducing SCE under the *in vitro* cultivation. Lymphocytes from smoking and nonsmoking subjects were cultivated in parallel in the presence of autologous serum (macro cultures) and fetal calf serum (micro cultures). It was found that the difference between nonsmokers and smokers is maintained even in the presence of fetal calf serum (Table 6.11). This finding suggests that the increased SCE frequency in smokers does indeed reflect some damage to the genetic material induced *in vivo*, as was also concluded by Hopkin and Evans (1980). It would also be of interest to know how soon after smoking an increase in SCE is detectable, and for how long the increase remains after a period of nonsmoking. A preliminary attempt to obtain some answers to these questions was made by

TABLE 6.9
The Frequency of SCE in Peripheral Lymphocytes of Various Reference Populations Divided by Smoking Habits

	Nonsmokers			Smokers			
		SCE/cell			SCE/cell		
Reference population[a]	No. of subjects	Group mean ± SD	Range	No. of subjects	Group mean ± SD	Range	Statistical analysis[b]
Blood donors	45	14.0 ± 3.0	7.8–20.2	37	15.7 ± 3.2	9.2–24.2	$P < 0.02$
Factory workers	50	12.7 ± 3.0	6.2–18.2	26	15.8 ± 4.1	8.4–24.1	$P < 0.001$
Psoriasis patients	10	14.3 ± 3.3	10.4–20.0	10	16.8 ± 2.4	13.8–22.0	$P < 0.1$
Melanoma patients	13	13.8 ± 3.1	10.1–19.8	9	17.4 ± 3.5	13.6–23.6	$P < 0.025$
Metronidazole patients	10	13.4 ± 2.9	9.7–20.6	9	18.8 ± 4.1	14.2–25.4	$P < 0.005$
All groups	128	13.5 ± 3.0	6.2–20.6	91	16.3 ± 3.6	8.4–25.4	$P < 0.001$

[a] All these groups are described in Sections III and IV of this Chapter. The patient groups were investigated before any recent medical therapy.
[b] t-Test for difference between nonsmokers and smokers within the reference populations.

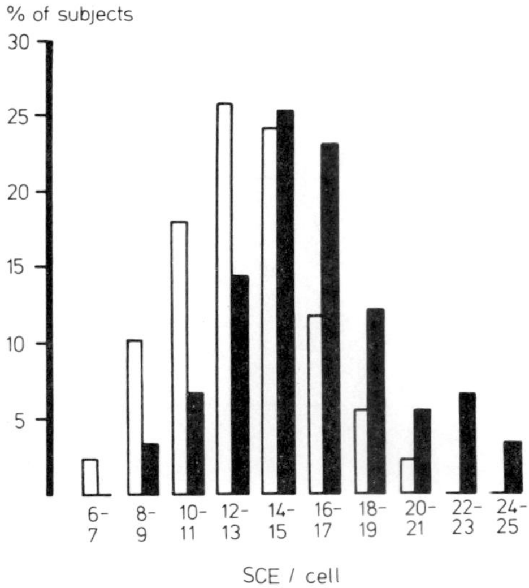

Fig. 6.5. Distribution of the individual SCE frequencies in (*open bars*) 128 nonsmokers and (*filled bars*) 91 smokers.

sampling blood from a single subject at various times during a day of smoking. The first analysis after 18 hours of nonsmoking showed a relatively low baseline value, which did not change immediately after the first cigarette. Six hours later, after six cigarettes, there was a small increase in the SCE frequency. After an additional 18 hours of nonsmoking the SCE frequency was again at the baseline level (Table 6.12). These results are not conclusive, but they suggest that the SCE response is closely related

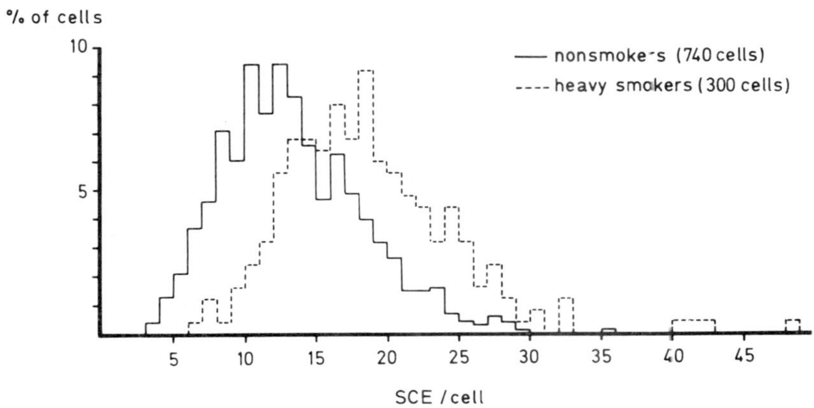

Fig. 6.6. Distribution of SCEs in cells from 27 nonsmokers and 15 heavy smokers.

TABLE 6.10
The frequency of SCE in lymphocytes from moderate (≤ 10 cigarettes/day) and heavy (> 10 cigarettes/day) smokers

Group of subjects	No. of subjects	SCE/cell		Statistical analysis[a]
		Group mean ± SD	Range	
Moderate smokers	32	15.7 ± 4.0	9.2–27.4	
Heavy smokers	32	17.9 ± 3.0	12.0–25.7	$P < 0.02$

[a]t-Test.

to the time of smoking. They also indicate that a relatively short smokingfree period could reduce the effect of previous smoking on the SCE frequency.

The results presented above show that there is a considerable overlap in the SCE frequencies between smokers and nonsmokers (Figure 6.5) and also between moderate and heavy smokers (Table 6.10). This may simply reflect different patterns of inhalation or the time between smoking and blood sampling (cf. Table 6.12). It could also be due to individual differences in the distribution of the active agents in the tobacco smoke from the bronchial tree by way of the mucosa into the blood. Constitutional differences in the metabolism of the active agents may cause additional variation as well. More information is needed about the cause of the great individual variation and about the time of appearence and duration of the increased SCE frequency in relation to smoking. Moreover, the SCE-inducing agents in tobacco smoke remain to be identified. Benzpyrene (Rüdiger et al., 1976; Schönwald et al., 1977) and cigarette smoke condensates (Hopkin and Evans, 1979) have been shown to induce SCE in human

TABLE 6.11
The Effect of Smoking Habits on the SCE Frequency in Lymphocytes Cultivated in Fetal Calf Serum

Subjects	No. of subjects	SCE/cell (mean ± SD)	
		Autologous serum	Fetal calf serum
Nonsmokers	7	12.4 ± 2.1	9.2 ± 0.9
Smokers	3	14.9 ± 1.7	11.0 ± 0.9

TABLE 6.12
The SCE Frequency in a Single Subject During One Day's Smoking

Time of blood sampling[a]	SCE/cell Mean ± SD	SCE/cell Range	No. of cells
Before smoking	11.8 ± 4.2	3–20	40
After 1 cigarette	11.4 ± 3.9	6–21	40
After 6 cigarettes	13.5 ± 4.2	5–24	40
Next morning	11.9 ± 4.8	5–24	34

[a]The donor is a regular smoker of 6–8 cigarettes per day. The first sample (before smoking) was taken in the morning after an 18 h period of no smoking. The second sample was taken 10 min after the first cigarette. The third sample was taken 6 h later, 10 min after the sixth cigarette. The last blood sample was taken the next morning after 18 h of no smoking.

lymphocytes *in vitro*. Very likely there are many factors in the cigarette smoke that can increase the SCE frequency *in vivo*.

The observation of the SCE-inducing effect of smoking is of obvious importance for the use of SCE in monitoring human populations. It allows one important source of variation in the baseline level of SCE to be eliminated, provided that control subjects, as well as test groups, are carefully interviewed with regard to smoking habits. Furthermore, the effects of tobacco smoking, which are well known in terms of medical and carcinogenic hazards, could be used as a reference for comparison with genotoxic effects of other types of environmental and occupational exposure.

D. Other Factors Influencing the Individual SCE Frequency

In parallel to SCE analysis in a number of subjects, information was collected about working and social conditions, medical treatment, hobbies, and private habits, including tobacco, coffee, and alcohol consumption. The subjects were randomly chosen among regular blood donors at the Karolinska Hospital, and apart from this they could be regarded as a representative sample of the general, healthy population. In the analysis of the questionnaires so far, smoking is the only single factor that has been found to have a clear effect on the SCE frequency. However, some preliminary observations concerning other factors are of interest. Table 6.13 shows the reference group divided according to smoking habits and sex. In the total male group there was a significant difference between smokers and nonsmokers ($P < 0.005$). No other significant differences were discovered within or between the nonsmoking and smoking groups with

TABLE 6.13
The SCE Frequency in Various Groups of a Reference Population

Group of subjects	Nonsmokers		Smokers	
	No. of subjects	SCE/cell (group mean ± SD)	No. of subjects	SCE/cell (group mean ± SD)
Males	33	13.5 ± 2.7	27	15.9 ± 3.4
Occupational exposure	13	14.3 ± 3.0	11	15.5 ± 3.7
Current infection	7	14.2 ± 3.8	5	15.5 ± 2.3
Females	30	15.7 ± 2.9	18	17.2 ± 4.4
Occupational exposure	11	15.8 ± 4.0	6	17.3 ± 5.3
Current infection	8	15.8 ± 2.9	6	19.9 ± 5.7
OC users	9	16.5 ± 2.1	7	20.3 ± 5.2

regard to allergic disability, occupational or habitual exposure to chemical agents, current infection (within the last 2 months), current medication, recent X-ray investigations, the use of alcohol, and so on. The variables shown in Table 6.13 were the only ones associated with a higher mean value than the total nonsmoking male population.

In the female group allergic reactions and the use of oral contraceptives (OC) were of interest. These features were common to several subjects in the nonsmoking and smoking group and were therefore not separated. Females positive for either or both of these conditions showed a higher, although not significant, increase in SCE frequency than the residual nonsmoking and smoking group of females (Table 6.13). These data are preliminary and are based on a small population. Nevertheless, they suggest that allergic condition and/or the use of oral contraceptives may be associated with an increased SCE frequency. This is not an entirely unexpected observation. Allergic conditions are frequently associated with changes in subpopulations of white blood cells, which may differ in the SCE frequency, as is discussed above (Section III-B). An increased level of SCE has recently been reported in oral contraceptive users by Bala Krishna Murthy and Prema (1979).

It is also worth noting that subjects stating that they are exposed to chemical agents of various kind in their daily work did not show a significant increase in the SCE frequency (Table 6.13). Thus occupational exposure in general does not have a pronounced influence on the SCE frequency. This problem is discussed further below (Section V).

IV. PATIENTS UNDER MEDICAL TREATMENT

The effect of a variety of chemicals and drugs on the SCE frequency in human lymphocytes *in vitro* has been studied in a number of reports (e.g., Craig-Holmes and Shaw, 1977; Gerner-Smidt and Friedrich, 1978; Csukás et al., 1979; Crossen, 1979; Littlefield et al., 1979; Lambert et al., 1979b).

There are comparatively few studies on human subjects exposed to radiotherapy or chemotherapy, although the evaluation of the SCE technique as an *in vivo* assay could benefit greatly from such studies. In addition to the defined dosage schedule in medical treatment, the clinical situation often offers the opportunity to analyze blood samples before, as well as at different times after, the treatment. This can provide information about the dose dependence and duration of changes in the SCE frequency.

The importance of studying the same patient before and after treatment is obvious in view of the great variation among individuals. A possible genotoxic effect of phototherapy against neonatal hyperbilirubinemia was suggested by Goyanes-Villaescusa et al. (1977) on the basis of their observation of a difference in SCE frequency between photoirradiated and nonirradiated groups of newborn infants. However, there is a great variation in the basal SCE frequency among healthy newborn infants, and a later study by Schwartz et al. (1978) did not reveal any difference in single individuals before as compared to after photoirradiation.

A. *PUVA Treatment*

PUVA treatment of severe psoriasis (a combination of oral 8-methoxypsoralen, 8-MOP, and long-wave ultraviolet light, UVA), has caused a great deal of concern because of the genetic hazards that may be involved (see Editorial, 1978). An increased risk for cutaneous carcinoma in PUVA-treated patients has also been reported (Stern et al., 1979). Although the genetic risk to germ cells of PUVA treatment seems to be small (Bridges, 1979), the apparent risk for somatic mutation has prompted a number of investigations on the frequency of SCE in lymphocytes from PUVA-treated patients (Wolff-Schreiner et al., 1977; Lambert et al., 1978b; Faed and Mourelatos, 1978; Brøgger et al., 1978). None of these studies showed an increase in the SCE frequency of PUVA-treated patients (cf. Table 6.14). In contrast, treatment of cells *in vitro* with 8-

TABLE 6.14
The SCE Frequency in Lymphocytes of Patients with Various Disorders Before and After Medical Treatment

Diagnosis and No. of patients		Type of treatment	SCE/cell (group mean ± SD)	
			Before treatment	After treatment
Psoriasis[a]	14	PUVA	15.8 ± 2.4	14.2 ± 4.3
Trichomoniasis or amoebiasis[b]	15	Metronidazole	16.6 ± 5.0	19.3 ± 6.6
Melanoma[c]	2	Actinomycin D	17.0 ± 0.6	18.0 ± 2.1
	8	DTIC	13.4 ± 2.2	14.2 ± 2.3
	5	CCNU	15.3 ± 4.9	35.4[d] ± 6.0
	7	Melphalan	19.8 ± 3.5	58.4[d] ± 10.7

[a]Data from Lambert et al., 1978b.

[b]Data from Lambert et al., 1979c.

[c]Data from Lambert et al., 1979d and e.

[d]Significantly different ($P < 0.01$) from pretreatment value according to Wilcoxon rank test.

MOP + UVA caused a dose-dependent increase of SCE even at a dose similar to that used in the clinic (Figure 6.7) (Carter et al., 1976; Lambert et al., 1978b). The explanation offered for this discrepancy between *in vitro* and *in vivo* results is that the lymphocytes *in vivo* are largely shielded from UVA, which penetrates through the epidermis to a small extent only.

B. *Metronidazole and Nalidixic Acid*

The genetic toxicity of metronidazole in bacterial test systems (e.g., Rosenkranz et al., 1976) has evoked some concern because of the widespread use of this drug for the treatment of protozoan and other anaerobic infections. It was therefore of interest to investigate subjects under therapy with metronidazole, to see if any genotoxic effect *in vivo* could be revealed by SCE analysis. Patients with vaginal trichomoniasis or amoebiasis were selected for the study. Blood samples were taken before initiation of therapy as well as on day 7 of daily intake of 0.6–4.2 g of metronidazole (Lambert et al., 1979c). No significant difference could be observed between the samples taken before and after treatment (Table 6.14). Studies on the effect of metronidazole *in vitro* have not given evidence that this drug induces DNA damage, chromosome aberrations, or SCE in human lymphocytes (Lambert et al., 1979b; Hartley-Asp, 1979). The discrepancy in the response to metronidazole between the bacterial and human cell may be due to different metabolic pathways. However, it is known that metronidazole has a radiosensitizing effect in mammalian

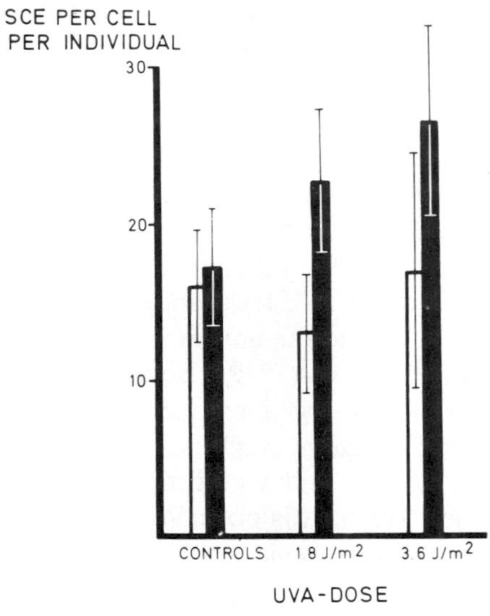

Fig. 6.7. Frequency of SCE in UVA-irradiated lymphocytes obtained from (*filled bars*) psoriasis patients 2 hours after oral administration of 8-methoxypsoralen (8-MOP) and (*open bars*) control patients receiving no 8-MOP. The bars indicate one standard deviation. Data from Lambert et al., 1978b.

cells that is utilized in cancer therapy. Mutagenic metabolites of metronidazole have been detected to the urine of treated human subjects (Speck et al., 1976). Thus the safety of this drug is still an open question.

An interesting study on children with urinary tract infection was recently reported by Kowalczyk (1980). He found a more than twofold increase of SCE in the lymphocytes of the children after treatment with nalidixic acid (Negram), a bacteriocidal agent known to inhibit DNA-gyrase and recombinational repair in *E. coli.* Nalidixic acid appears not to act directly on DNA. Its very significant effect on the SCE frequency *in vivo* could be due to interaction with recombinational functions mediated by DNA-gyrases. It is possible that further studies on the effect of nalidixic acid and other gyrase inhibitors will help to elucidate the mechanisms of sister chromatid exchange.

C. *Cancer Chemotherapy*

The best information about drugs causing an alteration in the SCE frequency in human lymphocytes after *in vivo* treatment so far has been obtained from cancer patients under cytostatic treatment (Table 6.14). A number of drugs have been investigated and the preliminary indication is that agents that bind to DNA or give rise to structural DNA damage increase the SCE frequency, while cytostatic agents interfering with the precursor supply for DNA synthesis have no such effect (Table 6.15). Interesting exceptions to this generalization are actinomycin D, bleomycin, and DTIC. These drugs are known to bind to DNA and/or cause chromosome aberrations *in vitro* but have not been shown to induce SCE *in vivo*. Explanations for this discrepancy could be that these drugs exert their primary cytostatic action at the clinical doses used by means other than interference with the DNA structure, or that the doses used may be too low to demonstrate an effect on the SCE frequency *in vivo*.

Results from several *in vitro* studies with human lymphocytes suggest that mono- and bifunctional alkylating agents such as mitomycin C, chlorambucil, thiotepa (Littlefield et al., 1979), and busulphan (Honeycombe, 1978) cause a marked increase in the SCE frequency, while the antimetabolites methotrexate and cytarabine and the antitumor antibiotic bleomycin have no such effect (Littlefield et al., 1979). Similar results have been reported in Chinese hamster cells. A large increase of SCE was found after *in vitro* treatment with alkylating agents and little or no effect, for example, with bleomycin, actinomycin D, and methotrexate (Perry and Evans, 1975; Banerjee and Benedict, 1979). It is possible that agents exerting their primary cytostatic action by metabolic inhibition of DNA

TABLE 6.15
Cancer-Chemotherapeutic Drugs Investigated for Their Effect on the Frequency of SCE in Human Lymphocytes After Clinical Treatment

Drug	Effect and its duration[a]	Reference
Actinomycin D	−	Lambert et al., 1979e
Adriamycin	+	Nevstad, 1978
	+ 12 weeks	Musilová et al., 1979
Bleomycin	−	Lambert et al., 1978c
Busulphan	+ 5 weeks	Musilová et al., 1979
CCNU	+ 8 weeks	Lambert et al., 1979d
Cyclophosphamide	+	Raposa, 1978
Cyclophosphamide	+ 16 weeks	Musilová et al., 1979
Cytosine arabinoside+ +thioguanine	−	Raposa, 1978
DTIC	−	Lambert et al., 1979d
5-Fluorouracil	−	Musilová et al., 1979
Melphalan	+ 4 weeks	Lambert et al., 1979e
Methotrexate	−	Lambert et al., 1978c
Mitomycin C	+ 1 week	Littlefield et al., 1980

[a]−: no demonstrable effect on the SCE frequency; +: a significant increase is found within 7 days after treatment. The duration is the longest time in weeks after termination of the treatment at which an increase in SCE is still observed. In the case of mitomycin C there was no indication of a reduction of the SCE frequency at the time indicated, and longer times were not studied.

synthesis or after the S phase have little or no capacity for SCE induction, while agents with a maximum effect during the G_0-G_1 phase or G_1-S transition are more potent SCE inducers. This interpretation of current data (Littlefield et al., 1979) is highly relevant to the use of SCE in the monitoring of human populations for exposure to unknown mutagens, since it implies that not all genotoxic agents yield positive results with the SCE technique.

Another important question concerns the duration *in vivo* of the SCE-inducing DNA damage in the peripheral lymphocytes. As is indicated in Table 6.15 the SCE frequency after cytostatic therapy returns to pretreatment values within a few months after cessation of the therapy. In the case of melphalan, it was shown that the usual clinical treatment interval of 4 weeks is not long enough to allow a restoration of the pretreatment SCE frequency. Instead there is a small increase of the SCE frequency during successive treatment courses (Figure 6.8) (Lambert et al., 1979e).

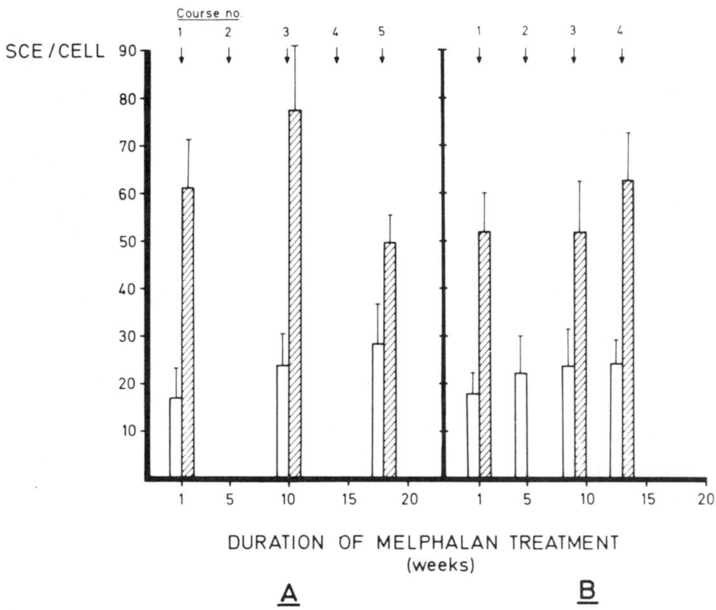

Fig. 6.8. SCE in the lymphocytes of two melanoma patients, (*A*) 76 and (*B*) 69 years old. Open bars are pretreatment values and striped bars represent day 1 after melphalan administration (1 mg/kg of body weight by i.v. infusion every 4 weeks). In course No. 5 of patient *A* the melphalan dose was reduced by 50% because of hematotoxic reactions after the previous course. From Lambert et al., 1979e.

It seems likely that this increase is due to an accumulation of melphalan-induced SCE-forming DNA damage in the circulating lymphocytes. On the other hand, it is evident that the majority of the DNA damage is removed or, alternatively, that the damaged cell population disappears during the 4 weeks after melphalan administration and is replaced by less damaged cells (Figure 6.8). Some information with regard to these alternatives was obtained from a study of the SCE distribution in the cell populations analyzed before treatment, on day 1 after treatment, and 4 weeks later, that is immediately before the second course. As is shown in Figure 6.9 almost all cells display an increase in the SCE frequency 1 day after melphalan administration. Four weeks later the SCE distribution is biphasic. One peak coincides with the pretreatment distribution, and another peak has a distribution overlapping that of day 1. It is not likely that this change in the distribution of SCE in single cells could be due entirely to a repopulation with undamaged cells. We therefore interpret these data

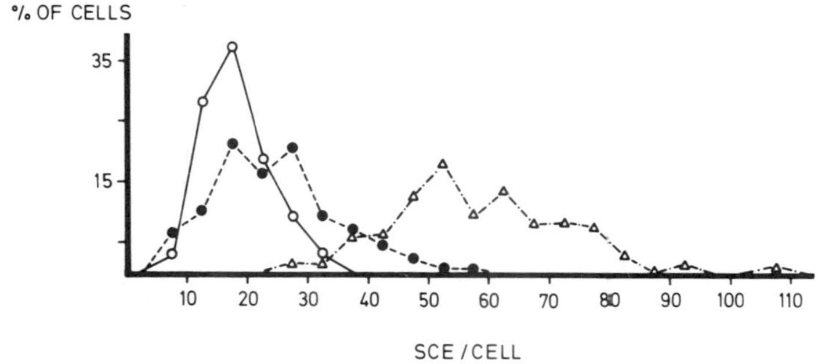

Fig. 6.9. Distribution of SCE in cells from four melanoma patients (○) before melphalan treatment, (△) on day 1 of melphalan administration, and (●) 4 weeks after the previous course. The dose was as stated in the legend to Figure 6.8. From Lambert et al., 1979e.

as indicating removal of SCE-forming damage, possibly by a slowly operating DNA repair process.

V. OCCUPATIONAL EXPOSURE

A. *Laboratory Workers*

It is often stated that SCE analysis appears to be a promising system for the monitoring of human populations with regard to genotoxic exposure. However, very few studies have actually put this suggestion to a test. We have previously reported on the SCE frequency in personnel from laboratories performing hormone analysis and organic chemical research (Funes-Cravioto et al., 1977). The occupational exposure in these subjects involved a great number of chemicals and organic solvents. The average individual SCE frequency was found to be significantly increased compared to current control values. However, nonsmoking laboratory workers were found to have a significantly higher SCE frequency than the nonsmoking controls, whereas no such difference was found among smokers (Table 6.16) (Lambert et al., 1978a). Later, an additional small group of laboratory workers occupied in the pharmaceutical industry was investigated. In this group there was a small, although not significant, increase in the SCE frequency among the 11 nonsmokers, but no increase in the 6 smokers compared to the relevant control values (Table 6.16, laboratory workers group b). Further studies are needed to evaluate these observations and to sort out the possible SCE-inducing agent(s) in the laboratory environment.

TABLE 6.16
SCE Frequencies in Various Groups of "Occupationally Exposed" Populations

	Nonsmokers		Smokers	
Group of subjects	No. of subjects	SCE/cell (group mean ± SD)	No. of subjects	SCE/cell (group mean ± SD)
Reference population, total[a]	63	13.9 ± 2.7	45	16.4 ± 3.8
Occupational exp.	24	15.0 ± 3.5	17	16.1 ± 4.3
Chemical factory workers[b]	41	12.5 ± 3.2	24	16.0 ± 4.3
Laboratory personnel[c]				
Group a	17	19.9 ± 4.6	18	18.4 ± 3.5
Group b	11	16.6 ± 2.2	6	15.3 ± 0.8
Ethylene oxide exposed[d]	2	16.1 ± 5.7	3	21.1 ± 6.9
Operating room personnel	6	13.0 ± 3.7	4	14.1 ± 2.6

[a]Data from Table 6.13 including males and females.

[b]Data from Table 6.9; nonexposed controls excluded.

[c]Data from Lambert et al., 1978a.

[d]Unpublished results of C. J. Calleman, L. Ehrenberg, and B. Lambert.

B. Operating Room Personnel

The mutagenicity of inhalation anesthetics was recently reviewed by Baden and Simmon (1980). According to their conclusions there is some uncertainty with regard to the mutagenic effect of nitrous oxide and halothane, whereas other frequently used anesthetics have shown negative results in several mutagenicity assays. However, several epidemiologic studies have reported an increased incidence of spontaneous abortion, malformed offspring, and cancer among operating room personnel (see Baden and Simmon, 1980). It was therefore considered of interest to study the SCE frequency in operating room personnel to see if any evidence of genotoxic exposure could be obtained.

A group of 10 anesthesiologists and nurse-anesthetists were selected for study. They were occupied in the department of orthopedic surgery and mainly used halothane. The blood samples were taken after 1 week of daily work. The mean SCE frequency of the operating room personnel was found to be very similar to that of the reference group (Table 6.16) and that of a matched control group of nonanesthesiology staff personnel from other surgical departments (not shown). In addition, 10 patients who were given at least 1 hour of halothane narcosis were found to have the same mean SCE frequency before and after the operation. Thus, neither acute exposure to narcotic doses of halothane in the patients, nor sustained occupational exposure in the operating room personnel was found to change the SCE frequency. These results are in accordance with the report of White et al. (1979), which did not show any increase in the SCE frequency in Chinese hamster ovary cells exposed for 1 hour to several anesthetics, including halothane.

C. Ethylene Oxide

Exposure to ethylene oxide (EO), a gaseous sterilant with mutagenic activity used in health care facilities and as an intermediate in chemical manufacturing, was shown to be associated with leukemia in a recent study (Hogstedt et al., 1979). We have studied the frequency of SCE in peripheral lymphocytes from five female workers occupied in a German sterilization plant. The rationale for carrying out this study on such a small number of persons was the unusually reliable information about the weekly exposure dose, which varied between 250 and 1560 ppm/hour in the subjects (Calleman et al., 1978). The average SCE frequency in the EO-exposed women was increased to a level similar to that in laboratory workers (Table 6.16). A positive, although not statistically significant correlation ($r = 0.7$) was found between the SCE frequency and the EO dose.

These results indicate that EO, very likely, is capable of inducing SCE in human lymphocytes *in vivo*.

D. Various Occupations and Types of Exposure

Mäki-Paakkanen et al. (1980) studied 32 rotogravure workers with daily exposure to toluene and found no difference in the SCE frequency compared to 15 unexposed control subjects. A study of 16 xylene-exposed workers in a dye industry showed no significant changes in the SCE frequency (Haglund et al., 1980).

Styrene-exposed workers in a plastic industry (Andersson et al., 1980) and 24 truck drivers exposed to motor combustion and fuels (K. Fredga, personal communication) were found to have SCE frequencies within or very close to the control range. The most significant difference in the SCE frequency observed in these studies was that between smokers and nonsmokers. Thus occupational exposure to several organic solvents has not been shown to be associated with changes in the SCE frequency. Moreover, benzene, toluene, and xylene have not been found to change the number of SCE in human lymphocytes *in vitro* (Gerner-Smidt and Friedrich, 1978).

We have recently obtained some preliminary results from an ongoing study of the SCE frequency in a population of employees in a chemical factory. One of the two groups investigated were workers occupied with chemical manufacturing of a number of pharmaceutical products. They were continuously exposed to various chemicals, raw materials, semimanufactured and finished pharmaceutical products, and several organic solvents that were handled in large quantity and partly in open work. The other group was laboratory personnel occupied with chemical analysis and quality control. In this group the exposure involved the same chemicals as in the previous group, but in much smaller quantities. In addition, a large variety of chemicals for analytical purposes were used. The work was carried out in hoods whenever possible. There was no difference in the average SCE frequency between the two groups. The total mean values are presented in Table 6.16. As shown, these chemical factory workers were found to have an SCE frequency very similar to that of the nonsmoking and smoking reference populations, respectively.

Recently, Brøgger and coworkers (1980) showed that a group of 11 nurses handling cytostatics in a cancer clinic had a small, but significantly increased, frequency of SCE in peripheral lymphocytes compared to a group of 10 female hospital clerks. This may not be a totally unexpected finding in view of the pronounced increase of SCE that can be observed in patients receiving these drugs (e.g., Lambert et al., 1979e).

VI. CONCLUSIONS

In this chapter we have discussed some problems concerning the use of SCE analysis in human lymphocyte cultures for monitoring subjects exposed to genotoxic agents *in vivo*. The SCE frequency is highly influenced by the conditions of lymphocyte cultivation. Most pronounced are the differences in SCE frequency between macro cultures and micro or whole blood cultures. Macro cultures yield a basal SCE frequency approximately 40% higher than that observed with the micro technique. It appears that the macro cultures also are more sensitive to methodological variation due to changes, for example, in BrdUrd concentration and type of serum used. It is therefore important to standardize the methods used as far as possible to permit a faithful comparison between groups of "exposed" and "nonexposed" human subjects.

The collection of reference and control data for the SCE frequency in "nonexposed" populations needs careful consideration. Cigarette smokers have a higher basal SCE frequency than nonsmokers. The control of smoking habits is therefore of great importance. Small differences between males and females may partly be attributed to the use of oral contraceptives. In addition, conditions affecting subpopulations and growth rates of peripheral lymphocytes, for example, allergy and vaccination may also influence the SCE frequency.

Studies on cancer patients under chemotherapy have demonstrated that several alkylating agents give rise to an increased SCE frequency in peripheral lymphocytes during a period of at least 4–16 weeks after drug administration. The duration of the effect indicates that it is caused by alterations in the genetic material of the cells and not by the persistence of the drug in the serum. However, it is also clear that several other drugs with a pronounced genotoxic activity do not increase the SCE frequency significantly over control values. When SCE analyses are used to monitor human populations with an occupational or environmental exposure of unknown specificity or composition, this diversity in the SCE response should be carefully considered.

Occupational exposure to chemicals in general appears not to be associated with an increased frequency of SCE. Several studies on workers exposed, for example, to inhalation anesthetics, toluene, xylene, motor combustion, and fuels have also not shown significant changes in the SCE frequency. So far only laboratory workers, nurses handling cytostatics, and possibly styrene-and ethylene-exposed personnel have been found to have an increase of the SCE frequency in the peripheral lymphocytes.

The conditions of exposure, the dose as well as the time of exposure in relation to sampling, should be taken into account when these data are interpreted. It is likely that an increase in the SCE frequency due to an acute genotoxic exposure will decline during a few weeks. On the other hand, repeated or chronic exposure could cause an accumulation of SCE-inducing DNA damage. The practical solution to this problem may be to carry out several SCE analyses in the single individual before, during, and after exposure, as in the studies of cancer patients. These studies have clearly shown that certain potent cytostatic drugs indeed give rise to an increase of SCE *in vivo*. Also the SCE-inducing effect of smoking indicates the potential utility of SCE analysis for the study of acute exposure to specific, genotoxic agents. However, more studies are needed to obtain the necessary information about which types of occupational exposure increase the SCE frequency in *vivo*. It is therefore too early to draw firm conclusions about the general utility of the SCE analysis for the purpose of monitoring occupationally exposed subjects for genetic damage due to the presence of chemicals in the environment.

REFERENCES

Alves, P. & Jonasson, J. (1978) New staining method for the detection of sister chromatid exchanges in BrdU-labelled chromosomes. *J. Cell Sci.* **32,** 185–195.

Andersson, H. C., Trauberg, E. Å., Uggla, A. H., & Zetterberg, G. (1980) Chromosomal aberrations and sister chromatid exchanges in lymphocytes of men occupationally exposed to styrene in a plastic boat factory. *Mutat. Res.* **73,** 387–401.

Baden, J. M. & Simmon, U. F. (1980) Mutagenic effects of inhalation anesthetics. *Mutat. Res.* **75,** 169–189.

Bala Krishna Murthy, P. (1979) Frequency of sister chromatid exchanges in cigarette smokers. *Hum. Genet.* **52,** 343–345.

Bala Krishna Murthy, P., & Prema, K. (979) Sister chromatid exchanges in oral contraceptive users. *Mutat. Res.* **68,** 149–152.

Banerjee, A. & Benedict, W. F. (1979) Production of sister chromatid exchanges by various cancer chemotherapeutic agents. *Cancer Res.* **39,** 797–799.

Bridges, B. A. (1979) An estimate of genetic risk from 8-methoxypsoralen photochemotherapy. *Hum. Genet.* **49,** 91–96.

Bridges, B. A., Clemmesen, J., & Sugimura, T. (1979) Cigarette smoking—does it carry a genetic risk? *Mutat. Res.* **65,** 71–81.

Brøgger, A., Waksvik, H., & Thune, P. (1978) Psoralen/UVA treatment and chromosomes. II. Analyses of psoriasis patients. *Arch. Dermatol. Res.* **261,** 287–294.

Brøgger, A., Klepp, O., & Waksvik, H. (1981) Chromosome analyses of nurses handling cytostatic agents. *Mutat. Res.*, **85,** 254.

Buckton, K. E. & Evans. H. J., Eds. (1973) *Methods for the analysis of Human Chromosome Aberrations,* WHO, Geneva.

Calleman, C. J., Ehrenberg, L., Jansson, B., Osterman-Golkar, S., Segerbäck, D., Svensson, K., & Wachtmeister, C. A. (1978) Monitoring and risk assessment by means of alkyl groups in haemoglobin in persons occupationally exposed to ethylene oxide. *J. Environ. Pathol. Toxicol.* **2,** 427–442.

Carter, D. M., Wolff, K., & Schnedl, W. (1976) 8-Methoxypsoralen and UVA promote sister chromatid exchanges. *J. Invest. Dermatol.* **67,** 548–551.

Craig-Holmes, A. P. & Shaw, M. W. (1977) Effects of six carcinogens on SCE frequency and cell kinetics in cultured human lymphocytes. *Mutat. Res.* **46,** 375–384.

Crossen, P. E. (1979) The effect of acridine compunds on sister-chromatid exchange formation in cultured human lymphocytes. *Mutat. Res.* **68,** 295–299.

Crossen, P. E., Drets, M. E., Arrighi, F. E., & Johnston, D. A. (1977) Analysis of the frequency and distribution of sister chromatid exchanges in cultured human lymphocytes. *Hum. Genet.* **35,** 345–352.

Csukás, I., Gungl, E., Fedorcsák, I., Vida, G., Antoni, F., Turtóczky, I., & Solymosy, F. (1979) Urethane and hydroxyurethane induce sister-chromatid exchanges in cultured human lymphocytes. *Mutat. Res.* **67,** 315–319.

Daoud, C., Shaw, M., & Craig-Holmes, A. (1976) Sister chromatid exchange frequency among normal individuals and breast cancer patients. *Mamm. Chromosomes Newslett.* **17,** 26.

Davidson, R. L., Kaufman, E. R., Dougherty, C. P., Ouellette, A. M., DiFolco, C. M. & Latt, S. A. (1980) Induction of sister chromatid exchanges by BUdR is largely independent of the BUdR content of DNA. *Nature* **284,** 74–76.

Editorial (1978) PUVA. *Bri. Med. J.* July 1, 2–3.

Faed, M. & Mourelatos, D. (1978) Sister chromatid exchanges in lymphocytes treated with 8-methoxypsoralen and exposed to long-wave ulraviolet light. In *Mutagen-Induced Chromosome Damage in Man* H. J. Evans and D. C., Loyd Eds.), Edinburgh University Press, pp. 216–220.

Funes-Cravioto, F., Zapata-Gayon, C., Kolmodin-Hedman, B., Lambert, B., Lindsten, J., Norberg, E., Nordenskjöld, M., Olin, R., & Swensson, Å. (1977) Chromosome aberrations and sister-chromatid exchange in workers in chemical laboratories and a rotoprinting factory and in children of women laboratory workers. *Lancet*, Aug. 13, **2,** 322–325.

Gerner-Smidt, P. & Friedrich, U. (1978) The mutagenic effect of benzene, toluene and xylene studied by the SCE-technique. *Mutat. Res.* **58,** 313–316.

Goyanes-Villaescusa, V. J., Ugarte, M., & Vazquez, A. (1977) Sister chromatid exchanges in babies treated by phototherapy. *Lancet,* Nov. 19, **2,** 1084–1085.

Haglund, U., Lundberg, I., & Zech, L. (1980) Chromosome aberrations and sister chromatid exchanges in Swedish paint industry workers. *Scand. J. Work, Environ. & Health,* **6,** 291–298.

Hartley-Asp, B. (1979) Mutagenicity of metronidazole. *Lancet,* Feb. 3, 275.

Hogstedt, C., Malmqvisit, N., & Wadman, B. (1979). Leukemia in workers exposed to ethylene oxide. *JAMA,* **241,** 1132–1133.

Hollander, D. H., Tockman, M. S., Liang, Y. W., Borgaonkar, D. S. & Frost, J. K. (1978) Sister chromatid exchanges in the peripheral blood of cigarette smokers and in lung cancer patients; and the effect of chemotherapy. *Hum. Genet.* **44,** 165–171.

Honeycombe, J. (1978) The effect of busulphan on the chromosomes of normal human lymphocytes. *Mutat. Res.* **57,** 35–49

Hopkin, J. M. & Evans, H. J. (1979; Cigarette smoke condensatees damage DNA in human lymphocytes. *Nature* **279,** 241–242.

Hopkin, J. M. & Evans, H. J. (1980) Cigarette smoke-induced DNA damage and lung cancer risk. *Nature* **283,** 388–390.

Kato, H. (1974). Spontaneous sister chromatid exchanges detected by a BUdR-labelling method. *Nature* **251,** 70–72.

Kato, H. & Sandberg, A.A. (1977). The effects of sera on sister chromatid exchanges *in vitro*. *Exp. Cell Res.* **109,** 445–448.

Knuutila, S., Mäki-Paakkanen, J., Kähkönen, M., & Hokkanen, E. (1978) An increased frequency of chromosomal changes and SCE's in cultured blood lymphocytes of 12 subjects vaccinated against smallpox. *Hum. Genet.* **41,** 89–96.

Kowalczyk, J. (1980) Sister chromatid exchanges in children treated with nalidixic acid. *Mutat. Res.* **77,** 371–375.

Lambert, B., Hansson, K., Lindsten, J., Sten, M., & Werelius, B. (1976) Bromodeoxyuridine-induced sister chromatid exchanges in human lymphocytes. *Hereditas* **83,** 163–174.

Lambert, B., Lindblad, A., Nordenskjöld, M., & Werelius, B. (1978a) Increased frequency of sister chromatid exchanges in cigarette smokers. *Hereditas* **88,** 147–149.

Lambert, B., Morad, M., Bredberg, A., Swanbeck, G., & Thyresson-Hök, M. (1978b) Sister chromatid exchanges in lymphocytes from psoriasis patients treated with 8-methoxypsoralen and longwave ultraviolet light. *Acta Dermato venereol.* **58,** 13–16.

Lambert, B., Ringborg, U., Harper, E., & Lindblad, A. (1978c) Sister chromatid exchanges in lymphocyte cultures of patients receiving chemotherapy for malignant disorders. *Cancer Treat Tep.* **62,** 1413–1419.

Lambert, B., Ehrnst, A., Hansson, K., Lindblad, A., Morad, M., & Werelius, B. (1979a). Sister chromatid exchange in peripheral lymphocytes of subjects vaccinated against measles. *Hum. Genet.* **50,** 291–296.

Lambert, B., Lindblad, A., & Ringborg, U. (1979b). Absence of genotoxic effects of metronidazole and two of its urinary metabolites on human lymphocytes *in vitro*. *Mutat. Res.* **67,** 281–287.

Lambert, B., Lindblad, A., Lindsten, J., Morad, M., Bengtsson, E., Faehlmann, M., Eneroth, G., & Marsk, L. (1979c) Genotoxic effects of metronidazole in human lymphocytes *in vitro* and *in vivo*. In *Metronidazole,* Royal Society of Medicine International Congress and Symposium Series No. 18 (I. Phillips and J. Collier, Eds.) Academic Press, London.

Lambert, B., Ringborg, U., & Lindblad, A. (1979d) Prolonged increase of sister chromatid exchanges in lymphocytes of melanoma patients after CCNU treatment. *Mutat. Res.* **59,** 295–300.

Lambert B., Ringborg, U., Lindblad, A., & Sten, M. (1979e) The effects of DTIC, melphalan, actinomycin D and CCNU on the frequency of sister chromatid exchanges in peripheral lymphocytes of melanoma patients. In *Adjuvant Therapy of Cancer II* (S. E. Jones, and S. E. Salmon, Eds.) Grune & Stratton, New York, pp. 56–61.

Latt, S. A. (1973) Microfluorometric detection of deoxyribonucleic acid replication in human metaphase chromosomes. *Proc. Natl. Acad. Sci. U.S.* **70,** 3395–3399.

Latt, S. A. (1974) Localization of sister chromatid exchanges in human chromosomes. *Science* **185,** 74–76.

Lindblad, A. & Lambert, B. (1981) Relation between sister chromatid exchange, cell proliferation and proportion of B and T cells in human lymphocyte cultures. *Hum. Genet.* **57,** 31–34.

Littlefield, L. G., Colyer, S. P., Sayer, A. M., & DuFrain, R. J. (1979) Sister chromatid exchanges in human lymphocytes exposed during G_0 to four classes of DNA-damaging chemicals. *Mutat. Res.* **67,** 259–269.

Littlefield, L. G., Colyer, S. P., & DuFrain, R. J. (1980). Comparison of sisterchromatid exchanges in human lymphocytes after G_0 exposure to mitomycin *in vivo* vs. *in vitro*. *Mutat. Res.* **69,** 191–197.

Mäki-Paakkanen, J., Husgafvel-Pursiainen, K., Kalliomäki, P.-L., Tuominen, J., & Sorsa, M. (1980) Toluene-exposed workers and chromosome aberrations. *J. Toxicol Environ. Health* **6,** 775–781.

Mazrimas, J. A. & Stetka, D.G. (1978). Direct evidence for the role of incorporated BUdR in the induction of sister chromatid exchanges. *Exp. Cell Res.* **117,** 23–30.

Morgan, W. F. & Crossen, P. E. (1977). The incidence of sister chromatid exchanges in cultured human lymphocytes. *Mutat. Res.* **42,** 305–312.

Musilová, J., Michalová, K. & Urban, J. (1979) Sister-chromatid exchanges and chromosomal breakage in patients treated with cytostatics. *Mutat. Res.* **67,** 289–294.

Nevstad, N.P. (1978) Sister chromatid exchanges and chromosomal aberrations induced in human lymphocytes by the cytostatic drug adriamycin *in vivo* and *in vivo*. *Mutat. Res.* **57,** 253–258.

Pedersen, C., Oláh, E., & Merrild, U. (1979) Sister chromatid exchanges in cultured peripheral lymphocytes from twins. *Hum. Genet.* **52,** 281–294.

Perry, P. & Evans, H.J. (1975) Cytological detection of mutagen–carcinogen exposure by sister chromatid exchange. *Nature* **258,** 121–125.

Perry, P. & Wolff, S. (1974) New Giemsa method for the differential staining or sister chromatids. *Nature* **251,** 156–158.

Raposa, T. (1978) Sister chromatid exchange studies for monitoring DNA damage and repair capacity after cytostatics *in vitro* and in lymphocytes of leukaemic patients under cytostatic therapy. *Mutat. Res.* **57,** 241–251.

Rosenkranz, H. S., Speck, W. T., & Stambaugh, J. E. (1976) Mutagenicity of metronidazole: Structure activity relationships. *Mutat. Res.* **38,** 203–206.

Rüdiger, H. W., Kohl, F. V., Mangels, W., v. Wichert, P., Bartram, C.R., Wöhler, W. & Passarge, E. (1976) Benzpyrene induced sister chromatid exchanges in cultured human lymphocytes. *Nature* **262,** 290–292.

Santesson, B., Lindahl-Kiessling, K., & Mattsson, A. (1979). SCE in B and T lymphocytes. Possible implications for Bloom's syndrome. *Clin. Genet.* **16,** 133–135.

Schönwald, A. D., Bartram, C. R. & Rüdiger, H. W. (1977) Benzpyrene-induced sister chromatid exchanges in lymphocytes of patients with lung cancer. *Hum. Genet.* **36,** 261–264.

Schwartz, A. L., Cole, F. S., Fiedorek, F., Mathhews, D., Paika, I., Frantz, I. D., & Latt, S.A. (1978) Effect of phototherapy on sister chromatid exchange in premature infants. *Lancet,* July 15, **2,** 157–158.

Snope, A. J. & Rary, J. M. (1979) Cell-cycle duration and sister-chromatid exchange frequency in cultured human lymphocytes. *Mutat. Res.* **63,** 345–349.

Speck, T., Stein, A. B., & Rosenkranz, H. S. (1976) Mutagenicity of metronidazole: Presence of several active metabolities in human urine. *J. Int. Cancer Inst.* **56,** 283–284.

Stern, R. S., Thibodeau, L. A., Kleinerman, R. A. Parnish, J. A., & Fitzpatrick, T. B. (1979) Risk of cutaneous carcinoma in patients treated with oral methoxsalen photochemotherapy for psoriasis, *N. Engl. J. Med.* **300,** 809–813.

White, A. E., Takehisa, S., Eger, E. I., Wolff, S., & Stevens, W.C. (1979) Sister chromatid exchanges induced by inhaled anesthetics. *Anesthesiology* **50,** 426–430.

Wolff, S. & Perry, P. (1974) Differential Giemsa staining of sister chromatids and the study of sister chromatid exchanges without autoradiography. *Chromosoma (Berl.)* **48,** 341–353.

Wolff-Schreiner, E. C., Carter, M., Schwarzacher, H. G., & Wolff, K. (1977). Sister chromatid exchanges in photochemotherapy. *J. Invest. Dermatol.* **69,** 387–391.

Chapter 7

Sister Chromatid Exchanges and Disease States in Man

H. J. EVANS
MEDICAL RESEARCH COUNCIL
CLINICAL AND POPULATION CYTOGENETICS UNIT
WESTERN GENERAL HOSPITAL
EDINBURGH, UNITED KINGDOM

I. Introduction 184
II. Incidence of SCE in cells taken from normal healthy individuals 185
III. The incidence of SCE in cells from people with inherited predispositions to chromosomal instabilities, defective DNA repair, and/or the development of neoplasms 187
  A. Bloom's syndrome 188
  B. Fanconi's anemia 191
  C. Ataxia telangiectasia 192
  D. Xeroderma pigmentosum 194
  E. Down's syndrome 197
  F. Cockayne's syndrome 198
  G. The Hutchinson-Gilford progeria syndrome 198
  H. Huntington's chorea 199
  I. Friedreich's ataxia 199
  J. Retinoblastoma 200
  K. Dyskeratosis congenita, Scoggins type 200
  L. Gardner's syndrome 201
  M. Rothmund-Thomson syndrome 201
  N. Other possible inherited DNA-repair defects 201
IV. The incidence of SCE in various acquired disease states 204
  A. Virus infections 205
  B. Neoplastic disease 207
  C. Ill-health associated with exposure to physical and chemical mutagens 208
    1. Exposure to physical agents 208
    2. Therapeutic exposure to chemical agents 210

3. Occupational exposure to chemical agents 211
4. Exposure to chemical agents by other means, including ingestion and inhalation. 212
V. Concluding comments 214

I. INTRODUCTION

A disease state may be defined as an abnormality in health resulting from an interaction between genotype and environment; the relative importance of genotype and environment will, of course, differ considerably in the case of a patient acquiring a broken limb during the course of a game of football and in that of a child inheriting a deficiency in a blood clotting factor and suffering from hemophilia. In this chapter I use the term disease state in its broadest sense and although there is emphasis on the influence of genotype in the induction of sister chromatid exchanges (SCE), the discussion is not restricted to inherited chromosome instabilities and defective DNA repair states.

The first, dramatic demonstration of an elevated incidence of SCE frequency associated with a disease state in man was the report by Chaganti et al. (1974) showing a more than ten-fold increase in SCE blood lymphocytes from a patient with the inherited autosomal recessive disease of Bloom's syndrome relative to cells from controls. This work followed as a natural consequence of the availability of simple techniques to visualise SCE and the knowledge that cells from Bloom's patients were characterized by a high spontaneous frequency of chromosome aberrations, including symmetrical exchanges between chromatids of homologous chromosomes (German, 1974a).

The association between gross chromosomal aberrations and SCE is discussed elsewhere in this volume (Wolff, 1981), and although the mechanisms involved in their development may be different, they may both reflect responses to lesions in DNA and/or to deficiencies or errors in repairing such lesions. Moreover, since SCEs are a consequence of a switching and exchange of DNA strands during the replication process, defects, anomalies, or even small perturbations in the normal processes of replication may alter the spontaneous incidence of SCE or the response of the cell following the introduction of DNA lesions by external agents. In very general terms, therefore, we might consider that the incidence of SCE depends on three factors: (*a*) the initial number and type of DNA lesions induced in a cell, which itself is a function of cell type, cell state, and phenotype with respect to metabolic activation, detoxification and so forth; (*b*) the potential for the cell to remove and correctly repair DNA lesions prior to undergoing DNA replication; and (*c*) the integrity of the normal DNA replication machinery and its efficiency in terms both of the coordination of the various processes involved and its ability to bypass lesions or gaps without necessitating a switching of polynucleotide strands. Each of these three factors is influenced by the genotype and, to varying degrees, by the cellular environment.

II. INCIDENCE OF SCE IN CELLS TAKEN FROM NORMAL HEALTHY INDIVIDUALS

The normal processes of DNA replication and repair involve the concerted actions of a range of enzymes including topoisomerases, gyrases, various nucleases, polymerases, and ligases and there are further batteries of enzyme systems involved in activating and/or detoxifying electrophilic, or potentially electrophilic, substances that may damage DNA. Thus, even on the incorrect assumption that all cells have an equal ability to absorb substances that could result in damage leading to SCE, the scores of genetic loci that might code for products that could affect SCE frequency, together with the known genetic heterogeneity of the human species, lead us to expect that both the spontaneous levels of SCE and the susceptibility to their induction following exposure to various agents might differ even among apparently healthy individuals. The possibility must also exist that increasing age may bring with it an increasing accumulation of DNA defects and protein alterations (Orgel, 1963, but see Harley et al., 1980) resulting in inefficiency of enzyme systems that would tend to increase SCE frequency. What evidence is there for these premises?

The original study of Galloway and Evans (1975) on peripheral blood lymphocyte chromosomes of seven adults and six newborns showed that the mean frequency of SCE differed by a factor of up to 2 within each group and that the mean incidence in newborn lymphocytes was twice that in lymphocytes from adults aged from 25–36 years. No differences were observed between the sexes. Later studies on a similar number of individuals (Crossen et al., 1977) revealed a similar up to two-threefold difference in mean frequencies between individuals. In some cases a significant difference in SCE incidence between an original culture and a later repeat sample from the same individual was noted, but, in general, the SCE levels in original samples were maintained in later repeats. Our later unpublished studies on lymphocytes from 104 individuals have confirmed that although there may be significant differences among SCE frequencies in blood cells of healthy individuals cultured at the same time, and under the same conditions, the range is not large and the incidence shows little dependence on age. It may, however, be influenced by exposure to various environmental factors, for example, cigarette smoke, virus infection, certain contraceptive pills (see below), and is heavily dependent on culture conditions including bromodeoxyuridine (BrdUrd) concentration, serum, and culture medium. For instance, in our own laboratory, lymphocytes cultured in 10 μM BrdUrd give mean SCE rates of 5.3/cell, whereas cells cultured in 100 μM BrdUrd give mean frequencies of 7.8 SCEs/cell.

In a summary of published data, Crossen et al. (1977) drew attention to a fivefold difference in mean SCE frequency in reported studies on lymphocyte chromosomes from normal individuals from a number of different laboratories, the mean incidence of SCE per cell ranging from 5.3 to 27.3, with most laboratories reporting values of around 6 SCEs/cell or around 14 SCEs/cell. These differences are, in part, attributable to differences in BrdUrd concentration used (see Wolff and Perry, 1974), but other culture factors may also be involved. For example, the SCE frequency depends on the type of culture medium used and is also influenced by serum type and concentration. The lowest frequencies are found when cells are cultured with autologous serum, substitution with fetal calf or human AB serum giving a more than 50% increase in exchange frequency (Ghosh and Nand, 1979). Similarly, cells cultured in media deficient in basic amino acids may show up to a threefold increase in SCE frequency relative to cells cultured under similar conditions but using complete medium (Schempp and Krone, 1979). These differences underline the fact that comparisons of results from laboratories using different culture techniques may be fraught with problems.

An evaluation of SCE incidence is only possible in cells that have undergone two rounds of replication in the presence of BrdUrd, so that the range of cell types that may be studied in man is limited. Information is, however, available on SCE incidence in cells from blood, bone marrow, skin, and various established cell lines. For example, Shiraishi et al. (1976) studied SCE frequencies in bone marrow and blood lymphocyte cells of two patients, one with Bloom's syndrome and the other with polycythemia vera, and noted that within each patient the two types of cells showed similar frequencies, although the incidence of SCE in the Bloom's patient's cells was 10 times as great as that in the polycythemic cells and in blood cells from normal individuals. Knuutila et al. (1978a) in their study of eight hematologically healthy patients reported that the SCE incidence in bone marrow cells cultured in Eagle's MEM medium with human AB serum was slightly lower (mean frequency of 3.9–6.0 SCEs/cell) than in blood lymphocytes from a different series of patients (mean frequency 7.9 SCEs/cell) and cultured under different conditions

(Knuutila et al., 1978b). Similar results have also been reported by Becher et al. (1979) on a series of six healthy individuals where the mean SCE in bone marrow cells was found to be around 4/cell, in contrast to the approximately 6/cell observed in blood lymphocytes. Evans and Vijayalaxmi (1980) studied the incidence of SCE in blood lymphocytes responding at different rates to the stimulatory action of phytohemagglutinin (PHA) and saw no change in SCE incidence in cells arriving at the second mitosis in culture from the second through to the ninth day in culture. The response of differentially developing T lymphocytes would therefore appear to be similar. However, Santesson et al. (1979) in a comparison of SCE incidence in B and T lymphocytes exposed to various levels of BrdUrd reported that B cells showed a lower response to the induction of SCE by BrdUrd than T cells. In these, and other, comparative studies different cell types have usually been cultured in different media and under different conditions so that *in vitro* differences in SCE incidence among different cells may by no means be a faithful reflection of *in vivo* differences among tissues. It may be relevant to note here, therefore, that a simultaneous *in vivo* assay of SCE incidences in murine alveolar macrophages and regenerating liver and bone marrow cells revealed no significant differences in mean frequencies among the three cell types (Conner et al., 1979).

In conclusion, we note that possible differences in SCE incidence may exist between tissues, but that any such differences are very small. Moreover, within-laboratory comparisons do reveal significant differences in SCE frequencies among normal healthy individuals, but these differences are not large and may at most encompass a range of a factor of two or three.

III. THE INCIDENCE OF SCE IN CELLS FROM PEOPLE WITH INHERITED PREDISPOSITIONS TO CHROMOSOMAL INSTABILITIES, DEFECTIVE DNA REPAIR, AND/OR THE DEVELOPMENT OF NEOPLASMS

The original demonstration by Cleaver (1968) that fibroblasts from patients with the autosomal recessive disease xeroderma pigmentosum were defective in their ability to repair UV-induced lesions in their DNA, highlighted the possibility that defective DNA repair might be an important etiological factor in UV-induced skin cancer. Prior to this discovery, a number of authors had established that three other autosomal recessive diseases in man that predisposed to an increased risk of malignancy, namely, Bloom's syndrome (German et al., 1965), Fanconi's anemia (Schroeder et al., 1964), and ataxia telangiectasia (Hecht et al., 1966;

Gropp and Flatz, 1967), were associated with markedly elevated frequencies of chromosome aberrations in blood lymphocytes and/or cultured fibroblasts. The association between inherited chromosomal instability and neoplasia, on the one hand, and between defective DNA repair and neoplasia, on the other, coupled with the obvious possibility that chromosomal instability might reflect basic errors in the processes of DNA replication and/or DNA repair, have led to considerable activity in this area. In parallel with the earlier and elegant studies on the use of bacterial mutants to dissect the processes involved in the repair of lesions induced by different mutagens (e.g., see Hanawalt and Setlow, 1975), much effort has been devoted to studying the response to mutagens/carcinogens of cells from individuals with various inherited disease states that are associated with a chromosomal instability and a predisposition to neoplastic disease, or from individuals suspected of having possible DNA repair defects. It is not the purpose of this chapter to review this work in the context of sensitivity to mutation, cell killing, and neoplastic transformation by carcinogen/mutagens, and the interested reader is referred to a number of recent and comprehensive review articles (Arlett and Lehmann, 1978; Friedberg et al., 1979; Paterson and Smith, 1979). Interest in the context of our present discussion is, of course, based on the fact that inherited genetic defects that diminish the ability to remove potential mutagenic or carcinogenic lesions in DNA, or that influence the efficiency of DNA replication, may result in an altered incidence of spontaneous SCEs or in an abnormal response to the induction of SCE by externally applied mutagens. Inherited human disease states that have been shown, or have been suspected, to fall into this category are discussed below.

A. Bloom's Syndrome (BS)

The original report by Chaganti et al. (1974) that the "spontaneous" SCE incidence in BS lymphocytes is some 10–15 times the incidence in lymphocytes from normal individuals has been considerably extended and repeatedly confirmed by various authors (see German et al., 1977). Moreover, it has been shown that this increase is not due to an increase in BrdUrd uptake or to an ultrasensitivity of BS cells to the induction of SCE by BrdUrd, since similar results are obtained when BS cells are labeled with [^{3}H]deoxycytidine to enable SCE to be resolved by autoradiography (Shiraishi et al., 1976). The enhanced incidence of SCE is not confined to lymphocytes of BS patients, but is also evident in bone marrow cells (Shiraishi et al., 1976) and cultured fibroblasts (Sperling et al., 1976), so that the factor(s) responsible are truly systemic. However, German et al. (1977) have reported that in several Bloom's patients although the majority of lymphocytes show elevated SCE frequencies, a substantial subpopulation of cells may show the normal lower levels. There are several possible interpretations of this finding, but no evidence to indicate the cause of this heterogeneity.

There have been various attempts to determine if Bloom's cells release a substance (or agent) into the plasma *in vivo* or culture medium *in vitro* that is capable of inducing aberrations or SCE. The culture of normal lymphocytes in medium containing serum from BS patient, and vice versa, was shown to have no effect in altering spontaneous aberration yields (Evans et al., 1978), but Tice et al. (1978a) claimed that the SCE frequency in normal fibroblasts and lymphocytes could be significantly increased when these cells were cultured in medium that had previously been used to culture BS fibroblast strain GM1492. The implication of this latter work was that the BS cell line secreted an agent or mutagen into the culture medium. The apparent mutagenicity of culture medium used to grow BS line GM1492 could not be confirmed in a repeat experiment (Ockey, 1979), which suggests that the results of Tice et al. simply could be due to differences in cell cycle kinetics brought about by culturing cells under different media conditions. Similarly, an elegant experiment by Schonberg and German (1980) that involved placing $HGPRT^-$ (Lesch-Nyhan) fibroblasts in contact-dependent metabolic coupling with BS fibroblasts by culturing in HAT medium (which would kill uncoupled $HGPRT^-$ cells) gave no increase in SCE frequency in the Lesch-Nyhan cells.

Other experiments on the cocultivation of BS cells with normal human cells (Bartram et al., 1979) and with Chinese hamster ovary (CHO) cells (van Buul et al., 1978) have also failed to confirm the report that Bloom's cells secrete an agent that increases the SCE rate in cocultivated normal cells. Moreover, these two studies demonstrated that cocultivation of human or CHO cells with BS cells results in a 20% reduction of SCE frequency in Bloom's cells if both cell types are in close contact, implying that the normal cells provide materials that are presumably lacking in BS cells and that are capable of mitigating the chromosomal manifestation of the BS mutation. These cocultivation experiments have been taken a stage further by Bryant et al. (1979), who examined the incidence of SCE in cell hybrids between BS fibroblasts and normal human fibroblasts. In this work the elevated SCE frequency of BS cells in BS/BS hybrids was characteristically maintained, but in BS/normal cell hybrids there was a complete correction of the mutant phenotype, the SCE frequency being down to normal control levels. Similar findings have recently been reported by Alhadeff et al. (1980) in their studies on BS (GM1492)/CHO hybrid cells. These hybridization experiments further refute the suggestion that BS cells secrete a mutagenic agent and are in line with the kind of result that might be expected for an enzymatic deficiency in BS cells and with the fact that BS behaves as a recessive trait that involves the impairment of control over chromosomal integrity and/or replication.

The molecular defect responsible for the BS phenotype has not yet been identified. Although some evidence suggests that the basic defect might involve perturbations of DNA replication and/or deficiencies of DNA repair, the reports are not always consistent and the majority reveal no differences between BS and normal cells. For example, the rate of DNA chain growth has been reported to be less in BS than in normal cells (Hand and German, 1975; Giannelli et al., 1977), but this diminished rate of DNA chain growth in Bloom's cells has since been claimed to be an artifact introduced by culture conditions (Ockey, 1979). The rate of replication repair in BS cells exposed to UV radiation is normal (Tice et al., 1978b; Cleaver, 1970), as is the rate of UV-induced dimer excision (Regan et al., 1973) and postreplication repair (Giannelli et al., 1977). Enzymological studies have not revealed any defects in apurinic endonuclease activity in Bloom's cells (Inoue et al., 1977) or in the levels of any of the DNA polymerases (Parker and Lieberman, 1977). Recently, however, Krepinsky et al. (1980) have reported that of nine BS fibroblast strains studied for their response to chromosome damage and cell killing by exposure to UV light, only one was found to be ultrasensitive to the induction of SCE, or chromosome aberrations, following UV exposure; four showed enhanced killing by UV and the remainder showed a normal response as measured by a range of criteria, including host cell reactivation. However, BS lymphocytes from five different patients all showed an enhanced sensitivity, relative to controls and heterozygotes, to the induction of SCE following exposure to the chemical mutagen ethyl methanesulfonate (EMS) (Krepinsky et al., 1979), but apparently no increased SCE response (relative to controls) has been reported for BS cells exposed to mitomycin C (MMC) (Shiraishi and Sandberg, 1978). In the studies with EMS one observation of interest was the finding that the enhanced SCE response to EMS was not evident in those subpopulations of lymphocytes in Bloom's patients that did not show the typical high spontaneous SCE levels. This presence of apparently normal lymphocytes in some Bloom's patients with respect to both spontaneous and induced SCE levels might imply that this recessive condition may be a regulatory mutation and it is tempting to consider that since an elevated spontaneous aberration level reflects an elevated spontaneous mutation rate, the normal cells in BS may represent mutant revertant clones. Nevertheless, at the present time no defect in a known DNA repair process has been characterized, or even demonstrated, for Bloom's syndrome cells, although one would expect that the enhanced SCE frequency that occurs at spontaneous levels and following exposure to EMS must presumably reflect some recessively inherited perturbation or deficiency in the DNA replication or repair machinery.

B. Fanconi's Anemia (FA)

The original demonstration by Schroeder et al. (1964) that lymphocytes of patients with FA showed high frequencies of chromosome aberrations has since been confirmed by many workers (see Schroeder and German, 1974), as has an often elevated, but less consistent, spontaneous aberration frequency in bone marrow cells and cultured skin fibroblasts from FA patients (see Friedberg et al., 1979). In contrast with the findings on BS cells, however, a number of studies on FA cells show that they have a normal spontaneous incidence of SCE (Chaganti et al., 1974; Latt et al., 1975; Hayashi and Schmid, 1975; Sperling et al., 1975; Kato and Stich, 1976).

The elevated frequency of spontaneous aberrations in FA cells is paralleled by their enhanced response to aberration induction by the DNA cross-linking agents nitrogen mustard (HN2), MMC, and long-wave UV plus 8-methoxypsoralen (Sasaki and Tonomura, 1973), although differences in the response have been reported among cell lines derived from different Fanconi patients (Sasaki, 1978). A similar enhanced sensitivity to aberration induction relative to controls is also observed for FA cells exposed to diepoxybutane (DEB), EMS (Auerbach and Wolman, 1976), thiotepa (TEPA) (Novotna et al., 1979) and isonicotinic acid hydrazide (Schroeder and Stahl-Mauge, 1979). Moreover, an enhanced response to the induction of aberrations by DEB has also been reported to be evident in FA heterozygotes (Auerbach and Wolman, 1978; Auerbach et al., 1978).

The sensitivity of FA cells to aberration induction by external mutagens is not paralleled by a similar increase in sensitivity to the induction of SCE by these agents. Indeed, Latt et al. (1975) originally reported that FA lymphocytes, but not fibroblasts, showed a lower than normal number of SCEs after exposure to MMC, a finding that has since been confirmed for MMC and EMS by some workers (Wolff and Bodycote quoted by Galloway, 1977), but not by others (Novotna et al., 1979).

The enhanced response of FA cells to aberration induction by DNA cross-linking agents and the importance of DNA interstrand cross-links was emphasized by the demonstration of Sasaki (1975) that the sensitivity of FA cells to the monofunctional derivative of MMC (dicarbamol MMC) was no greater than normal. Fujiwara et al. (1977) further showed that different strains of FA fibroblast cells had different abilities to remove cross-links from their DNA and that this was paralleled by the differing responses of these cell strains to the induction of aberrations by MMC. This finding clearly implicates interstrand cross-links as being important lesions in the formation of chromosome aberrations but implies that the induction of cross-links and their persistence may be unimportant in the formation of SCEs.

The reduced ability of certain FA lines to remove interstrand DNA cross-links appears to be the major DNA repair defect noted in FA cells. The rate of DNA chain growth and replicon size appear to be normal (Hand, 1977); no defect has been noted in the excision of UV dimers (Hirsch-Kauffmann et al., 1978) or in the induction of unscheduled DNA synthesis after exposure to UV light, MMC, or HN2 (Fujiwara et al., 1977); there is no detectable defect in postreplication repair after UV (Lehmann et al., 1977); the rate of excision of acetylaminofluorene-induced lesions is normal (Amacher and Lieberman, 1977), as is the rejoining of X- or gamma-ray induced single strand breaks (Poon et al., 1974). The levels of apurinic endonucleases (Teebor and Duker, 1975) and of DNA polymerases (Parker and Lieberman, 1977) are also normal. Remsen and Cerutti (1976) have, however, reported that two of four FA cell lines tested had a reduced ability to excise gamma-ray-induced pyrimidine base damage. This difference in response among these four cell lines, together with the earlier reported findings of Fujiwara et al. on the variability among cell lines, emphasizes once again the genetic heterogeneity of response among cells derived from different Fanconi patients. Finally, we should note that Hirsch-Kauffmann et al. (1978) have reported a significant decrease in DNA ligase activity in lymphocytes and fibroblasts of one FA patient, with the ligase activity of a heterozygous parent being intermediate between normals and the homozygote. In this connection, however, it should be noted that both the spontaneous and MMC-induced aberration frequency in controls and the Fanconi heterozygotes were the same, whereas the homozygotes showed the typical enhanced response.

In conclusion, and with regard to the presence and induction of SCE in FA cells, it would appear that aside from a lowered response of FA lymphocytes to the induction of SCE by MMC, FA cells show normal spontaneous SCE levels and normal SCE responses to the few mutagens studied. The reduced ability of FA cells to remove cross-links in DNA and their diminished ligase activity therefore do not appear to be important with regard to the induction of SCE.

C. *Ataxia Telangiectasia (AT)*

The observations of large numbers of spontaneous chromosome aberrations in blood lymphocytes of AT patients (Hecht et al., 1966; Gropp and

Flatz, 1967) and in cultured AT skin fibroblasts (Oxford et al., 1975) have been confirmed by numerous authors (see Galloway, 1977; Hecht and McCaw, 1977). Cells from AT patients also show an enhanced response to cell killing (Taylor et al., 1975; Paterson and Smith, 1979) and induction of aberrations (Higurashi and Conen, 1973; Taylor et al., 1976) by ionizing radiation, implying a possible inherent DNA repair defect. This implication has been reinforced by the demonstration that AT cells are also unduly sensitive to killing following exposure to methyl methanesulfonate (MMS) and MMC (Hoar and Sargent, 1976), although not to UV light (Arlett, 1977). More recently, Paterson et al. (1979) have also reported that cultured fibroblasts from AT heterozygotes show an increased sensitivity to cell killing by gamma rays and a reduced ability to perform DNA repair replication. A high incidence of X-ray-induced chromosomal aberrations, including an unusually large number of chromatid-type changes observed in X-irradiated G_1 lymphocytes from AT patients, was suggested to reflect a possible delay in repair of single-strand breaks and a defect in the rejoining of double-strand breaks (Taylor et al., 1976; Taylor, 1978), but see below.

Despite the high background of spontaneous aberrations, the spontaneous incidence of SCE in the lymphocytes of a number of AT patients was shown to be normal (Chaganti et al., 1974; Galloway and Evans, 1975; Bartram et al., 1976; Galloway, 1977). Moreover, in studies on lymphocytes obtained from four AT patients and exposed to X rays, EMS, MMC, and adriamycin (AM) *in vitro,* Galloway (1977) showed that in all cases the response was the same as was observed in cells from normal individuals, although studies by others had indicated that some AT cell lines were sensitive to cell killing by MMC (see above) and others were not (Arlett and Harcourt, 1978). Although there has been no indication from any published work to date of any unusual incidence of SCE in AT cells, in view of the heterogeneity of response to the cell-killing effects of different mutagens among AT lines derived from different patients, and because of the limited number of studies that have so far been carried out on SCE in AT cells, a brief resume of the work on DNA repair in these cells may be relevant. The interested reader is referred to recent comprehensive reviews for further details (Kraemer, 1977; Paterson and Smith, 1979).

A number of studies have indicated that AT cells have a normal capacity to repair both single- (Taylor et al., 1975; Vincent et al., 1975) and double-strand (Lehmann and Stevens, 1977) breaks; to have a normal DNA chain growth (Hand, 1977); and to have normal levels of alpha, beta, and gamma DNA polymerases (Bertazzoni et al., 1978). A specific DNA repair defect was, however, established and characterized by Paterson and colleagues (Paterson et al., 1976 and see Paterson and Smith, 1979) and involved a defect in excising base damage induced by ionizing radiation, particularly under anoxic conditions. This defect has been assumed to reflect a lack of endonucleases responsible for excision at the damaged sites, but their presence has recently been established in AT cell extracts (Sheridan and Huang, 1979) and it is suggested that the defect(s) may not be due to a lack of nucleolytic function, but rather to a deficiency in the recognition functions of the nucleases. Paterson and his colleagues have reported that approximately one-half of all the AT strains studied display reduced levels of DNA repair, and these have been labeled as exr^- strains in contrast to the exr^+ strains that are proficient in gamma-ray-induced DNA repair synthesis. It is reported that exposure to MMS gives normal levels of repair in all strains, but exposure to *N*-methyl-N′-nitro-*N*-nitrosoguanidine (MMNG), like gamma rays, elicits normal repair only in the exr^+ strains (Scudiero, 1978). Moreover, exr^- AT strains fall into at least two complementation groups so that the responses are clearly complex (Paterson and Smith, 1979).

Insofar as the induction of SCE is concerned, to date AT cells have shown normal responses in all tests so far conducted. However, in view of the intriguing heterogeneity of the syndrome—at least in terms of the response of cells in culture—further studies on the different strains characterized appear to be worthwhile.

D. Xeroderma Pigmentosum (XP)

The original observation of Cleaver (1968, 1969), later confirmed by Setlow et al. (1969), on the response to UV light of cultured skin fibroblast cells from XP patients was the first report of a defective DNA repair mechanism in man. Moreover, since the inherited defect in repairing UV-damaged DNA was associated with a disease characterized by ultrasensitivity to exposure to sunlight and to the development of light-induced skin cancers, the findings stimulated a considerable amount of work over the following decade. At the present time it is evident that there may be a multiplicity of different repair defects in unrelated cases of XP. In all cases, however, there is no increase in the spontaneous frequencies of chromosomal aberrations or of SCE, but there may be differing responses to the induction of aberrations and of SCE by ultraviolet light and chemical mutagens, which may be related to differences in the exact nature of the DNA repair defect in the various strains of XP.

All XP cells except the so-called XP variant (Cleaver, 1970) are extremely sensitive to the lethal effects of ultraviolet light, but not X rays (Cleaver and Bootsma, 1975), and the major DNA repair defect in classical XP is a reduced ability to excise UV photoproducts; in other words, excision repair is reduced or absent in these cells. Cell fusion studies on classical XP cell strains have demonstrated the existence of at least seven complementation groups labeled A–G (Bootsma, 1978). These complementation groups are defined by their genetic interaction, but it should be noted that there is heterogeneity in the levels of repair synthesis within groups (de Weerd-Kastelein et al., 1976) and the innate levels of repair synthesis of the different complementation groups (usually estimated as relative levels of UV-induced unscheduled DNA synthesis) do not show a simple numerical correlation with the sensitivity of these groups to the cell-killing effects of UV (Andrews et al., 1976). De Weerd-Kastelein et al. (1977) studied the induction of SCEs by UV-light in a range of XP fibroblast strains and Cheng et al. (1978) reported on similar experiments on XP lymphoid cell lines. Both groups found that most excision-defective cells responded to UV radiation with enhanced frequencies of SCE compared to wild-type cell strains, and excision-proficient cells had the normal frequency of UV-induced SCE. These findings can in general be interpreted as an indication that unexcised UV damage is involved in the production of SCE. However, as is observed for the endpoint of cell killing, there are some exceptions to the general rule and there is no simple one-to-one numerical relationship between the efficiency of a cell type in carrying out unscheduled DNA synthesis and its sensitivity to SCE induction.

Individuals diagnosed as XP variants have clinical XP but normal excision repair; XP variant cells, however, are UV sensitive, are defective in postreplication repair (Lehmann et al., 1975) and also have very low levels of photoreactivating enzyme (Sutherland and Oliver, 1975). Classical XP cell strains may also show a somewhat lower than normal level of postreplication repair and of photoreactivating enzyme (Sutherland, 1978). A further and important difference between the XP variant and classical XP cell strains is the fact that in the latter, exposure to UV light results in an above normal frequency of chromosome aberrations (Parrington et al., 1971) and also of SCE (Wolff et al., 1975; Marshall and Scott, 1976; Perry et al., 1978 and see above). However, in contrast, exposure of XP variant cells to UV light does not result in an increase in SCE frequency over and above that expected from normal cells (de Weerd-Kastelein et al., 1977). This implies that defective postreplication repair is unimportant in relation to the induction of SCE.

Excision repair is often considered to be of two sorts: "short patch repair" ("X-ray-like"), with three or four nucleotides being inserted and the repair being of brief duration (up to 1 hour) and predominant after exposure of cells to, for example, ionizing radiations, and "long-patch repair" ("UV-like"), up to 100 or so nucleotides inserted and protracted (up to 18–20 hours), which is predominant after exposure of cells to UV light and certain alkylating agents (Regan and Setlow, 1974a). XP cells are thought to be efficient for short-patch, but not long-patch, repair; in other words they are able to repair damage by ionizing radiations and certain chemical agents reasonably efficiently, but are inefficient in repairing damage induced, for example, by chemical agents that form large adducts with the DNA molecules (San et al., 1977). This, however, is a rather simplistic view since, as we noted earlier, although a major defect in XP cells is in excision repair, there are defects observed in postreplication repair and photoreactivation and indeed also in base excision, where Kuhnlein and colleagues (Kuhnlein et al., 1978) have reported the absence of one of two forms of apurinic endonuclease in all XP strains of complementation group D studied. Goth-Goldstein (1977) also reported that although XP cells could efficiently excise N7-methylguanine following exposure to ethylnitrosourea (ENU) or methylnitrosourea (MNU), they were inefficient in excising O6-methylguanine. On the other hand, Pedrini et al. (1974) reported that the total DNA polymerase, DNA ligase, polynucleotide kinase, and two endonucleases were normal in cells from nine XP patients, and Parker and Lieberman (1977) report normal activities of the DNA polymerases alpha, beta, and gamma. Thus the evidence favors the hypothesis that the major defect in classical XP cells occurs at the level of DNA incision, incision being defined as recognition of damage followed by endonucleolytic breakage. Confirmation of a defective incision process was provided by the convincing demonstrations of Tanaka et al. (1975, 1977), who introduced the *E. coli* phage T4 UV-specific endonuclease (endonuclease V) into XP cells and showed that these cells could then repair UV damage in the normal way. It has since been shown that extracts of XP cells can remove UV-induced dimers in isolated DNA, but are not able to remove the dimers in chromatin (Mortelmans et al., 1976; Lieberman et al., 1979), implicating a defect of access or the importance of some cofactor.

Damage caused by "UV-like" chemicals, such as 4-nitroquinoline 1-oxide (4NQO), is not repaired as efficiently in XP cells as in normal cells (Regan and Setlow, 1974b; Sasaki, 1973) and this is associated with a decreased level of unscheduled DNA synthesis (Stich and San, 1971), an increased level of cell killing (Stich et al., 1973), and an increased frequency of chromosomal aberrations with this mutagen (Stich and San, 1973). Damage due to "X-ray-like" chemicals, however, such as EMS and MNNG, can be handled by the XP excision repair systems (Cleaver, 1971) and results in a similar level of unscheduled DNA synthesis, cell killing, and chromosome aberrations to normal cells (Sasaki, 1973). However, studies by Wolff et al. (1977) have shown that transformed XP fibroblasts give abnormally high SCE levels following exposure to a wide range of chemical mutagens, including both X-ray-like (DMS, MMS, EMS, MNNG and ENU) and UV-like (4NQO, MMC) agents, substantiating the already considerable evidence that the mechanism of formation of SCE cannot be directly equated with the processes involved in unscheduled DNA synthesis or in the development of chromosome aberrations.

The enhanced SCE induction in the virus-transformed XP cell line exposed to EMS was not confirmed by Heddle (quoted by Arlett and Lehmann, 1978), who could find no enhanced response to SCE induction in three untransformed strains of cells of complementation group A. However, results somewhat similar to those of Wolff et al. were obtained by Perry et al. (1978), who showed that lymphocytes from a single XP patient showed an enhanced response to SCE induction following exposure to EMS, 4NQO, and HN2. These latter findings imply a clear dissociation between SCE induction on the one hand and chromosome aberration induction and lethality on the other, and further studies in this area are warranted.

E. Down's Syndrome (DS)

Classical DS, trisomy 21, has repeatedly been shown (see Jackson et al., 1968) to be associated with an increased incidence of leukemia and, like FA fibroblasts, DS cells have an increased sensitivity to transformation by SV40 virus (Todaro and Martin, 1967). These observations, taken together with the findings that lymphocytes from DS patients showed a consistent and significant ultrasensitivity to the induction of chromosome aberrations by X rays (Kucerova, 1967; Sasaki et al., 1970; Evans and Adams, 1973; Holmberg, 1974) and that DS fibroblast cell lines showed a higher sensitivity than normal to the induction of aberrations by 7,12-dimethylbenz(*a*)anthracene (DMBA) (O'Brien et al., 1971) and MNU (Kaina et al., 1977), have stimulated efforts to examine the response of these cells to the induction of SCE and to probe for possible defects in DNA repair.

O'Brien et al. (1971) found normal levels of repair synthesis in DS cells after exposure to UV light and 4NQO, so that the cells were normal for UV-like repair. In contrast, Lambert et al. (1976) reported an approximately 30% reduction in DNA repair synthesis in UV-exposed blood lymphocytes from 12 DS patients. However, there is no evidence for any increased spontaneous aberration frequency in cultured DS fibroblasts or in DS lymphocytes from patients not exposed to mutagens, and the spontaneous incidence of SCE in DS blood lymphocytes is normal (Lezana et al., 1977; Kucerova and Polikova, 1978). Studies on DS lymphocytes exposed to MMC have revealed no increased response to the induction of either aberrations or SCE (Kucerova and Polikova, 1978).

F. Cockayne's Syndrome (CS)

Cockayne's syndrome, an autosomal recessive disease with its dwarfism, mental retardation, premature aging and, in most cases, sun sensitivity (MacDonald et al., 1960), seemed a likely prospect for yet a further DNA repair-deficient syndrome, and Schmickel et al. (1977) first reported that fibroblasts from CS patients are also ultrasensitive to killing by UV light. In contrast to XP, however, these authors found that CS cells were proficient in both excision and postreplication DNA repair and that DNA polymerase activity was normal (Wade and Chu, 1978). The ultrasensitivity of CS cells to cell killing by UV light has been confirmed on four CS cell lines by Marshall et al. (1980), and these authors have also demonstrated that these cell lines are ultrasensitive to the induction of SCE by UV light, although only a single fixation-sampling time was used, so that further work may be necessary to confirm this finding. In a previous study by Cheng et al. (1978) it had been noted that the UV-induced SCE frequencies in two lymphoblastoid cell lines from CS patients were elevated relative to controls, although the differences were not significant. At the present time there is no known DNA repair defect in this syndrome and despite the UV sensitivity of CS cells in terms of both SCE and cell killing, there is not a close relationship between these two endpoints. In the work of Marshall et al. (1980) the four CS strains studied fell into two clear groups for UV cell killing but gave a homogeneous response to SCE induction. This absence of an absolute correlation between SCE sensitivity and cell killing parallels the observations previously made on XP cells (de Weerd-Kastelein, 1977; Cheng et al., 1978).

G. The Hutchison-Gilford Progeria Syndrome (HG)

The HG progeria syndrome is one of several progeroid (premature aging) diseases of man (deBusk, 1972). There is conflicting evidence on whether HG cells are able to repair single-strand DNA breaks induced by X- or gamma-irradiation (Epstein et al., 1974; cf. Regan and Setlow, 1974b) and there is no published information on the incidence of SCE or of chromosome aberrations following exposure to chemical mutagens.

H. Huntington's Chorea (HC)

Neurological abnormalities are associated with the heritable repair-deficient syndromes XP, AT, and CS and it has often been considered that heritable diseases characterized by neurological degeneration might be likely prospects for conditions associated with inherited defects in DNA repair processes. Andrews et al. (1976) noted that the degree of severity of neurological abnormalities in XP were directly related to the ability to repair UV-damaged DNA and suggested that the lack of DNA repair was causally related to neurological degeneration. XP patients with neurological abnormalities fall into complementation groups A, B, and D and neuronal degeneration is progressive and, in some ways, shows some parallels with the progressive neurological degeneration in AT (Kraemer, 1977). It is this background that led Arlett and Muriel (1979) and Moshell et al. (1980) to examine the response of cells from HC patients to the lethal effects of X rays. HC is a fatal late-onset autosomal dominantly inherited condition, with a prevalence of around 1 in 10,000 in North Europe, and is characterized by primary neuronal degeneration. Arlett and Muriel (1979) stated in a brief report that four fibroblast HC cell strains were more sensitive than normal cells to the killing effects of gamma rays, as were cells from three "at risk" individuals (possible HC carriers). Moshell et al. (1980) provided some supporting evidence in demonstrating a 30% reduction in viability of four HC lymphoblastoid cell lines, relative to controls, after X-irradiation, but normal sensitivity following exposure to UV. There is no published information on the sensitivity of HC cells to the induction of chromosome aberrations or SCE, but in our own laboratory we have been unable to demonstrate any difference in the incidence of X-ray-induced chromosome aberrations in blood lymphocytes from four HC patients yet find a small, but consistent, increased response to the induction of SCE by MMC and EMS. These data will be published elsewhere (Evans et al., unpublished work), but we note that the spontaneous levels of SCE and aberrations in HC lymphocytes do not differ from normal.

I. Friedreich's Ataxia (FrA)

FrA is a rare autosomal recessive condition with spinocerebellar degeneration that is considered to be distinct from AT although it shows obvious similarities (see McKusick, 1976). Moreover, the neurological degeneration in FrA is to some extent paralleled by the spinocerebellar changes that occur in XP, and these associations prompted Lewis et al. (1979) to examine the response of FrA cells to exposure to gamma-radiation. Skin fibroblast cultures from a single FrA patient were gamma-irradiated and were shown to be hypersensitive to cell killing, the hypersensitivity being less extreme than that recorded for AT cells. Preliminary results on one other fibroblast FrA cell strain exposed to X-rays have also indicated an increased sensitivity over normal (Moshell et al., 1980).

In our own laboratory we have found a significant increased sensitivity of blood lymphocyte chromosomes of two related patients with an inherited autosomal dominant form of ataxia associated with retinitis pigmentosum and choroidal sclerosis (apparently not Refsum's syndrome) to the induction of aberrations by X rays and to the induction of SCE by both MMC and EMS (Evans et al., unpublished work). Note that there is no increased response of AT cells to SCE induction by MMC or EMS and there is no information available on possible DNA-repair defects in ataxia of the Friedreich type.

J. Retinoblastoma (Rb)

In a proportion of cases with the malignant eye tumor retinoblastoma, and, more particularly, in those patients presenting with bilateral retinoblastoma, the predisposition to the development of the tumor is inherited and in a small proportion of these inherited cases there is a constitutional deletion of part of the long arm of chromosome 13 (see Knudson et al., 1976). There is some evidence that the incidence of malignant tumors developing within tissues exposed to radiotherapy for Rb is higher than expected (Sagerman et al., 1969) and this, coupled with the heritability of some Rb cases, led Weichselbaum et al. (1977, 1978a,b) to examine the response of Rb cells to X rays. Their findings showed that fibroblast cell lines from patients with the hereditary form of Rb were more sensitive than normal cells to the killing effects of X rays, but the response of cells from patients with spontaneous Rb did not differ from normal. There is no published information on the induction of aberrations or SCE in Rb cells, nor on their response to chemical mutagens, and a report that patients with Rb have a higher than normal frequency of spontaneous aberrations in their lymphocytes (Czeizel et al., 1974) has not been substantiated (Knight et al., 1979).

K. Dyskeratosis Congenita, Scoggins Type (DC)

This condition is manifested as a rare autosomal dominant genodermatosis with pronounced reticular hyperpigmentation of the skin and hematological and immunological similarities with FA. The hematological similarity with FA is further emphasized by the apparent increased levels of spontaneous chromosome aberrations (Burgdorf et al., 1977a). Early studies on DC fibroblasts revealed normal repair replication following exposure to UV light (Cleaver, 1970), but Carter et al. (1978) have recently shown that DC cells are inefficient in removing psoralen-induced cross-links in their DNA and are hypersensitive to the induction of SCEs by psoralen and long-wavelength UV light. Whether the cells are ultrasensitive to SCE induction by other cross-linking agents or, indeed, to monofunctional alkylating agents is not known.

L. *Gardner's Syndrome (GS)*

GS is one of the dominant hereditary adenomatosis with an associated high risk of developing intestinal cancer. Sasaki (1978) showed that fibroblast cell lines from patients with this syndrome, and from those with other forms of familial polyposis coli, were more sensitive than normal to the killing effects of MMC. More recently, Little et al. (1980) studied the response of cell strains from two GS patients, confirming their higher sensitivity to cell killing by MMC, but also demonstrating that they were hypersensitive to UV—to the same extent as XP cells of complementation group C—and also, perhaps surprisingly, to X-radiation. As with the UV-sensitive Cockayne's syndrome cells, GS cells showed normal UV excision repair. As yet there is no published information on chromosome aberrations and SCE induction in this syndrome and whether GS cells show a similar response to CS cells in their sensitivity to the induction of SCE remains to be seen.

M. *Rothmund-Thomson Syndrome (RT)*

RT is a rare inherited autosomal recessive dermatosis with telangiectasia and cataract, sun sensitivity, and a predisposition to cancer. Early studies on cells from this syndrome revealed no abnormalities in DNA repair following exposure to UV light (Cleaver, 1970), but a recent brief report (Smith and Paterson, 1979) describes ultrasensitivity of RT cells to the killing effects of both X-irradiation and near-UV light but a normal response to far-UV (254 nm). There is no information on chromosome damage.

N. *Other Possible Inherited DNA-Repair Defects*

Various other inherited conditions that predispose to increased sensitivity to mutagens have been reported, but in most of these cases the information available is rather limited. There are various reports that patients with the basal cell nevus syndrome show an enhanced reaction to radiotherapy, but there are no studies on cells from these patients (Harnden and Taylor, 1978). Patients with solar actinic keratosis, a premalignant lesion of the epidermis characterized by sun sensitivity, have been reported to show a 30% lower than normal level of unscheduled DNA synthesis in UV-irradiated lymphocytes (Lambert et al. 1976), but no further data are available. Cells from patients with the rare sun-sensitive autosomal dominant diseases of acanthosis nigrocans and Darier's keratosis follicularis were originally reported by Cleaver (1970) to have reduced repair replication following exposure to UV light, but no further studies appear to have been undertaken. A cell line from a light-sensitive child (line 11961) has been shown to be highly sensitive to cell killing by UV light and EMS but not by MMS or gamma rays (Arlett and Lehmann, 1978). Excision of UV-induced pyrimidine dimers and post-replication repair after UV exposure are reported to be normal for 11961 cells (Lehmann et al., 1977) and it was considered that the patient studied might have Cockayne's syndrome; however Scott and Marshall (1978) report no hypersensitivity to the induction of SCE by UV in these cells.

Finally, we should not forget that there is a wide range of different inherited conditions that predispose to the development of cancers, many of which may have nothing whatsoever to do with abnormalities of DNA repair and few of which have, in any case, been examined from this viewpoint and even fewer for their response to the induction of SCE following exposure to carcinogens/mutagens. These include conditions that may involve a constitutional chromosome anomaly, for example, Wilm's tumor with aniridia (Riccardi et al., 1978; Franke et al., 1979), inherited renal cell carcinoma (Cohen et al., 1979), and other autosomal dominants (e.g., neurofibromatosis and tuberous sclerosis) or recessives (e.g., Chediak-Higashi syndrome) and the interested reader is referred to Mulvihill et al. (1977) and to the listings in McKusick (1976).

The original discovery by Cleaver that UV-sensitive XP cells were defective in repairing UV-induced DNA damage was stimulated by knowledge of bacterial mutants that owed their enhanced UV-sensitivity to defective DNA repair processes. The later discovery of other defective DNA repair mutants in man has in no way been surprising, and indeed the notion that there would be a variety of defective DNA repair mutants in eukaryotes, including our own species, similar to those that have been described previously in bacteria has been remarked on earlier (Evans, 1967). In the light of this knowledge, and considering possible consequences of mutations influencing DNA repair in man, we may divide some of the most obvious inherited human disease states that might contain mutants of interest into the following five categories:

1. Sun-sensitive conditions that are associated with one or more of the other conditions listed below.
2. Degenerative diseases due to premature cell death following an accumulation of DNA damage, particularly in nonreplaceable or slowly replaceable tissues, such as neuronal degeneration or certain muscular dystrophies.
3. Diseases related to an insufficiency of the production of end cells in proliferating systems due to an accumulation of genetic defects (particularly lethal mutations), for example, bone marrow, skin, and gut epithelium. Inefficiencies in systems that require high rates of cell proliferation to maintain good health may result in an insufficient production of cells of the immune response system (resulting in increased infection, diminished immune surveillance), premature aging of skin, and, in conjunction with (2) above, premature aging of all body systems.
4. A predisposition to the development of cancers ("cancer families") might be expected to occur as a consequence of an expected increased somatic mutation rate associated with some forms of defective DNA repair (also possible augmentation by deficient immune systems).
5. Sterility; DNA repair processes are involved in the normal events of meiosis in the gonads and defects in these processes lead to meiotic breakdown in plants and animals and presumably also in man.

The above list is, not unnaturally, all embracing and from the preceding pages it is evident that identified heritable defective DNA repair syndromes fall into one or more of the above categories. We should stress, however, that many of the disease states considered represent the extremes and in view of the implied multiple allelic nature of at least two of the known repair-deficient syndromes, XP and AT, the existence of a range of inefficiencies in any given type of repair process might be expected, from alleles having barely discernible effects to those resulting in serious disabilities. Similarly, since many loci must be involved in DNA housekeeping (there are, for instance, at least 50 loci coding for DNA repair functions in bakers' yeast; Haynes et al., 1978), the spectrum of efficiencies for maintaining the integrity of our DNAs must be quite considerable. Thus, although many DNA-repair mutations are lethal in early stages of development, we must expect a range of efficiencies in dealing with damaged DNAs even in apparently normal healthy populations.

Why study SCE in the context of inherited predisposition to chromosomal instability, defective DNA repair, and neoplastic transformation? Our interest essentially has been two pronged. First, since the analysis of SCE is rapid and often technically much simpler to undertake than certain of the assays for DNA-repair deficiencies, and since it offers a further parameter for assay, it has a role as a means for detecting abnormal responses to the induction of DNA damage by certain kinds of mutagens/carcinogens. We must recognize that this role has limitations and some of these are evident in our discussions of the various syndromes listed above. Second, the study of SCE induction in cells of mutants defective in some aspects of DNA replication and/or repair may provide a powerful means for unraveling the mechanisms involved in the formation of SCE. We are obviously quite a long way from understanding the nature of the repair defects in many of the syndromes listed above and in only a few instances has the induction of SCE been thoroughly studied. Nevertheless, this approach is invaluable and has, for instance, already pointed to the noninvolvement of postreplication repair in the process of formation of SCEs and to the apparent irrelevance of defective ligase activity in this context and has cast doubt on the importance of some particular DNA lesions, for example, interstrand cross-links, as causal factors in exchange. The data acquired thus far are, however, scanty and much further work is necessary.

IV. THE INCIDENCE OF SCE IN VARIOUS ACQUIRED DISEASE STATES

To the purist, few, if any, disease states can be attributed solely to environmental causes, since most, if not all, disease states involve some element of heritable susceptibility or predisposition on the part of the affected individual. The relative importance of the genotype depends on the type of disease and the type of environment in question, so that the definition of an "acquired disease," in contrast to a "heritable disease," is to a great extent a loose convenience. In the context of our discussion on SCEs, our interest is in acquired disease states that might result from, or be associated with, changes in the cell nucleus, and in particular in chromosomal structure, or in nuclear phenomena relevant to cell proliferation. Under this heading, diseases that may result in, or be associated with, exposure to chromosome-damaging agents (including biological, physical, and chemical mutagens), neoplastic diseases, and other conditions in which there is evidence for an unusual incidence of chromosome aberrations are all of special relevance.

A. *Virus Infections*

The fact that certain virus infections of mammalian cells can result in visible chromosome damage was first demonstrated by Hampar and Ellison (1961) in their study on the effects of infection of cultured Chinese hamster cells by the herpes simplex virus. In the following year Nichols et al. (1962) reported that similar viral-induced chromosome aberrations were to be found in blood lymphocytes of children infected with measles virus and later reports describe chromosome abnormalities in lymphocytes of patients suffering from smallpox, mumps (Aula, 1963; Grippenberg, 1965), and poliomyelitis (Bartsch et al., 1967) or following vaccination with vaccinia virus (Schuler et al., 1968) and in lymphocytes exposed to some of these viruses *in vitro* (Nichols et al., 1965a,b; Aula and Nichols, 1968). Interest in this area was especially aroused since infection by a range of DNA and RNA viruses known to be oncogenic in mammalian cells was often seen to be associated with the induction of chromosome structural changes. Since these early studies a wide range of DNA or RNA viruses that can induce chromosome damage in infected human cells has been well documented (Harnden, 1974; Nichols, 1974), although conflicting results have often been reported and a multiplicity of factors may be involved in influencing the cellular, or chromosomal, response. Nevertheless, it is clear that virus infection *in vitro* or *in vivo* can result in gross chromosomal damage, that certain viruses are oncogenic, that chromosome aberrations may be associated with neoplastic change, and that conditions or agents that result in aberrations may also induce SCE. It was no surprise, therefore, when Wolff et al. (1975) reported a slightly increased level of SCE in SV40 transformed human diploid cells relative to untransformed cells, and slightly more pronounced increases, up to around 50%, were later reported in mouse embryo cell lines infected with the Rauscher leukemia virus (Brown and Crossen, 1976). In contrast, Kato and Sandberg (1977) observed no increases in SCE in human diploid fibroblasts exposed to Herpes simplex virus 1 (HSV 1) and HSV 2 under conditions where both viruses produced considerable amounts of chromosome damage in the early phases of infection. These authors also reported similar findings for cells exposed to adenovirus type 2 and to the Rous sarcoma virus, underlining once again the lack of association between aberration induction and SCE formation.

More recently, Nichols et al. (1978) undertook a detailed study of SCE frequencies in human diploid fibroblast cultures following the addition of SV40 virus. They noted that the SCE frequency remained the same in controls as in infected cultures up until the time the latter became tumor antigen positive. When the tumor antigen became expressed, the SCE frequency increased up to threefold and was accompanied by increases in chromosome aberrations.

The fact that many chemical mutagens that promote the formation of SCEs are also known to be potent in inducing the production of infectious virus in rodent cells transformed by DNA viruses, or harboring the pro virus of C-type oncoviruses, led Kaplan et al. (1978) to show that exposure of SV40-transformed hamster kidney cells to MMC and EMS resulted in the coinduction of SV40 infectious virus and SCE formation, with parallel dose response curves for these two endpoints. Later studies (Zamanski et al., 1980) revealed similar findings of coinduction of virus and SCE formation following exposure of hamster kidney cells to UV light or BrdUrd. Details of the data obtained in these experiments, however, indicate that the increased SCE levels were observed in many cases at very low virus titers, so that the appearance of SCEs did not seem to be caused by the concomitant induction of virus. Moreover, exposure of UV-irradiated cells to caffeine enhanced virus induction by an order of magnitude, but had little or no effect on SCE incidence. The relationship between lesions and mechanisms resulting in virus induction and SCEs is therefore obscure and it remains to be determined whether there are any common steps involved prior to the divergence of pathways resulting in virus synthesis on the one hand and SCE formation on the other.

The positive findings of increased SCE levels following virus infection *in vitro* have also been obtained in *in vivo* studies. Kurvink et al. (1978a) reported small but significant increases in SCE frequencies at various times after infection in a number of cases in a series of 17 patients suffering from infections with herpes simplex, cold/flu (unspecified), or infectious hepatitis virus. Knuutila et al. (1978c) reported similar SCE increases in lymphocytes sampled at the estimated time of viremia in 12 patients vaccinated against smallpox, and parallel results were noted in three individuals vaccinated with an attenuated measles virus (Lambert et al., 1979a). In our own laboratory we studied lymphocytes from 50 individuals immunized with an attenuated poliomyelitis virus vaccine and a similar number of individuals immunized with an attenuated or inactivated influenza virus (Stephenson, Hainey, and Evans, unpublished data). In all cases, we noted a significant increase in SCE, with a maximum increase of around 50% on the sixth or seventh day following immunization.

The above *in vitro* and *in vivo* findings clearly show that viral infection can result in the formation of SCE, but the mechanisms resulting in these increases are, however, obscure and may well differ in different cases. It is possible that lesions may be induced in host cell DNA because of viral integration or excision or as a result of viral replication; alternatively, the

SCE increases may reflect disturbances in the normally smoothly integrated processes that are involved in DNA replication so that defects could result from concomitant viral replication competing with host cell DNA replication. In the case of some of the small increases observed following vaccination *in vivo,* and particularly where inactive virus has been used, a further possibility is that these changes may reflect changes in the efficiency of cells for BrdUrd uptake, or incorporation, or the immunization process may result in a change in the type of cell sampled. At present we are unable to distinguish among any of these possibilities.

B. Neoplastic Disease

The association of chromosomal structural changes with the development, and in some cases apparently the initiation, of neoplasms (see German, 1974b; Evans, 1980) has stimulated studies on SCE incidence particularly in leukemias and lymphomas, where relatively ready access is available to proliferating materials that can be cultured over two cells cycles to give harlequin-stained chromosomes for SCE analysis. The first relevant report was that of Kurvink et al. (1978b), who showed that whereas the mean frequency of SCEs in lymphocytes of 40 healthy control individuals was 6.1/cell, the frequency in 13 untreated patients with malignant lymphoma—both Hodgkins and non-Hodgkins—was significantly higher at 12.7/cell. Thirteen patients were then studied after chemotherapy treatment alone, and 17 after radiotherapy and subsequent chemotherapy. In 7 other patients treated with radiotherapy alone, no increase in SCE was noted, which was consistent with earlier *in vitro* studies showing the inefficiency of ionizing radiations in inducing SCE (Perry and Evans, 1975). Following chemotherapy with cyclophosphamide or adriamycin plus bleomycin, SCE frequencies were shown to increase and to vary depending on the time of sampling over the days or weeks following treatment. It was also noted in this study that patients with active disease, whether recently treated or not, had higher SCE scores than patients in remission.

Sandberg and colleagues (see Shiraishi and Sandberg, 1980) have studied the SCE incidence in bone marrow cells in cases of acute and chronic leukemias. Their findings in untreated patients indicate a normal range of SCE incidence in leukemia cells, identified by aneuploidy or in the case of chronic myeloid leukemia (CML) by the presence of the Philadelphia chromosome. SCE values above normal were observed in leukemic cells of CML patients in the blastic phase and in most patients following chemotherapy. The workers conclude that leukemic cells are not characterized by an inherently high incidence of SCE, regardless of the type of leukemia, and the higher than normal frequency of SCE in circulating lymphocytes of some patients with lymphoma or leukemia are considered to be secondary consequences of the neoplasia, or of the viral infections to which these patients are particularly prone.

Finally, it is relevant to refer to the studies of Knuutila et al. (1978b) on patients with megaloblastic anemia, a nonmalignant condition in which bone marrow and cultured lymphocytes may show elevated incidences of chromosomal aberrations (Lawler et al., 1971; Krogh Jensen, 1977) that are often considered to be due to impaired DNA synthesis resulting from folic acid deficiency. In one study, on peripheral blood lymphocytes and bone marrow cells from eight patients, it was shown that some of the bone marrow cells had a tenfold increase in SCE incidence over normal levels, and bone marrow cells also showed chromosomal aberrations, although no such aberrations were present in blood lymphocytes. The average SCE frequency in bone marrow cells was some 100% greater than in controls, and a similar increase was observed in lymphocytes, although no lymphocytes were found to have the very high numbers of SCEs seen in some of the bone marrow cells.

Our conclusion from the above studies is that there is no evidence that neoplastic transformation per se is reflected in a dramatic change in SCE incidence, but the SCE frequency may increase as a consequence of disturbed replication, as evidenced by the anemia studies, by possible secondary viral infections associated with the malignancy, and, of course, by the mutagenic chemicals used as anticancer agents in treatment.

C. Ill-Health Associated With Exposure to Physical and Chemical Mutagens

1. Exposure to Physical Agents

The original *in vitro* studies on harlequin-stained chromosomes in cells exposed to X rays showed that X-irradiation was very inefficient in inducing SCEs (Perry and Evans, 1975) and these observations have been confirmed in a variety of later *in vitro* and *in vivo* studies. In contrast, exposure to nonionizing radiations and, as we see above, in particular to short-wavelength UV, has been amply demonstrated to result in large increases in SCE in cells irradiated *in vivo*. Doubtless, similar increases must occur *in vivo,* but because of the lack of penetration these occur in cells proliferating in the superficial layers of the body and there is no evidence of increased SCE levels in blood lymphocytes of people having had a significant erythema following a sun-tanning session on the beach.

We refer above to the fact that exposure of cells *in vitro* to 8-methoxypsoralen (8-MOP) and then to long-wave UV light (350 nm) results in the formation of SCEs, since in the presence of UV 8-MOP forms monoadducts with pyrimidines and also DNA cross-links in a ratio of around 3:1 (Cole, 1971). Since exposure to whole-body irradiation with long-wave UV light following oral intake of 8-MOP is a very effective treatment for patients with psoriasis (Parrish et al., 1974), a common, distressing, but nonmalignant, skin condition, it was important to establish whether this therapy regimen produces a significant chromosome damaging effect in treated patients. Initial *in vitro* studies showed that exposure of isolated peripheral blood lymphocytes to 8-MOP alone did not influence SCE frequency, but when followed by exposure to 350 nm UV it resulted in a threefold increase in SCE (Faed and Mourelatos, 1978). However, in studies on lymphocytes from a series of patients sampled after a combined therapy with 8-MOP and long-wavelength UV, no increases in SCE were observed, presumably because the blood cells were largely shielded from the ultra violet light (Faed and Mourelatos, 1978; Brøgger et al., 1978).

Exposure to light from fluorescent tubes radiating in the visible part of the spectrum has been shown to induce DNA damage (Tyrell et al., 1974) and mutations (Bradley and Sharkey, 1977) in a variety of mammalian cells irradiated *in vitro*. Moreover, Monticone and Schneider (1979) have shown that similar exposures to cultured human fibroblasts result in marked increases in SCE. Studies using filters have shown that the damaging wavelengths are in the near UV, 300–390 nm, and that most of the biological effects are consequences of the formation of toxic photoproducts in the culture medium and, to a lesser extent, within the cells. Phototherapy with similar fluorescent light sources is also employed in the treatment of neonatal hyperbilirubinaemia, where the technique involves giving affected babies intensive whole-body exposure over a 48–72 hours period. Goyanes-Villaescusa et al. (1977) have reported that such exposures result in a significant increase in SCE in lymphocytes from blood samples taken from treated babies relative to the SCE incidence in untreated but similarly afflicted infants, but that the increase is small. However, these findings were firmly refuted by two independent and thorough studies by Schwartz et al., (1978) and Hatcher et al., (1979) who observed no differences in SCE levels between blood cells of a series of control and phototherapy-exposed infants.

Exposure of cells to ultrasound had previously been believed to result in the formation of chromosome aberrations (Macintosh and Davey, 1972), although later extensive studies in this field failed to substantiate the original claim (Buckton and Baker, 1972; Watts et al., 1972). Recently, however, Liebskind et al. (1979) have reported that *in vitro* exposure of blood lymphocytes or lymphoblastoid cell lines for 30 min to 2.5–5.0

mW/cm^2 of diagnostic ultrasound resulted in a significant 30% increase in SCE frequency in a series of experiments. The same exposures had been shown previously by these authors to produce DNA damage and to induce unscheduled DNA synthesis, but the doses employed in all these studies would be considerably in excess of those received by a fetus during diagnostic exposure to ultrasound.

2. *Therapeutic Exposure to Chemical Agents*

In our earlier discussion on SCE associated with neoplasia, we comment on the fact that at least some of the increased SCE frequencies reported to be associated with the leukemias and lymphomas are a consequence of the actions of the chemotherapeutic agents used to treat these diseases. Since the original report by Perry and Evans (1975) of increased frequencies of SCEs following exposure of cancer patients to the drug adriamycin, there have been many reports of elevated SCE incidences in blood lymphocytes of patients exposed to a range of anticancer drugs. These include adriamycin (Perry and Evans, 1975; Musilova et al., 1979; Nevstad, 1978), cyclophosphamide (Raposa, 1978; Musilova et al., 1979; Natarajan et al., 1978); busulfan (Musilova et al., 1979); melphalan (Lambert et al., 1978a), and various standard cocktails for example CCNU (Natarajan et al., 1978; Lambert et al., 1979b). On the other hand, no increased SCE incidences were observed in cells from patients treated with certain other drugs, including 5-fluorouracil (Perry and Evans, 1975; Musilova et al., 1979) and cytosine arabinoside (Raposa, 1978). Where elevation of SCE incidence has been observed, the amount of increase is highly dependent on the nature of the drug used, its concentration/dose, whether given in single or multiple exposures, and the time of sampling of blood lymphocytes after the treatment has been completed. In general, the higher the dose, the longer the persistence of the raised SCE levels, and in most instances the SCE frequency returns to normal levels within 3 weeks of treatment. However, with some drugs, for example CCNU, elevated SCE levels may be observed for some months following discontinuation of treatment.

Elevated SCE incidences in cancer patients receiving cytostatic drugs is not surprising, since many of these cytostatics, including those of microbial origin, have been shown to be potent inducers of chromosomal aberrations. A surprising finding, however, is the report by Burgdorf et al. (1977b), who found a three fold increase in SCE frequencies relative to controls in six patients treated with potassium arsenite more than 24 years prior to blood sampling and SCE analysis. All the patients showed clinical symptoms of arsenic treatment, including skin cancers. Nordensen et al.

(1979), however, failed to confirm these findings in eight psoriatic patients treated with arsenic 15 years earlier, although they reported increased frequencies of chromosomal aberrations in these patients. Even more surprising is a report by Krishna Murthy and Prema (1979) of a significant 30% increase in SCEs in peripheral blood lymphocytes of women taking oral contraceptives. Similar findings had been noted independently in our laboratory (Hopkin, unpublished work), but whether this small, but consistent, increase is due to a change in the population of leucocytes in women on the pill or to an effect on membranes and on BrdUrd uptake is not known.

3. Occupational Exposure to Chemical Agents

The realization that many individuals may be exposed during the course of their work to possible mutagenic/carcinogenic chemicals, has provided a spur to the testing of a range of compounds in *in vitro* testing systems, as well as to various studies to detect the presence of chromosome aberrations in blood cells of exposed individuals (see Evans, 1975; Hollaender, 1976; Evans and Lloyd, 1978). The range of compounds of interest is wide and includes insecticides, pesticides, fungicides, heavy metals, various processing chemicals, organic solvents, plasticizers, anesthetic gases, and so on. In some instances, markedly elevated frequencies of chromosomal aberrations have been observed in blood cells of workers exposed, for example, to benzene (e.g. Forni et al., 1971), vinyl chloride monomer (e.g. Purchase et al., 1975), epichlorhydrin (Kucerova et al., 1977; Picciano, 1979; Sram et al., 1980), and styrene (Meretoja et al., 1977). Parellel *in vitro* studies on human cells in culture and *in vivo* studies on laboratory animals, as referred to elsewhere in this book, have demonstrated that many chemical agents encountered in everyday life may be mutagenic and may induce SCEs. The simplicity of the SCE technique has stimulated a number of studies on SCE incidence in blood cells of individuals exposed to known mutagens, and recent reports have described elevated SCE levels (two to three times control levels) in blood cells of workers exposed to vinyl chloride monomer (Kucerova et al., 1979) and in those of laboratory technicians exposed to organic solvents (Funes-Cravioto et al., 1978). A major problem in assaying for SCEs in workers exposed to suspected mutagens follows from the fact that the lesions that result in SCE, or the chemicals that are responsible, must be present in the blood cells at the time of sampling, so that the detection of raised SCE levels following an acute exposure to a mutagen many weeks previously may not be possible, (cf. the time studies on cancer patients treated with cytostatics and referred to earlier). Thus it may be relatively easy to demonstrate an effect of epichlorhydrin (White, 1980) or the insecticides Captan and Dichlorvos (Tezuka et al., 1980) in inducing SCE *in vitro,* but unless blood samples are taken shortly after an *in vivo* exposure, or unless individuals are exposed chronically to significant dose levels, SCE incidence is not a useful guide to detect or monitor *in vivo* exposures. Nevertheless, the presence of high levels of SCE in apparently healthy individuals is a cause for concern and the first suspicion would be that these reflect an exposure to a mutagenic substance.

4. *Exposure to Chemical Agents by Other Means, Including Ingestion and Inhalation*

The range of environmental mutagens/carcinogens that nowadays form part of the normal environment is large (Nagao et al., 1978), and substances such as the food additives 2-(2-furyl)-3-(5-nitro-2-furyl) acrylamide, or AF2 (Tonomura and Sasaki, 1973), at one time extensively used in Japan, and sunset yellow (Abe and Sasaki, 1977), synthetic hair dyes (Perry and Searle, 1977), fluorescent brighteners (Abe and Sasaki, 1977), the flame retardant tris (2,3-dibromopropyl) phosphate (Tris-BP) (Nakanishi and Schneider, 1979), aflatoxin Bl from the fungal contaminant of groundnuts (Takehisa and Wolff, 1977; Vijayalaxmi et al., 1979) and even such substances as saccharin (Abe and Sasaki, 1977; Wolff and Rodin, 1978) and vitamin C (Galloway and Painter, 1979) have all been shown to induce SCE and/or aberrations in *in vitro* tests or in animal *in vivo* studies (see Takehisa, 1981). In contrast, there is no information on the effects of these substances in inducing SCE *in vivo* in man, although it seems likely that such effects must occur. Two important much used, and abused, substances that have shown positive results in *in vivo* studies in man, however, are alcohol and cigarette tobacco smoke.

The demonstration (Yamasaki and Ames, 1977) that the urine of cigarette smokers, but not nonsmokers, contains materials that are mutagenic to bacteria, stimulated interest in the possible effects of inhaled cigarette smoke products on human chromosomes *in vivo.* Obe and Herha (1978) originally reported that cultured peripheral blood lymphocytes taken from heavy cigarette smokers showed a marked elevation of chromosome aberration frequency relative to those of light or non-smokers. This work was followed by a report by Lambert et al. (1978b), who studied SCEs in lymphocytes from groups of smokers and nonsmokers and found small, but consistent and significant, elevations of SCE frequencies in cigarette smokers. In contrast, Hollander et al. (1978) found no such effect in chronic smokers (including lung cancer patients) compared with healthy nonsmoking controls, a result that has also been reported by others (Ardito et al., 1980; Crossen and Morgan, 1980). In our own laboratory we have observed both no difference in SCE frequency between healthy cigarette smokers and nonsmokers in one study on 104 individuals (Hainey, Vijayalaxmi and Evans, unpublished data), although the majority of these smokers were light smokers, and a significant effect in a separate series of experiments (Hopkin and Evans, 1980). This latter work confirms the earlier report of Lambert et al. in demonstrating a small, but nevertheless consistent, excess of SCEs in blood lymphoctes of healthy individuals and lung cancer patients smoking, and inhaling from, more than 20 cigarettes/day. In addition, we noted that smoking lung cancer patients showed slightly higher yields than smoking controls and that these differences were maintained, and indeed enhanced, in blood lymphocyte cells exposed to cigarette smoke condensates *in vitro* (Hopkin and Evans, 1980). A similar small effect of cigarette smoking on SCE incidence has recently been reported by Krishna Murthy (1979), who found that while moderate smoking had a significant effect on SCE frequency, light smoking (less than 10 cigarettes/day) had no effect. It would seem from these studies, therefore, that inhaling cigarette smokers who smoke more than 20 cigarettes/day show small, but consistent, elevations of SCE incidences in their blood cells, as well as increased frequencies of chromosome aberrations.

Cigarette smoke condensates are powerful inducers of SCE in human lymphocytes *in vitro* (Hopkin and Evans, 1979, 1980) and the component that induces SCE in tobacco tars is not the polycyclic aromatic hydrocarbon benzpyrene, or any other hydrocarbon whose activation is blocked in the presence of alpha-naphthoflavone (Hopkin and Perry, 1980). Mizusaki et al. (1977) have shown that harman and norharman, products of the pyrolosis of tryptophan, are present in tobacco tar and indeed in other burnt organics, at concentrations 1000 times more than that of benzpyrene. Moreover, these agents and other pyrolysates, for example, that of glutamic acid, have been shown to be powerful comutagens, or in some cases are themselves mutagenic (Matsumoto et al., 1978). It seems possible, therefore, that the effects of cigarette smoke *in vivo* in man may well reflect the actions of pyrolyzed amino acids rather than of the polycyclic aromatic hydrocarbons.

The facts that alcohol appears to be teratogenic (Hanson et al., 1976) and carcinogenic (McMichael, 1978) in man and that early studies on plant root-tip cells indicated that exposure to alcohol could result in chromosomal aberrations (Rieger and Michaelis, 1970) prompted Obe and Herha (1975) to examine lymphocyte chromosomes in chronic alcoholics and to demonstrate enhanced frequencies of chromosomal aberrations in cultured lymphocytes from these individuals. From later *in vitro* studies

(Ristow and Obe, 1978 and see review by Obe and Ristow, 1979) it emerged that alcohol (ethanol) per se is ineffective in inducing chromosomal aberrations, but that its first metabolic product, acetaldehyde, is a potent inducer of both chromosome aberrations and SCE. Moreover, human lymphocytes cultured in the presence of ethanol at concentrations equivalent to those found in blood after moderate to heavy consumption, show a 50% increase in SCE incidence relative to unexposed cells (Alvarez et al., 1980). In this context it may be relevant to note that recent findings of Goedde and colleagues (Goedde et al., 1979; Harada et al., 1980) on autopsy livers of Japanese suggest that the initial alcohol sensitivity that is quite common in individuals of mongoloid origin is due to delayed oxidation of acetaldehyde. There is no published information on SCE levels in individuals having the atypical alcohol dehydrogenase, and lacking the enzyme that has a high affinity for acetaldehyde, that results in the high levels of acetaldehyde in bloods from these individuals. Finally, it is also relevant to note that the urine of cirrhotic patients who had stopped drinking and smoking has recently been shown (Gelbart and Sontag, 1980) to be mutagenic to bacteria. If this is a consequence of a diminished hepatic detoxification, then it would be of interest to examine the chromosomes of such patients.

V. CONCLUDING COMMENTS

Our discussion of SCE in relation to human disease has encompassed a range of often seemingly disparate studies, some of which are seen to be rewarding and others of which may sometimes appear to amount to no more than "stamp collecting." In all cases, however, the simple technology involved in analyzing SCE incidence and the fact that in the majority of cases these exchange events mirror abnormalities in replication or indeed DNA structure have been paramount factors in the reported studies. We have seen that SCE incidence can provide a useful guide to exposure to mutagens and, moreover, appears to be a useful parameter for study in the context of mutagen testing (see Perry, 1980). However, the mechanisms involved in the formation of SCE are still unclear, although there are obvious departures from pathways that result in aberration formation, and yet many similarities in terms of positive common responses to a wide range of mutagens. A better understanding of the lesions responsible for, and mechanisms involved in, the formation of SCEs and of the many factors that may influence the response of the cells both *in vivo* and *in vitro* are essential requirements. Such an understanding can undoubtedly be enhanced by a study of SCE in disease states and this will in turn throw more light on what the presence of these exchange events might mean in terms of individual ill health.

References

Abe, S. & Sasaki, M. (1977) Chromosome aberrations and sister chromatid exchanges in Chinese hamster cells exposed to various chemicals. *J. Natl. Cancer Inst.* **58,** 1635–1641.

Alhadeff, B., Velivasakis, M., Pagan-Charry, I., Wright, W. C., & Siniscalco, M. (1980) High rate of sister chromatid exchanges of Bloom's syndrome chromosomes is corrected in rodent human somatic cell hybrids. *Cytogenet. Cell Genet.* **27,** 8–23.

Alvarez, M. R., Cimino, L. E., Cory, M. J., & Gordon, R. E. (1980) Ethanol induction of sister chromatid exchanges in human cells *in vitro. Cytogenet. Cell Genet.* **27,** 66–69.

Amacher, D. E. & Lieberman, M. W. (1977) Removal of acetylaminofluorene from the DNA of control and repair-deficient human fibroblasts. *Biochem. Biophys. Res. Commun.* **74,** 285–290.

Andrews, A. D., Barrett, S. F., & Robbins, J. H. (1976) Relation of DNA repair processes to pathological ageing of the nervous system in xeroderma pigmentosum. *Lancet* **1,** 1318–1320.

Ardito, G., Lamberti, L., Ansaldi, E., & Ponzetto, P. (1980) Sister chromatid exchanges in cigarette-smoking human females and their newborns. *Mutat. Res.* **78,** 209–212.

Arlett, C. F. (1977) Lethal response to DNA damaging agents in a variety of human fibroblast cell strains. *Mutat. Res. 46, 106.*

Artlett, C. F. & Harcourt, S. A. (1978) Cell killing and mutagenesis in repair-defective human cells. In *DNA Repair Mechanisms* P. C., Hanawalt, E. C. Friedberg, and C. F. Fox, Eds.) Academic Press, New York, pp. 633–636.

Artlett, C. F. & Lehmann, A. R. (1978) Human disorders showing increased sensitivity to the induction of genetic damage. *Annu. Rev. Genet.* **12,** 95–115.

Artlett, C. F. & Muriel, W. J. (1979) Radiosensitivity in Huntington's chorea cell strains: a possible pre-clinical diagnosis. *Heredity* **42,** 276.

Auerbach, A. D. & Wolman, S. R. (1976) Susceptibility of Fanconi's anaemia fibroblasts to chromosome damage by carcinogens. *Nature (Lond.)* **261,** 494–496.

Auerbach, A. D. & Wolman, S. R. (1978) Carcinogen-induced chromosome breakage in Fanconi's anaemia heterozygous cells. *Nature (Lond.)* **271,** 69–71.

Auerbach, A. D., Warburton, D., & Chaganti, R. S. K. (1978) Pre-natal detection of the Fanconi's anaemia gene by cytogenetic methods. *J. Supramol. Struct.,* Suppl. 2, 84.

Aula, P. (1963) Chromosome breaks in leukocytes of chickenpox patients. *Hereditas* **49,** 451–453.

Aula, P. & Nichols, W. W. (1968) Lysosomes and virus induced chromosome breakage. Exp. *Cell Res.* **51,** 595–601.

Bartram, C. R., Koske-Westphal, T., & Passarge, E. (1976) Chromatid exchanges in ataxia telangiectasia, Bloom syndrome, Werner syndrome, and xeroderma pigmentosum. *Ann. Hum. Genet. Lond.* **40,** 79–86.

Bartram, C. R., Rüdiger, H. W., & Passarge, E. (1979) Frequency of sister chromatid exchanges in Bloom syndrome fibroblasts reduced by cocultivation with normal cells. *Hum. Genet.* **46,** 331–334.

Bartsch, H. D., Habermehl, K.-O., & Diefenthal, W. (1967) Chromosomal damage after infection with poliomyelitis virus. *Exp. Cell Res.* **48,** 671–675.

Becher, R., Schmidt, C. G., Theis, G., & Hossfeld, D. K. (1979) The rate of sister chromatid exchange in normal human bone marrow cells. *Hum. Genet.* **50,** 213–216.

Bertazzoni, U., Scovassi, A. I., Stefanini, M., Giulotto, E., Spadari, S., & Pedrini, A. M. (1978) DNA polymerases α, β and γ in inherited diseases affecting DNA repair. *Nucleic Acids Res.* **5,** 2189–2196.

Bootsma, D. (1978) *Xeroderma pigmentosum.* In *DNA Repair Mechanisms* (P. C. Hanawalt, E. C. Friedberg, and C. F. Fox, Eds.), Academic Press, New York, pp. 589–601.

Bradley, M. W. & Sharkey, N. A. (1977) Mutagenicity and toxicity of visible fluorescent light to cultured mammalian cells. *Nature (Lond.)* **266,** 724–726.

Brøgger, A., Waksvik, H., & Thune, P. (1978) No evidence for chromosome damage in psoriasis patients treated with psoralen and long-wave ultra violet light. In *Mutagen-induced Chromosome Damage in Man* (H. J. Evans and D. C. Lloyd, Eds.), University Press, Edinburgh, pp. 221–226.

Brown, R. L. & Crossen, P. E. (1976) Increased incidence of sister chromatid exchanges in Rauscher leukemia virus infected mouse embryo fibroblasts. *Exp. Cell Res.* **103,** 418–420.

Bryant, E. M., Hoehn, H., & Martin, G. M. (1979) Normalisation of sister chromatid exchange frequencies in Bloom's syndrome by euploid cell hybridisation. *Nature (Lond.)* **279,** 795–796.

Buckton, K. E. & Baker, N. V. (1972) An investigation into possible chromosome damaging effects of ultrasound on human blood cells. *Br. J. Radiol.* **45,** 340–342.

Burgdorf, W., Kurvink, K., & Cervenka, J. (1977a) Sister chromatid exchange in dyskeratosis congenita lymphocytes. *J. Med. Genet.* **14,** 256–257.

Burgdorf, W., Kurvink, K., & Cervenka, J. (1977b) Elevated sister chromatid exchange rate in lymphocytes of subjects treated with arsenic. *Hum. Genet.* **36,** 69–72.

Carter, D. M., Gaynor, A., & McGuire, J. (1978) Sister chromatid exchanges in *dyskeratosis congenita* after exposure to trimethyl psoralen and UV light. In *DNA Repair Mechanisms* (P. C. Hanawalt, E. C. Friedberg, and C. F. Fox, Eds.), Academic Press, New York, pp. 671–674.

Chaganti, R. S. K., Schonberg, S., & German, J. (1974) A manyfold increase in sister chromatid exchanges in Bloom's syndrome lymphocytes. *Proc. Natl. Acad. Sci. U.S.* **71,** 4508–4512.

Cheng, W. –S., Tarone, R. E., Andrews, A. D., Whang-Peng, J.S., & Robbins, J. H. (1978) Ultraviolet light-induced sister chromatid exchanges in xeroderma pigmentosum and in Cockayne's syndrome lymphocyte cell lines. *Cancer Res.* **38,** 1601–1609.

Cleaver, J. E. (1968) Defective repair replication of DNA in xeroderma pigmentosum. *Nature (Lond.)* **218,** 652–656.

Cleaver, J. E. (1969) Xeroderma pigmentosum: a human disease in which an initial stage of DNA repair is defective. *Proc. Natl. Acad. Sci. U.S.* **63,** 428–435.

Cleaver, J. E. (1970) DNA damage and repair in light-sensitive human skin disease. *J. Invest. Dermatol.* **54,** 181–195.

Cleaver, J. E. (1971) Repair of alkylation damage in ultraviolet-sensitive (xeroderma pigmentosum) human cells. *Mutat. Res.* **12,** 453–462.

Cleaver, J. E. & Bootsma, D. (1975) Xeroderma pigmentosum: biochemical and genetic characteristics. *Annu. Rev. Genet.* **9,** 19–38.

Cohen, A. J., Li, F. P., Berg, S., Marchetto, D. J., Tsai, S., Jacobs, S. C., & Brown, R. S. (1979) Hereditary renal-cell carcinoma associated with a chromosomal translocation. *N. Engl. J. Med.* **301,** 592–595.

Cole, R. S. (1971) Psoralen monoadducts and interstrand cross-links in DNA. *Biochim. Biophys. Acta* **254,** 30–39.

Conner, M. K., Alarie, Y., & Dombroski, R. L. (1979) Sister chromatid exchange in murine alveolar macrophages, regenerating liver and bone marrow cells—a simultaneous multicellular *in vivo* assay. *Chromosoma (Berl.)* **74,** 51–55.

Crossen, P. E. & Morgan, W. F. (1980) Sister chromatid exchange in cigarette smokers. *Hum. Genet.* **53,** 425–426.

Crossen, P. E., Drets, M. E., Arrighi, F. E., & Johnston, D. A. (1977) Analysis of the frequency and distribution of sister chromatid, exchanges in cultured human lymphocytes. *Hum. Genet.* **35,** 345–352.

Czeizel, A., Csosz, L., Gardonyi, J., Remenar, L., & Ruziscka, P. (1974) Chromosome studies in twelve patients with retinoblastoma. *Humangenetik* **22,** 159–166.

deBusk, F. L. (1972) The Hutchinson-Gilford progeria syndrome. *J. Pediat.* **80,** 697–724.

deWeerd-Kastelein, E. A., Keijzer, W., Sabour, M., Parrington, J. M., & Bootsma, D. (1976) A xeroderma pigmentosum patient having a high residual activity of unscheduled DNA synthesis after UV is assigned to complementation group A. *Mutat. Res.* **37,** 307–312.

deWeerd-Kastelein, E. A., Keijzer, W., Rainaldi, G., & Bootsma, D. (1977) Induction of sister chromatid exchanges in xeroderma pigmentosum cells after exposure to ultraviolet light. *Mutat. Res.* **45,** 253–261.

Epstein, J., Williams, J. R., & Little, J. B. (1974) Rate of DNA repair in progeric and normal human fibroblasts. *Biochem. Biophys. Res. Commun.* **59,** 850–857.

Evans, H. J. (1967) Repair and recovery at chromosome and cellular levels: similarities and differences. In *Recovery and Repair Mechanisms in Radiobiology,* Brookhaven Symposium in Biology, No. 20, pp. 111–133.

Evans, H. J. (1975) Environmental agents and chromosome damage in somatic cells of man: *in vivo* studies. In *Chromosome Variations in Human Evolution* (A., Boyce, Ed.) Taylor and Francis, London, pp. 63–82.

Evans, H. J. (1980) Genes, chromosomes and neoplasia: a summary. In *Genes, Chromosomes and Neoplasia* (F. E. Arrhigi, P. N. Rao, and E. Stubblefield, Eds.), Raven Press, New York, pp. 511–527.

Evans, H. J. & Adams, A. (1973) X-ray induced chromosome aberrations in human lymphocytes irradiated *in vitro*: the influence of exposure conditions, genotype and age on aberration yields. In *Advances in Radiation Research, Vol. I: Biology and Medicine* (J. F. Duplan and A. I. Chapiro, Eds.), Gordon and Breach, New York, pp. 335–348.

Evans, H. J. & Lloyd, D. C., (Eds.) (1978) *Mutagen-induced Chromosome Damage in Man,* University Press, Edinburgh.

Evans, H. J. & Vijayalaxmi (1980) Storage enhances chromosome damage after exposure of human leukocytes to mitomycin C. *Nature (Lond.)* **284,** 370–372.

Evans, H. J., Adams, A. C., Clarkson, J. M., & German, J. (1978) Chromosome aberrations and unscheduled DNA synthesis in X- and UV-irradiated lymphocytes from a boy with Bloom's syndrome and a man with xeroderma pigmentosum. *Cytogenet. Cell Genet.* **20,** 124–140.

Faed, M. & Mourelatos, D. (1978) Sister chromatid exchanges in lymphocytes treated with 8-methoxypsoralen and exposed to longwave ultraviolet light. In *Mutagen-induced Chromosome Damage in Man* (H. J. Evans and D. C. Lloyd, Eds.), University Press, Edinburgh, pp. 216–220.

Forni, A. M., Cappellini, A., Pacifido, E., & Vigliana, E. (1971) Chromosome changes and their evolution in subjects with past exposure to benzene. *Arch. Environ. Health* **23,** 385–391.

Francke, U., Holmes, L. B., Atkins, L., & Riccardi, V. M. (1979) Aniridia-Wilms' tumor association: evidence for specific deletion of 11p13. *Cytogenet. Cell Genet.* **24,** 185–192.

Friedberg, E. C., Ahmann, U. K., & Williams, J. I. (1979) Human diseases associated with defective DNA repair. In *Advances in Radiation Biology,* Vol. 8, Academic Press, New York, pp. 85–174.

Fujiwara, Y., Tatsumi, M., & Sasaki, M. S. (1977) Cross-link repair in human cells and its possible defect in Fanconi's anemia cells. *J. Mol. Biol.* **113,** 635–649.

Funes-Cravioto, F., Zapata-Gayon, C., Kolmodin-Hedman, B., Lambert, B., Lindsten, J., Norberg, E., Nordenskjöld, M., Olin, R., & Swensson, Å. (1978) Chromosome aberrations and sister chromatid exchange in laboratory and factory workers and their children. In *Mutagen-induced Chromosome Damage in Man* (H. J. Evans, and D. C. Lloyd, Eds.), University Press, Edinburgh, pp. 275–289.

Galloway, S. M. (1977) Ataxia telangiectasia: the effects of chemical mutagens and X-rays on sister chromatid exchanges in blood lymphocytes. *Mutat. Res.* **45,** 343–349.

Galloway, S. M. & Evans, H. J. (1975) Sister chromatid exchanges in human chromosomes from normal individuals and patients with ataxia telangiectasia. *Cytogenet. Cell Genet.* **15,** 17–29.

Galloway, S. M. & Painter, R. B. (1979) Vitamin C is positive in the DNA synthesis inhibition and sister-chromatid exchange tests. *Mutat. Res.* **60,** 321–327.

Gelbart, S. M. & Sontag, S. J. (1980) Mutagenic urine in cirrhosis. *Lancet* **1,** 894–896.

German, J. (1974a) Bloom's syndrome. II. The prototype of human genetic disorders predisposing to chromosomal instability and cancer. In *Chromosomes and Cancer* (J. German, Ed.), Wiley, New York, pp. 601–617.

German, J., (Ed.) (1974b) *Chromosomes and Cancer,* Wiley, New York.

German, J., Archibald, R., & Bloom, D. (1965) Chromosomal breakage in a rare and probably genetically determined syndrome in man. *Science* **148,** 506–507.

German, J., Schonberg, S., Louie, E., & Chaganti, R. S. K. (1977) Bloom's syndrome. IV. Sister-chromatid exchanges in lymphocytes. *Am. J. Hum. Genet.* **29,** 248–255.

Ghosh, P. K. & Nand, R. (1979) Reduced frequency of sister chromatid exchanges in human lymphocytes cultured with autologous serum. *Hum. Genet.* **51,** 167–170.

Giannelli, F., Benson, P. F., Pawsey, S. A., & Polani, P. E. (1977) Ultraviolet light sensitivity and delayed DNA-chain maturation in Bloom's syndrome fibroblasts. *Nature (Lond.)* **265,** 466–469.

Goedde, H. W., Harada, S., & Agarwal, D. P. (1979) Racial differences in alcohol sensitivity: a new hypothesis. *Hum. Genet.* **51,** 331–334.

Goth-Goldstein, R. (1977) Repair of DNA damaged by alkylating carcinogens is defective in xeroderma pigmentosum-derived fibroblasts. *Nature (Lond.)* **267,** 81–82.

Goyanes-Villascusa, V. J., Ugarte, M., & Vazquez, A. (1977) Sister chromatid exchange in babies treated by phototherapy. *Lancet* **2,** 1084–1085.

Grippenberg, U. (1965) Chromosome studies on some virus infections. *Hereditas* **54,** 1–18.

Gropp, A. & Flatz, G. (1967) Chromosome breakage and blastic transformation of the lymphocytes in ataxia-telangiectasia. *Humangenetik* **5,** 77–79.

Hampar, B. & Ellison, S. A. (1961) Chromosomal aberrations induced by animal virus. *Nature (Lond.)* **192,** 145–147.

Hanawalt, P. C. & Setlow, R. B. (Eds.) (1975) *Molecular Mechanisms for Repair of DNA,* Plenum Press, New York.

Hand, R. (1977) Human DNA replication: fiber autoradiographic analysis of diploid cells from normal adults and from Fanconi's anemia and ataxia telangiectasia. *Hum. Genet.* **37,** 55–64.

Hand, R. & German, J. (1975) A retarded rate of DNA chain growth in Bloom's syndrome. *Proc. Natl. Acad. Sci. U.S.* **72,** 758–762.

Hanson, J., Jones, K., & Smith, D. (1976) Fetal alcohol syndrome: experience with 41 patients. *J. Am. Med. Assn.* **235,** 1458–1460.

Harada, S., Misawa, S., Agarwal, D. P., & Goedde, H. W. (1980) Liver alcohol dehydrogenase and aldehyde dehydrogenase in the Japanese: Iozyme variation and its possible role in alcohol intoxication. *Am. J. Hum. Genet.* **32,** 8–15.

Harley, C. B., Pollard, J. W., Chamberlain, J. W., Stanners, C. P. & Goldstein, S. (1980) Protein synthetic errors do not increase during aging of cultured human fibroblasts. *Proc. Natl. Acad. Sci. U.S.* **77,** 1885–1889.

Harnden, D. G. (1974) Viruses, chromosomes, and tumours; the interaction between viruses and chromosomes. In *Chromosomes and Cancer* (J. German, Ed.), Wiley, New York, pp. 151–190.

Harnden, D. G. & Taylor, A. M. R. (1978) The effects of radiation on the chromosomes of patients susceptible to cancer. In *Mutagen-induced Chromosome Damage in Man* (H. J. Evans and D. C. Lloyd, Eds.), University Press, Edinburgh, pp. 52–61.

Hatcher, N. H., Risemberg, H. M., Powers, M. M., & Hook, E. B. (1979) Sister-chromatid exchange and phototherapy. *Mutat. Res.* **60,** 401–403.

Hayashi, K. & Schmid, W. (1975) The rate of sister chromatid exchanges parallel to spontaneous chromosome breakage in Fanconi's anemia and to trenimon-induced aberrations in human lymphocytes and fibroblasts. *Humangenetik* **29,** 201–206.

Haynes, R. H., Prakash, L., Resnick, M. A., Cox, B. S., Moustacchi, E., & Boyd, J. B. (1978) Workshop summary: DNA repair in lower eucaryotes. In *DNA Repair Mechanisms* (P. C. Hanawalt, E. C. Friedberg, and C. F. Fox, Eds.), Academic Press, New York, pp. 405–411.

Hecht, F. & McCaw, B. K. (1977) Chromosome instability syndromes. In *Genetics of Human Cancer* J. J. Mulvihill, R. W. Miller, and J. F. Fraumeni, Eds.) Raven, New York, pp. 105–123.

Hecht, F., Koler, R. D., Rigas, D. A., Dahnke, G. S., Case, M. P., Tisdale, V., & Miller, R. W. (1966) Leukaemia and lymphocytes in ataxia-telangiectasia. *Lancet* **2,** 1193.

Higurashi, M. & Conen, P. (1973) *In vitro* chromosomal radiosensitivity in "chromosomal breakage syndromes." *Cancer* **32,** 380–383.

Hirsch-Kauffmann, M., Schweiger, M., Wagner, E. F., & Sperling, K. (1978) Deficiency of DNA ligase activity in Fanconi's anemia. *Hum. Genet.* **45,** 25–32.

Hoar, D. I. & Sargent, P. (1976) Chemical mutagen hyper-sensitivity in ataxia telangiectasia. *Nature (Lond.)* **261,** 590–592.

Hollaender, A., Ed. (1976) *Chemical Mutagens: Principles and Methods for their Detection,* Vol. 4, Plenum Press, New York.

Hollander, D. H., Tockman, M. S., Liang, Y. W., Borgaonkar, D. S., & Frost, J. K. (1978) Sister chromatid exchanges in the peripheral blood of cigarette smokers and in lung cancer patients; and the effect of chemotherapy. *Hum. Genet.* **44,** 165–171.

Holmberg, M. (1974) No interaction between ultraviolet and X irradiation on chromosome aberrations in cells with trisomy 21. *Nature (Lond.)* **249,** 448–449.

Hopkin, J. M. & Evans, H. J. (1979) Cigarette smoke condensates damage DNA in human lymphocytes. *Nature (Lond.)* **279,** 241–242.

Hopkin, J. M. & Evans, H. J. (1980) Cigarette smoke induced DNA damage and lung cancer risks. *Nature (Lond.)* **283,** 388–390.

Hopkin, J. M. & Perry, P. E. (1980) Benzo (*a*) pyrene does not contribute to the SCEs induced by cigarette smoke condensate. *Mutat. Res.* **77,** 377–381.

Inoue, T., Hirano, K., Yokoiyama, A., Kada, T., & Kato, H. (1977) Dna repair enzymes in ataxia telangiectasia and Bloom's syndrome fibroblasts. *Biochim. Biophys. Acta* **479,** 497–500.

Jackson, E. W., Turner, J. H., Klauber, M. R., & Norris, F. D. (1968) Down's syndrome: variation of leukaemia occurence in institutionalised populations. *J. Chron. Dis.* **21,** 247–253.

Kaina, B., Waller, H., Waller, M., & Rieger, R. (1977) The action of *N*-methyl-*N*-nitrosourea on non-established human cell lines *in vitro.* I. Cell cycle inhibition and aberration induction in diploid and Down's fibroblast. *Mutat. Res.* **43,** 387–400.

Kaplan, J. C., Zamansky, G. B., Black, P. H., & Latt, S. A. (1978) Parallel induction of sister chromatid exchanges and infectious virus from SV40-transformed cells by alkylating agents. *Nature (Lond.)* **271,** 662–663.

Kato, H. & Sandberg, A. A. (1977) Effects of herpes simplex virus on sister chromatid exchange and chromosome abnormalities in human diploid fibroblasts. *Exp. Cell Res.* **109,** 423–427.

Kato, H. & Stich, H. F. (1976) Sister chromatid exchanges in ageing and repair-deficient human fibroblasts. *Nature (Lond.)* 260, 447–448.

Knight, L. A., Gardner, H. A., & Gallie, B. L. (1979) Absence of chromosome breakage in patients with retinoblastoma. *Hum. Genet.* **51,** 73–78.

Knudson, A. G., Meadows, A. T., Nichols, W. W., & Hill, R. (1976) Chromosomal deletion and retinoblastoma. *N. Engl. J. Med.* **295,** 1120–1123.

Knuutila, S., Helminen, E., Vuopio, P., & de la Chapelle, A. (1978a) Sister chromatid exchanges in human bone marrow cells. I. Control subjects and patients with leukaemia. *Hereditas* **88,** 189–196.

Knuutila, S., Helminen, E., Vuopio, P., & de la Chapelle, A. (1978b) Increased sister chromosome exchange in megaloblastic anaemia: studies on bone marrow cells and lymphocytes. *Hereditas* **89,** 175–181.

Knuutila, S., Maki-Paakkanen, J., Kahkonen, M., & Hokkanen, E. (1978c) An increased frequency of chromosomal changes and SCE's in cultured blood lymphocytes from 12 subjects vaccinated against smallpox. *Hum. Genet.* **41,** 89–96.

Matsumoto, T., Yoshida, D., Mizusaki, S., & Okamoto, H. (1978) Mutagenicities of the pyrolyzates of peptides and proteins. *Mutat. Res.* **56,** 281–288.

Meretoja, T., Vainio, H., Sorsa, M., & Härkönen, H. (1977) Occupational styrene exposure and chromosomal aberrations. *Mutat. Res.* **56,** 193–197.

Mizusaki, S., Takashima, T., & Tomaru, K. (1977) Factors affecting mutagenic activity of cigarette smoke condensate in *Salmonella typhimurium* TA-1358. *Mutat. Res.* **48,** 29–36.

Monticone, R. E. & Schneider, E. L. (1979) Induction of sister chromatid exchanges in human cells by fluorescent light. *Mutat. Res.* **59,** 215–221.

Mortelmans, K., Friedberg, E. C., Slor, H., Thomas, G. & Cleaver, J. E. (1976) Defective thymine dimer excision by cell-free extracts of xeroderma pigmentosum cells. *Proc. Natl. Acad. Sci. U.S.* **73,** 2757–2761.

Moshell, A. N., Barret, S. F., Tarone, R. E., & Robbins, J. H. (1980) Radiosensitivity in Huntington's disease: implications for pathogenesis and presymptomatic diagnosis. *Lancet* **1,** 9–11.

Mulvihill, J. J., Miller, R. W., & Fraumeni, J. F., (Eds.) (1977) *Genetics of Human Cancer,* Raven Press, New York.

Musilova, J., Michalova, K. & Urban, J. (1979) Sister-chromatid exchanges and chromosomal breakage in patients treated with cystostatics. *Mutat. Res.* **67,** 289–294.

Nagao, M., Sugimura, T., & Matsushima, T. (1978) Environmental mutagens and carcinogens. *Annu. Rev. Genet.* **12,** 117–159.

Nakanishi, Y. Schneider, E. L. (1979) *In vivo* sister-chromatid exchange: a sensitive measure of DNA damage. *Mutat. Res.* **60,** 329–337.

Natarajan, A. T., Van Buul, P. P. W., & Raposa, T. (1978) An evaluation of the use of peripheral blood lymphocyte systems for assessing cytological effects induced *in vivo* by chemical mutagens. In *Mutagen-induced Chromosome Damage in Man* (H. J. Evans and D. C. Lloyd, Eds.) University Press, Edinburgh, pp. 268–274.

Nevstad, N. P. (1978) Sister chromatid exchanges and chromosomal aberrations induced in human lymphocytes by the cytostatic drug adriamycin *in vivo* and *in vitro*. *Mutat. Res.* **57,** 253–258.

Nichols, W. W. (1974) Viruses and chromosomes. In *The Cell Nucleus* (H. Busch, Ed.), Vol. 2 Academic Press, New York, pp. 437–458.

Nichols, W. W., Levan, A., Hall, B., & Ostergren, G. (1962) Measles-associated chromosome breakage. Preliminary communication. *Hereditas* **48,** 367–370.

Nichols, W. W., Levan, A., Aula, P., & Norrby, E. (1965a) Chromosome damage associated with measles virus *in vitro*. *Hereditas* **54,** 101–118.

Nichols, W. W., Levan, A., Heneen, W. K., & Peluse, M. (1965b) Synergism of the Schmidt-Ruppin of the strain of the Rous sarcoma virus and cytidine triphosphate in the induction of chromosome breaks in human cultured leucocytes. *Hereditas* **54,** 213–236.

Nichols, W. W., Bradt, C. I., Toji, L.H., Godley, M. & Segawa, M. (1978) Induction of sister chromatid exchanges by transformation with Simian virus 40. *Cancer Res.* **38,** 960–964.

Nordenson, I., Salmonsson, S., Brun, E., & Beckman, G. (1979) Chromosome aberrations in psoriatic patients treated with arsenic. *Hum. Genet.* **48,** 1–6.

Novotna, B., Goetz, P., & Surkova, N. I. (1979) Effects of alkylating agents on lymphocytes from controls and from patients with Fanconi's anemia. *Hum. Genet.* **49,** 41–50.

Obe, G. & Herha, J. (1975) Chromosomal damage in chronic alcohol users. *Humangenetik* **29,** 191–200.

Obe, G. & Herha, J. (1978) Chromosomal aberrations in heavy smokers. *Hum. Genet.* **41,** 259–263.

Obe, G. & Ristow, H. (1979) Mutagenic, cancerogenic and teratogenic effects of alcohol. *Mutat. Res.* **65,** 229–259.

O'Brien, R. L., Poon, P., Kline, E., & Parker, J.W. (1971) Susceptibility of chromosomes from patients with Down's syndrome to 7,12-dimethylbenz(*a*)anthracene-induced aberrations *in vitro*. *Int. J. Cancer* **8,** 202–210.

Ockey, C. H. (1979) Quantitative replicon analysis of DNA synthesis in cancer-prone conditions and the defects in Bloom's syndrome. *J. Cell Sci.* **40,** 125–144.

Orgel, L. E. (1963) The maintenance of the accuracy of protein synthesis and its relevance to ageing. *Proc. Natl. Acad. Sci. U.S.* **49,** 517–521.

Oxford, J. M., Harnden, D. G., Parrington, J. M., & Delhanty, J. D. A. (1975) Specific chromosome aberrations in ataxia telangiectasia. *J. Med. Genet.* **12,** 251–262.

Parker, V. P. & Lieberman, M. W. (1977) Levels of DNA polymerases α, β and γ in control and repair-deficient human diploid fibroblasts. *Nucleic Acids Res.* **4,** 2029–2037.

Parrington, J. M., Delhanty, J. D. A., & Baden, H. P. (1971) Unscheduled DNA synthesis, u.v.-induced chromosome aberrations and SV_{40} transformation in cultured cells from xeroderma pigmentosum. *Ann. Hum. Genet. Lond.* **35,** 149–160.

Parrish, J. A., Fitzpatrick, T. B., Tanenbaum, L., & Pathak, M. A. (1974) Photochemotherapy of psoriasis with oral methoxsalen and longwave ultraviolet light. *N. Engl. J. Med.* **291,** 1207–1211.

Paterson, M. C. & Smith, P. J. (1979) Ataxia telangiectasia: an inherited human disorder involving hypersensitivity to ionising radiation and related DNA-damaging chemicals. *Annu. Rev. Genet.* **13,** 291–318.

Paterson, M. C., Smith, B. P., Lohman, P. H. M., Anderson, A. K., & Fishman, L. (1976) Defective excision repair of γ-ray-damaged DNA in human (ataxia telangiectasia) fibroblasts. *Nature (Lond.)* **260,** 444–447.

Paterson, M. C., Anderson, A. K., Smith, B. P., & Smith, P. J. (1979) Enhanced radiosensitivity of cultured fibroblasts from ataxia telangiectasia heterozygotes manifested by defective colony-forming ability and reduced DNA repair replication after hypoxic γ-irradiation. *Cancer Res.* **39,** 3725–3734.

Pedrini, A. M., Dalpra, L., Ciarrocchi, G., Pedrali Noy, G. C. F., Spadari, S., Nuzzo, F., & Falaschi, A. (1974) Levels of some enzymes acting on DNA in xeroderma pigmentosum. *Nucleic Acids Res.* **1,** 193–202.

Perry, P. E. (1980) Chemical mutagens and sister-chromatid exchange. In *Chemical Mutagens: Principles and Methods for their Detection,* Vol. 6 (F. J. de Serres and A. Hollaender, Eds.) Plenum, New York, pp. 1–39.

Perry, P. & Evans, H. J. (1975) Cytological detection of mutagen-carcinogen exposure by sister chromatid exchange. *Nature (Lond.)* **258,** 121–125.

Perry, P. E. & Searle, C. E. (1977) Induction of sister chromatid exchanges in Chinese hamsters cells by the hair dye constituents 2-nitro-*p*-phenylenediamine and 4-nitro-*o*-phenylenediamine. *Mutat. Res.* **56,** 207–210.

Perry, P. E., Jager, M., & Evans, H. J. (1978) Mutagen-induced sister chromatid exchanges in xeroderma pigmentosum and normal lymphocytes. In *Mutation-induced Chromosome Damage in Man* (H. J. Evans and D. C. Lloyd, Eds.), University Press, Edinburgh, pp. 201–207.

Picciano, D. (1979) Cytogenetic investigation of occupational exposure to epichlorohydrin. *Mutat. Res.* **66,** 169–173.

Poon, P. K., O'Brien, R. L., & Parker, J. W. (1974) Defective DNA repair in Fanconi's anaemia. *Nature (Lond.)* **250,** 223–225.

Purchase, I. F. H., Richardson, C. R., & Anderson, D. (1975) Chromosomal and dominant lethal effects of vinyl chloride. *Lancet* **2,** 410–411.

Raposa, T. (1978) Sister chromatid exchange studies for monitoring DNA damage and repair capacity after cytostatic therapy. *Mutat. Res.* **57,** 241–251.

Regan, J. D. & Setlow, R. B. (1974a) Two forms of repair in the DNA of human cells damaged by chemical carcinogens and mutagens. *Cancer Res.* **34,** 3318–3325.

Regan, J. D. & Setlow, R. B. (1974b) DNA repair in human progeroid cells. *Biochem. Biophys. Res. Commun.* **59,** 858–864.

Regan, J. D., Setlow, R. B., Carrier, W. I. & Lee, W. H. (1973) Molecular events following the ultra-violet irradiation of human cells from ultraviolet-sensitive individuals. In *Advances in Radiation Research.* Vol. 1, *Biology and Medicine* J. E. Duplan and A. I. Chapiro, Eds.), Gordon and Breach, New York, pp. 119–126.

Remsen, J. F. & Cerutti, P. A. (1976) Deficiency of gamma-ray excision repair in skin fibroblasts from patients with Fanconi's anemia. *Proc. Natl. Acad. Sci. U.S.* **73,** 2419–2423.

Riccardi, V. M., Sujansky, E., Smith, A. C., & Francke, U. (1978) Chromosomal imbalance in the aniridia-Wilms' tumor association: 11p interstitial deletion. *Pediatrics* **61,** 604–610.

Rieger, R. & Michaelis, A. (1970) Intrachromosomal clustering of chromatid aberrations induced by maleic hydrazyde and ethyl alcohol: studies with a new karyotype of *Vicia faba. Mutat. Res.* **10,** 162–164.

Ristow, H. & Obe, G. (1978) Acetaldehyde induces cross-links in DNA and causes sister-chromatid exchanges in human cells. *Mutat. Res.* **58,** 115–119.

Sagerman, R. H., Cassady, J. R., Tretter, P., & Ellsworth, R. M. (1969) Radiation-induced neoplasia following external beam therapy for children with retinoblastoma. *Am. J. Roentgenol.* **105,** 529–535.

San, R. H. C., Stitch, W., & Stich, H. F. (1977) Differential sensitivity of *xeroderma pigmentosum* cells of different repair capacities toward the chromosome breaking action of carcinogens and mutagens. *Int. J. Cancer* **20,** 181–187.

Santesson, B., Lindahl-Kiessling, K., & Mattsson, A. (1979) SCE in B and T lymphocytes. Possible implications for Bloom's syndrome. *Clin. Genet* **16,** 133–135.

Sasaki, M. S. (1973) DNA repair capacity and susceptibility to chromosome breakage in xeroderma pigmentosum cells. *Mutat. Res.* **20,** 291–293.

Sasaki, M. S. (1975) Is Fanconi's anaemia defective in a process essential to the repair of DNA cross links? *Nature (Lond.)* **257,** 501–503.

Sasaki, M. S. (1978) Faconi's anemia: a condition possibly associated with a defective DNA repair. In *DNA Repair Mechanisms*)P. C. Hanawalt, E. C.Friedberg, & C. F. Fox, Eds.), Academic Press, New York, pp. 675–684.

Sasaki, M. S. & Tonomura, A. (1973) A high susceptibility of Fanconi's anemia to chromosome breakage by DNA cross-linking agents. *Cancer Res.* **33,** 1829–1836.

Sasaki, M. S., Tonomura, A., & Matsubara, S. (1970) Chromosome constitution and its bearing on the chromosomal radiosensitivity in man. *Mutat. Res.* **10,** 617–633.

Schempp, W. & Krone, W. (1979) Deficiency of arginine and lysine causes increase in the frequency of sister chromatid exchanges. *Hum. Genet.* **51,** 315–318.

Schmickel, R. D., Chu, E. H. Y., Trusko, J. E., & Chang, C. C. (1977) Cockayne syndrome: a cellular sensitivity to ultraviolet light. *Pediatrics* **60,** 135–139.

Schonberg, S. & German, J. (1980) Sister chromatid exchange in cells metabolically coupled to Bloom's syndrome cells. *Nature (Lond.)* **284,** 72–74.

Schroeder, T. M. & German, J. (1974) Bloom's syndrome and Fanconi's anemia: demonstration of two distinctive patterns of chromosome disruption and rearrangement. *Humangenetik* **25,** 299–306.

Schroeder, T. M. & Stahl-Mauge, C. (1979) Mutagenic effects of isonicotinic acid hydracide in Fanconi's anemia. *Hum. Genet.* **52,** 309–321.

Schroeder, T. M., Anschütz, F., & Knopp, A. (1964) Spontane Chromosomenaberrationen bei familiärer Panmyelopathie. *Humangenetik* **1,** 194–196.

Schuler, D., Hervei, S., Gacs, G., Kirchner, M., & Szathmary, J. (1968) The influence of vaccinia viruses on human chromosomes in vivo and in vitro. *Humangenetik* **6,** 55–60.

Schwartz, A. L., Cole, F. S., Fiedorek, F., Matthews, D., Frantz, I. D., Raika, I., & Latt, S. A. (1978) Phototherapy does not increase the sister chromatid exchange frequency in premature infants. *Paediatric Res.* **12,** 534 (abstract).

Scott, D. & Marshall, R. R. (1978) An investigation of UV-induced chromosome changes in relation to lethality in normal and UV-sensitive human cells. In *Mutagen-induced Chromosome Damage in Man* H. J. Evans and D. C. Lloyd, Eds.), University Press, Edinburgh, pp. 129–141.

Scudiero, D. A. (1978) Repair deficiency in n-methyl-n′-nitro-n-nitrosoguanidine treated ataxia telangiectasia fibroblasts. In *DNA Repair Mechanisms* (P. C. Hanawalt, E. C. Friedberg and C. F. Fox, Eds.), Academic Press, New York, pp. 655–658.

Setlow, R. B., Regan, J. D., German, J. & Carrier, W. L. (1969) Evidence that xeroderma pigmentosum cells do not perform the first step in the repair of ultraviolet damage to their DNA. *Proc. Natl. Acad. Sci. USA* **64,** 1035–1041.

Sheridan, R. B. & Huang, P. C. (1979) Endonucleotytic activity for γ–irradiated DNA in normal and ataxia telangiectasia fibroblast cell extracts. *Mutat. Res.* **63,** 387–392.

Shiraishi, Y. & Sandberg, A. A. (1978) Effects of mitomycin C on sister chromatid exchange in normal and Bloom's syndrome cells. *Mutat. Res.* **49,** 233–238.

Shiraishi, Y. & Sandberg, A. A. (1980) Sister chromatid exchange in human chromosomes, including observations in neoplasia. *Cancer Genet. Cytogenet.* **1,** 363–380.

Shiraishi, Y., Freeman, A. I., & Sandberg, A. A. (1976) Increased sister chromatid exchange in bone marrow and blood cells from Bloom's syndrome. *Cytogenet. Cell Genet.,* **17,** 162–173.

Smith, P. J. & Paterson, M. C. (1979) Sensitivity to near UV light in cultured fibroblasts from various human syndromes. *Proc. Am. Assn. Cancer Res.* **20,** 88.

Sperling, K., Wegner, R.-D., Riehm, H., & Obe, G. (1975) Frequency and distribution of sister chromatid exchanges in a case of Fanconi's anemia. *Humangenetik* **27,** 227–230.

Sperling, K., Goll, U., Kunze, J. Lüdtke, E.-K., Tolksdorf, M., & Obe, G. (1976) Cytogenetic investigations in a new case of Bloom's syndrome. *Hum. Genet.* **31,** 47–52.

Sram, R. J., Zudova, Z., & Kuleshov, M. P. (1980) Cytogenetic analysis of peripheral lymphocytes in workers occupationally exposed to epichlorohydrin. *Mutat. Res.* **70,** 115–120.

Stitch, H. F. & San, R. H. C. (1971) Reduced DNA repair synthesis in xeroderma pigmentosum cells exposed to the oncogenic 4-nitroquinoline l-oxide and 4-hydroxyaminoquinoline l-oxide. *Mutat. Res.* **13,** 279–282.

Stitch, H. F. & San, R. H. C. (1973) DNA repair synthesis and survival of repair deficient human cells exposed to the K-region epoxide of benz(a)anthracene. *Proc. Soc. Exp. Biol. Med.* **142,** 155–188.

Stitch, H. J., San, R. H. C., & Kawazoe, Y. (1973 Increased sensitivity of xeroderma pigmentosum cells to some chemical carcinogens and mutagens. *Mutat. Res.* **17,** 127–137.

Sutherland, B. M. (1978) Enzymatic photoreactivation of DNA. In *DNA Repair Mechanisms* (P. C. Hanawalt, E. C. Friedberg, and C. F. Fox, Eds.), Academic Press, New York, pp. 113–122.

Sutherland, B. M. & Oliver, R. (1975) Low levels of photoreactivating enzyme in xeroderma pigmentosum variants. *Nature (Lond.)* **257,** 132–134.

Takehisa, S. (1981) Induction of sister chromatid exchanges by chemical agents. In *Sister Chromatid Exchange* (S. Wolff, Ed.), Wiley-Interscience, New York, pp. 87–147.

Takehisa, S. & Wolff, S. (1977) Induction of sister chromatid exchanges in Chinese hamster cells by carcinogenic mutagens requiring metabolic activation. *Mutat. Res.* **45,** 263–270.

Tanaka, K., Sekiguchi, M., & Okada, Y. (1975) Restoration of ultraviolet-induced unscheduled DNA synthesis of xeroderma pigmentosum cells by the concomitant treatment with bacteriophage T4 endonuclease V and HVJ (Sendai virus). *Proc. Natl. Acad. Sci. U.S.* **72,** 4071–4075.

Tanaka, K., Hayakawa, H., Sekiguchi, M., & Okada, Y. (1977) Specific action of T4 endonuclease V on damaged DNA in xeroderma pigmentosum cells *in vivo*. *Proc. Natl. Acad. Sci. U.S.* **74,** 2958–2962.

Taylor, A. M. R. (1978) Unrepaired DNA strand breaks in irradiated ataxia telangiectasia lymphocytes suggested from cytogenetic observations. *Mutat. Res.* **40,** 407–418.

Taylor, A. M. R., Harnden, D. G., Arlett, C. F., Harcourt, S. A., Lehmann, A. R., Stevens, S., & Bridges, B. A. (1975) Ataxia telangiectasia: a human mutation with abnormal radiation sensitivity. *Nature (Lond.)* **258,** 427–429.

Taylor, A. M. R., Metcalfe, J. A., Oxford, J. M., & Harnden, D. G. (1976) Is chromatid-type damage in ataxia telangiectasia after irradiation at G_0 a consequence of defective repair? *Nature (Lond.)* **260,** 441–443.

Teebor, G. W. & Duker, N. J. (1975) Human endonuclease activity for DNA apurinic sites. *Nature (Lond.)* **258,** 544–547.

Tezuka, H., Ando, N., Suzuki, R., Terahata, M., Moriya, M., & Shirasu, Y. (1980) Sister-chromatid exchanges and chromosomal aberrations in cultured Chinese hamster cells treated with pesticides positive in microbial reversion assay. *Mutat. Res.* **78,** 177–191.

Tice, R., Windler, G., & Rary, J. M. (1978a) Effect of cocultivation on sister chromatid exchange frequencies in Bloom's syndrome and normal fibroblast cells. *Nature (Lond.)* **273,** 538–540.

Tice, R. R., Rary, J. M., & Bender, M. A. (1978b) An investigation of DNA repair potential in Bloom's syndrome. In *DNA Repair Mechanisms* (P. C. Hanawalt, E. C. Friedberg, and C. F. Fox, Eds.), Academic Press, New York, pp. 659–662.

Todaro, G. J. & Martin, G. M. (1967) Increased susceptibility of Down's syndrome fibroblasts to transformation by SV. *Proc. Soc. Exp. Biol. Med.* **124,** 1232–1236.

Tonomura, A. & Sasaki, M. S. (1973) Chromosome aberrations and DNA repair synthesis in cultured human cells exposed to nitrofurans. *Jap. J. Genet.* **48,** 291–294.

Tyrell, R. M., Ley, R. D., & Webb, R. B. (1974) Induction of single-strand breaks (alkali-labile bonds) in bacterial and phage DNA by near UV (365 nm) radiation. *Photochem. Photobiol.* **20,** 395–398.

van Buul, P. P. W., Natarajan, A. T., & Verdegaal-Immerzeel, E. A. M. (1978) Suppression of the frequencies of sister chromatid exchanges in Bloom's syndrome fibroblasts by co-cultivation with Chinese hamster cells. *Hum. Genet.* **44,** 187–189.

Vijayalaxmi, Thomson, E., & Evans, H. J. (1979) Induction of sister-chromatid exchanges in human lymphocytes and Chinese hamster cells exposed to aflatoxin B_1 and *N*-methyl-*N*-nitrosourea. *Mutat. Res.* **67,** 47–53.

Vincent, R. A., Sheridan, R. B., & Huang, P. C. (1975) DNA strand breakage repair in ataxia telangiectasia fibroblast-like cells. *Mutat. Res.,* **33,** 357–366.

Wade, M. H. & Chu, E. H. Y. (1978) Effects of DNA damaging agents on cultured fibroblasts derived from patients with Cockayne syndrome. In *DNA Repair Mechanisms* (P. C. Hanawalt, E. C. Friedberg, and C. F. Fox, Eds.), Academic Press, New York, pp. 667–670.

Watts, P. L., Hall, A. J., & Fleming, J. E. E. (1972) Ultrasound and chromosome damage. *Br. J. Radiol.* **45,** 335–339.

Weichselbaum, R. R., Nove, J., & Little, J. B. (1977) Skin fibroblasts from a D-deletion type retinoblastoma patient are abnormally X-ray sensitive. *Nature (Lond.)* **266,** 726–727.

Weichselbaum, R. R., Nove, J., & Little, J. B. (1978a) Deficient recovery from potentially lethal radiation damage in ataxia telangiectasia and xeroderma pigmentosum. *Nature (Lond.)* **271,** 261–262.

Weichselbaum, R. R., Nove, J., & Little, J. B. (1978b) X-ray sensitivity of diploid fibroblasts from patients with hereditary or sporadic retinoblastoma. *Proc. Natl. Acad. Sci. U.S.* **75,** 3962–3964.

White, A. D. (1980) *In vitro* induction of SCE in human lymphocytes by epichlorohydrin with and without metabolic activation. *Mutat. Res.* **78,** 171–176.

Wolff, S. (1980) Chromosome aberrations, sister chromatid exchanges, and the lesions that cause them. In *Sister Chromatid Exchange* (S. Wolff, Ed.), Wiley-Interscience, New York, pp. 41–57.

Wolff, S. & Perry, P. (1974) Differential Giemsa staining of sister chromatids and the study of sister chromatid exchanges without autoradiography. *Chromosoma (Berl.)* **48,** 341–353.

Wolff, S. & Rodin, B. (1978) Saccharin-induced sister chromatid exchanges in Chinese hamster and human cells. *Science* **200,** 543–545.

Wolff, S., Bodycote, J., Thomas, G. H., & Cleaver, J. E. (1975) Sister chromatid exchange in xeroderma pigmentosum cells that are defective in DNA excision repair or post-replication repair. *Genetics* **81,** 349–355.

Wolff, S., Rodin, B., & Cleaver, J. E. (1977) Sister chromatid exchanges induced by mutagenic carcinogens in normal and xeroderma pigmentosum cells. *Nature (Lond.)* **265,** 347–349.

Yamasaki, E. & Ames, B. N. (1977) Concentration of mutagens from urine by adsorption with the nonpolar resin XAD-2: cigarette smokers have mutagenic urine. *Proc. Natl. Acad. Sci. U.S.* **74,** 3555–3559.

Zamanski, G. G., Latt, S. A., Kaplan, J. C., Kleinman, L. F., Dougherty, C., & Black, P. H. (1980) The co-induction of sister chromatid exchanges and virus synthesis in mammalian cells. *Exp. Cell. Res.* **126,** 473–477.

Chapter 8

In Vivo Methods for Detecting Sister Chromatid Exchange

E. L. SCHNEIDER
NATIONAL INSTITUTE ON AGING
NATIONAL INSTITUTES OF HEALTH
BETHESDA, MARYLAND

D. KRAM
EPA
TS-792
WASHINGTON, D.C.

I. Introduction 230
II. Methodology 231
A. Rodents 231
1. Multiple injections 231
2. Continuous infusion of BrdUrd 232
3. Tablet implantation 233
4. Delayed release charcoal–BrdUrd 233
5. *In vivo/in vitro* combinations 233
B. *In ovo* BrdUrd differential chromatid labeling of chicken embryos 234
C. BrdUrd differential chromatid labeling of fish chromosomes 234
D. BrdUrd differential chromatid labeling of *Drosophila* 234
E. BrdUrd differential chromatid labeling of plant chromosomes 234
III. Comparisons between SCE analysis and other *in vivo* systems 235
IV. Applications 235
A. Screening for mutagens and carcinogens 235
1. Differential routes of exposure 236
2. *In utero* SCE analyses 236
3. Different tissue response 237
B. The spontaneous nature of sister chromatid exchange 237

C. Examination of the mechanisms of meiosis 238
D. Investigations of cell cycle kinetics 238
E. Tumor cell studies 239
F. Studies of aging cell populations 239
V. Concluding comments 239

I. INTRODUCTION

In vivo analyses of sister chromatid exchanges (SCEs) offer many advantages and relatively few disadvantages as compared to *in vitro* SCE determinations. While *in vitro* SCE analyses may be simpler and slightly more rapid, they have several potential problems. First, most cells are grown in medium containing serum and therefore, in undefined medium. It has been demonstrated that serum batch can affect baseline SCE levels (Kato and Sandberg 1977). In addition, media components such as riboflavin, phenol red, and tetracycline may be photoactivated by fluorescent lights to form reactive intermediates that induce SCE (Monticone and Schneider, 1979). Finally, it is very difficult to conduct experiments *in vitro* in total darkness and Ikushima and Wolff (1974) have demonstrated that light is an effective inducer of SCE in bromodeoxyuridine-(BrdUrd) substituted chromosomes.

In vivo SCE analysis provides an environment that stimulates human *in vivo* exposures. A variety of somatic tissues can be examined *in vivo*, including cells that are normally quiescent (liver cells) and germ cell precursors. Furthermore, varied routes of test chemical exposure can be examined. Thus environmental exposure to a mutagen can be more closely approximated by *in vivo* SCE analyses.

Another important aspect of *in vivo* SCE analyses is that the activation and/or inactivation of mutagens and carcinogens can be examined more readily than *in vitro* analyses. By analyzing cells from different tissues one can investigate differential activation of mutagens and thus examine the mechanisms for tissue-specific development of tumors.

Finally, *in vivo* SCE analysis may provide a particularly useful approach to questions concerning, for example, the spontaneity of SCE. In addition, the effects of aging on SCE induction can be readily examined in a setting that in many ways approximates normal human aging.

Initial *in vivo* investigations of SCE were performed by Bloom and Hsu (1975) in chick embryos. Other laboratories rapidly followed up this application and demonstrated SCE *in vivo* in rodent cell populations (Allen and Latt, 1976; Vogel and Bauknecht, 1976; Schneider et al., 1976). *In vivo*

SCE analyses have since been performed in a variety of other mammals, as well as insects, fish, and plants.

In this chapter methodologies for *in vivo* SCE analyses are discussed, as are some of their applications. In addition, the sensitivities of *in vivo* SCE analyses are compared with other systems with respect to mutagen/carcinogen detection.

II. METHODOLOGY

A. *Rodents*

Several important cellular parameters must be determined prior to examining mammalian cell populations *in vivo*. First, one must have some knowledge of the cell replication rates of the tissues to be examined in order to obtain significant numbers of second replication cycle cells for examination. While bone marrow cells may have replication rates approaching 8 hours (Schneider et al., 1977), the replication rates for spermatogonia are substantially longer (Allen and Latt, 1976a,b). Second, sufficient BrdUrd must reach the target tissue to permit differential staining of second replication cycle cells. Third, these concentrations of BrdUrd must not severely inhibit cellular replication, since this could lead to insufficient numbers of second replication cycle metaphase cells for SCE analysis. The replication rates of certain cell populations, such as fetal cells, may be quite sensitive to BrdUrd concentrations (Kram et al., 1979a). Finally, one has the choice of exposing cells to either one or two cycles of BrdUrd. Both approaches have been successfully employed *in vivo* (Allen et al., 1977; Schneider et al., 1976).

1. Multiple Injections

Two laboratories independently developed *in vivo* labeling of mammalian chromosomes by hourly intraperitoneal injections of BrdUrd, followed by a period of observation, and then a final injection of colchicine 3–5 hours prior to the sacrifice of the animal (Allen and Latt, 1976a,b; Vogel and Bauknecht, 1976). The total numbers of injections ranged from 6 to 14 and the animals were simultaneously treated with either fluorodeoxyuridine or deoxycytidine to reduce the toxicity of the BrdUrd. These techniques led to successful differential labeling of mouse bone marrow cells and spermatogonia (Allen and Latt, 1976a,b; Vogel and Bauknecht, 1976). The difficulties with this approach are tediousness and the trauma of multiple injections of the same animal.

2. *Continuous Infusion of BrdUrd*

Our laboratory examined continuous infusion of BrdUrd at the same time as we attempted multiple injections. It was clear that, in our hands, continuous infusion of BrdUrd produced more consistent results. For the restraint of the animals, we modified Bollman restrainers to permit access to food and water while the animals remained relatively immobile. After intravenous infusion was commenced. BrdUrd, the drug to be tested, and the Colcemid could all be administered without additional traumatization. Twenty-three and 27 gauge needles were used for intravenous infusion of rats and mice, respectively. A Watson-Marlow 10 channel constant speed infusion pump was utilized to deliver exact amounts of BrdUrd solution to the animals. By altering the autoanalyzer tubing that is used in these pumps, we could obtain flow rates between 2 and 50 ml/day. We found that doses of BrdUrd between 10 and 50 mg/kg wt/hour produced consistent results for sister chromatid exchange analysis in both rodent species.

The advantages of this technique are that the repeated trauma of multiple injections is avoided and the delivered doses of BrdUrd are individually tailored for each animal. The reproducibility of this approach is shown in Figure 8.1, where a typical dose response to mitomycin C in mouse bone marrow cells is presented. Regression coefficients obtained by the *in vivo* infusion of BrdUrd surpassed those obtained *in vitro* and were usually between 0.95 and 0.99.

The most obvious disadvantage of this approach is that animals without tails are difficult to examine with this methodology. However, subcutaneously implanted minipumps might permit *in vivo* labeling of these animals. Another potential problem is that rats require mild anesthesia to permit their placement into restrainers. This problem is not encountered in mice.

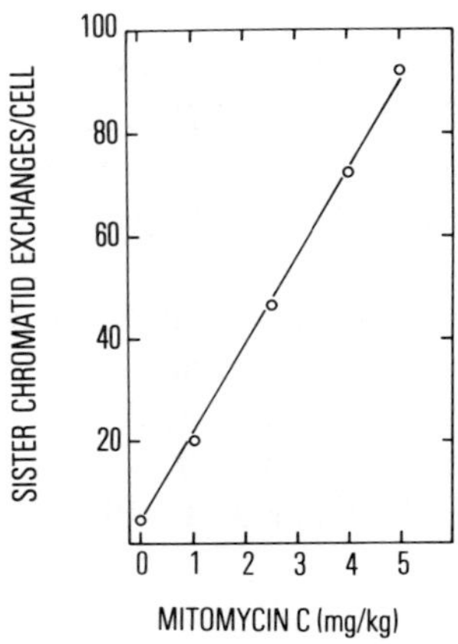

Fig. 8-1. Induction of SCE *in vivo* in mouse bone marrow cells as a function of BrdUrd concentration.

3. *Tablet Implantation*

Another approach to avoiding the multiple traumatization of experimental animals is the implantation of BrdUrd tablets. Allen et al. (1977, 1978) demonstrated that this technique can be performed easily and leads to differential labeling of mammalian chromatids in laboratory mice. BrdUrd tablets are prepared with a simple benchtop press (Parr pellet press equipped with a 0.178 in. diameter punch and die, Parr Instrument Company). Each tablet is composed of 55 mg of pure BrdUrd powder and is protected from light and stored below 0°C until use. The tablets are implanted in the lower, lateral abdominal region of mice that have been anesthetized with either methoxyflurane or ether.

The advantages of this technique are its simplicity and the lack of a requirement for restraining animals. It is thus applicable to a wide variety of laboratory animals. The potential difficulties include the requirements for anesthesia. the necessity for repeated trauma of injections of the test drug and mitotic inhibitor, and possible variations of baseline SCE levels due to the use of fixed BrdUrd doses in animals of different weight.

4. *Delayed Release Charcoal–BrdUrd*

The most recent approach to the *in vivo* BrdUrd labeling of rodent cell populations is the intraperitoneal injection of a slurry of BrdUrd adsorbed to charcoal. Kanda and Kato (1979) were able, with this approach, to obtain differential chromatid labeling of mouse bone marrow metaphase cells. While this technique should theoretically have all the advantages of tablet implantation, it has the disadvantage of causing considerable peritoneal irritation, resulting in peritoneal edema.

5. In vivo/in vitro *Combinations*

Another approach that permits *in vivo* exposure of cell populations to test chemicals involves the administration of these compounds *in vivo* followed by the *in vitro* cultivation of cells derived from these animals (Stetka and Wolff, 1976). This approach has been successfully used in rabbits, where the administration of the compound intraperitoneally was followed by the removal and culture of peripheral lymphocytes. It has the advantage of permitting exposure of cell populations *in vivo* and multiple samplings of the same laboratory animal. The disadvantages of this approach are those related to the artifacts that may be introduced by *in vitro* cell culture.

B. In Ovo *BrdUrd Differential Chromatid Labeling of Chicken Embryos*

The first *in vivo* BrdUrd differential chromatid labeling was accomplished by Bloom and Hsu (1975). These investigators injected BrdUrd into chicken eggs, which were then incubated for 3 days. Colcemid was injected prior to sacrifice. This procedure did not appear to induce significant numbers of morphologically abnormal embryos and produced consistent differential staining at BrdUrd concentrations above 12.5 mg/embryo. In subsequent reports, Bloom has demonstrated the efficacy of this approach for screening mutagens and carcinogens (Bloom, 1978, 1979).

C. BrdUrd Differential Chromatid Labeling of Fish Chromosomes

BrdUrd differential chromatid staining was first accomplished in fish by Kligerman and Bloom (1976) in adult mudminnows. These fish were injected with 0.5 mg of BrdUrd, followed 2–10 days later with an injection of colcemid. Metaphase chromatids with differential labeling were obtained in intestines, kidney, gills, and scales.

D. BrdUrd Differential Chromatid Labeling of Drosophila

Differential chromatid labeling was first achieved in *Drosophila* cells by incubating whole ganglia with surrounding imaginal discs in BrdUrd-containing culture medium for two cycles of DNA replication (Weinberg, 1977).

Recently, direct *in vivo* BrdUrd labeling of *Drosophila* chromosomes was achieved by feeding (See Chapter 10). The abundant genetic information available on *Drosophila* may provide a useful approach to examining the mechanisms of SCE.

E. BrdUrd Differential Chromatid Labeling of Plant Chromosomes

Sister chromatid exchanges were first observed in metaphase cells derived from the root tips of the plant *Vicia faba* (Taylor et al., 1957). The roots were incubated for one cell cycle in the presence of tritiated thymidine and were chased for a second cell cycle in unlabeled medium. Differential labeling of sister chromatids was seen using autoradiography.

BrdUrd differential labeling has since been successfully applied to this same system (Kihlman and Kornberg, 1975; Scheid, 1976). Schvartzman and Cortes (1977) and have used the BrdUrd technology to perform SCE analysis in root meristems of the plant *Allium cepa*. Cell cycle times average approximately 12 hours and labeling simply involves immersing root tips in tap water containing BrdUrd. This system can be easily applied to the screening of compounds that may be mutagenic to plant cells.

III. COMPARISONS BETWEEN SCE ANALYSIS AND OTHER *IN VIVO* SYSTEMS

The first comparison between SCE analysis and classical cytogenetics was performed *in vitro* using transformed Chinese hamster fibroblasts (Perry and Evans, 1975). This study demonstrated that the concentrations of clastogens that caused a doubling in the baseline SCE level were from 10 to 200 times lower than those necessary to induce a statistically significant increase in the frequency of chromosomal abberations.

This relationship was further studied *in vivo* in C57Bl/6J mice using a series of 16 chemicals and X-irradiation (Nakanishi and Schnieder, 1979). All chemicals tested induced many more SCEs than chromosomal aberrations, with the ratio of the two events ranging from 255 to 11. The only exception was X-irradiation, which induced a higher frequency of chromosomal breaks than SCE.

In a similar study the minimum doses of cyclophosphamide required to induce significant elevations of SCE and chromosomal aberration frequencies were compared in Chinese hamster bone marrow cells (Roszinski-Kocher and Rohrborn, 1979). While a dose of 1 mg/kg resulted in a statistically significant increase in SCE, 13.3 mg/kg was required to elevate the frequency of chromosomal aberrations.

The induction of SCE *in vivo* was also compared with the micronucleus assay using six known mutagens (Bauknecht et al., 1977). This quantitative comparison revealed that SCEs were induced at doses one-tenth to one-hundredth of those required for the detection of micronuclei.

In comparing the sensitivities of several *in vivo* techniques, Marquardt and Bayer (1977) observed that the lowest effective dose of EMS required to elevate SCE was 1 mg/kg. This compared with 25 mg/kg for the micronucleus test, 10–20 mg/kg for dominant lethals, 40 mg/kg for heritable translocations, and 50 mg/kg in the spot test.

IV. APPLICATIONS

A. *Screening for Mutagens and Carcinogens*

In vitro SCE have been demonstrated to be sensitive indicators of cellular exposure to mutagens and carcinogens (Perry and Evans, 1975; Abe and Sasaki, 1977; Popescu et al., 1977). A wide variety of carcinogens and mutagens have been tested in various *in vivo* systems (Nakanishi and Schneider, 1979; Bloom, 1978; Bauknecht et al., 1977). The results of these studies indicate that the majority of agents induce SCE in both systems. However, there are some discrepancies between *in vivo* and *in vitro* results. The obvious exceptions are those agents that require metabolic activation and do not produce SCE *in vitro* without the addition of metabolic activators but consistently induce SCE *in vivo*. Cyclophosphamide is a good example of this type of compound. Some agents that induce SCE only *in vitro* may be prohibitively toxic *in vivo*. An example of this type of drug is proflavin, which induced SCE *in vitro* but did not induce significant frequencies of SCE *in vivo* (Nakanishi and Schneider, 1979). However, proflavin did severely inhibit cellular replication kinetics *in vivo*, which was detected with the BrdUrd differential staining technique. Another interesting exception is the hair dye nitrophenylenediamine, NPPD. Although this agent clearly induces SCEs *in vitro* in a variety of cell types, it did not induce SCE *in vivo* (Bynum et al., in preparation). While this may be partially due to the severe inhibition of cellular replication kinetics, there is the intriguing possibility that this agent is rapidly inactivated *in vivo*. This latter possibility is important in evaluating the potential mutagenic risk of a compound.

1. Differential Routes of Exposure

One important aspect of screening for potential mutagens and carcinogens is to simulate as closely as possible human exposure to these agents. The *in vivo* analysis of SCE can be employed with a wide variety of exposure routes. Potential routes of exposure include oral, dermal and inhalation. The latter two routes have been examined and have been shown to be effective in inducing SCE with several compounds. Of particular interest is the report of Tice et al. (1980), who demonstrated that inhalation of benezene by laboratory animals resulted in the induction of SCE. Thus exposure by inhalation of various chemical compounds can be simulated in laboratory animals and the results of potential DNA damage can be obtained rapidly through *in vivo* SCE analyses.

2. In Utero *SCE Analyses*

It is particularly important that *in utero* exposure to carcinogenic and mutagenic compounds be accurately assessed. One approach to this goal is to examine SCE induction *in utero* in an appropriate laboratory animal model. Several investigators have achieved the BrdUrd differential labeling of fetal rodent chromosomes (Kram et al., 1979a; Basler, 1979). *In vivo* SCE analyses can thus be compared after exposure to test compounds in maternal and fetal tissues. Our studies have revealed that certain mutagens and/or carcinogens may have varying abilities to cross the placental barrier. In addition, these studies demonstrated that certain compounds cause differential induction of SCE at various stages of gestation. Thus *in utero* SCE analyses may also provide an indication of the carcinogenic and/or mutagenic potential of compounds at different stages of fetal development.

3. *Different Tissue Response*

It is clear that certain compounds manifest their carcinogenic potential on specific organs or organ systems. Therefore, it is important to identify those tissues that might have a differential susceptibility to these compounds. The advantage of SCE analyses is that a variety of normal cell types can be examined for the effect of diverse classes of mutagens.

B. *The Spontaneous Nature of Sister Chromatid Exchange*

While it is clear that SCEs are induced by a wide variety of agents, there is doubt over whether they occur in nature. It is difficult to examine the spontaneous nature of SCE, since the agents utilized for their detection are known inducers of SCE (Taylor et al., 1957; Kato, 1974). However, one can approach this question by analyzing an SCE dose-response to BrdUrd. The *in vivo* route has several advantages over the *in vitro* examinations of the spontaneous nature of SCE. First, there is no exposure to fluorescent light, which is a known inducer of SCE (Monticone and Schneider, 1979). Second, background levels of SCE are considerably lower *in vivo* than *in vitro* (Kram et al., 1979). Finally, one does not need tissue culture media that may contain compounds that damage DNA.

We have conducted BrdUrd dose-response experiments in both rats (Tice et al., 1976) and mice (Kram et al., 1979b) to examine this question. In both species, SCE were induced by BrdUrd *in vivo* in bone marrow cells. While it was impossible to rule out the possibility that all the observed SCEs were induced by BrdUrd, our data did suggest that a low level of SCE occurs in nature.

To measure the spontaneous level of SCE without the use of BrdUrd or tritiated thymidine, Gatti et al. (1979) looked at the frequency of double dicentric rings in cell lines of *Drosophila* neuroblasts. These rings result from the occurrence of SCE in a stable circular chromosome carried in two separate cell lines. The workers noted that the frequency of these "spontaneous" events was comparable to that seen in BrdUrd-substituted rod chromosomes in one cell line and higher in the second.

C. Examination of the Mechanisms of Meiosis

The BrdUrd differential chromatid staining technique can be applied *in vivo* to examine meiotic, as well as mitotic, cell populations. This was first accomplished by Allen and Latt (1978), who demonstrated BrdUrd differential staining of spermatogonial X and Y chromosomes. With a combination of *in vivo* and *in vitro* BrdUrd labeling Polani et al. (1979) were able to differentially label all the meiotic chromosomes of the female mouse. They demonstrated that crossover exchange of chromatids does indeed occur at chiasmata. This technique further opens the field for the examination of the effect of various agents on meiotic cells.

D. Investigations of Cell Cycle Kinetics

The number of cycles of DNA replication that a cell has undergone in the presence of BrdUrd can be discerned readily by the staining pattern of metaphase chromosomes. This is shown in Figure 8.2. Very little experience is required to distinguish a first replication cycle cell from a second, third, and subsequent cycle cell. Thus cell replication studies can be readily conducted in cell populations exposed to BrdUrd *in vitro* (Tice et al., 1976), as well as *in vivo* (Schneider et al, 1976). While other investigators have examined cell replication kinetics *in vivo* utilizing tritiated

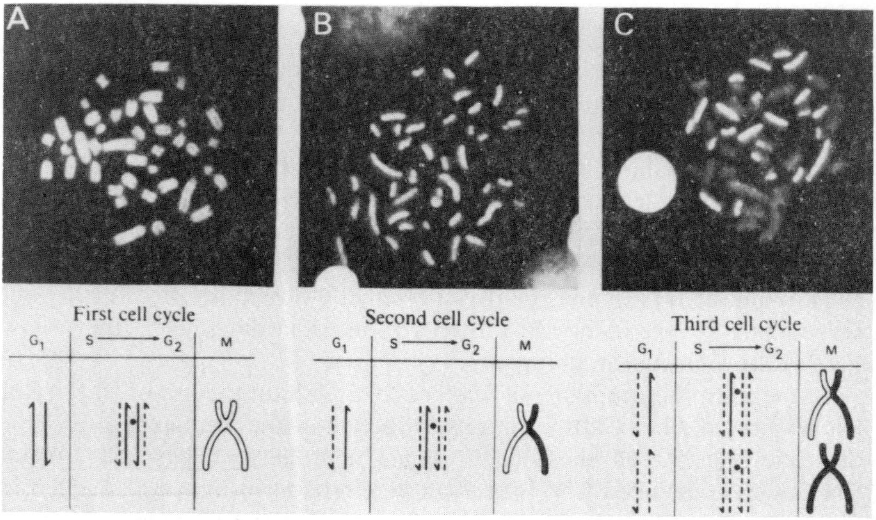

Fig. 8.2. Schema demonstrating that the replication cycle of a cell can be identified by the differential chromatid staining pattern at metaphase.

thymidine labeling, the BrdUrd approach has several advantages. It can discern not only that a cell has replicated, but also the number of previous replications. In addition, we have demonstrated that cell cycle kinetics can be determined using BrdUrd concentrations that do not inhibit cellular replication (Tice et al., 1976; Schneider et al., 1977).

E. *Tumor Cell Studies*

In vivo BrdUrd differential chromatid staining can be applied to examining tumor cell populations in their *in vivo* environment (Schneider et al., 1982). The advantage of this approach is that the effect of cancer chemotherapeutic agents can be examined concurrently in normal and tumor cell populations.

F. *Studies of Aging Cell Populations*

The effect of aging on the response of cells to DNA damage has been the subject of many investigations (Price et al., 1965; Wheeler and Lett, 1975; Mattevi and Salzano, 1975). Most studies have been conducted using the Hayflick model of *in vitro* aging using cultures of human diploid cells. We have examined SCE induction *in vitro* and have found a decline in SCE induction in late passage cells and cells from aged human donors (Schneider and Gilman, 1978; Schneider and Monticone, 1978). In addition, we have employed the *in vivo* BrdUrd techniques to examine aging cell populations in mice and rats. Our results indicate that cell populations *in vivo* also have diminished SCE induction as a foundation of aging (Kram et al., 1978; Nakanishi and Schneider, 1979). Thus the *in vivo* approach clearly confirmed our findings *in vitro*. We also have examined the effect of postembryonic development on SCE frequencies and found that SCE induction increases with development, plateaus during maturity, and diminishes with aging (Nakanishi and Schneider, 1979).

V. CONCLUDING COMMENTS

The application of the BrdUrd differential chromatid staining techniques *in vivo* for the examination of sister chromatid exchanges and cellular replication opens a new realm for the screening of environmental mutagens and carcinogens. It is clear that *in vivo* exposure and *in vivo* analysis more closely approximate human exposure to these agents. The development of several techniques for the administration of BrdUrd makes *in vivo* SCE analysis almost as simple as *in vitro* studies. In addition, since the cell populations are replicating at rapid rates, analysis generally can be completed in 24 hours.

In vivo techniques have already been applied to the examination of several important areas ranging from investigations of the nature of meiosis to the effect of aging on cell populations.

References

Abe, S. & Sasaki, M. (1977) Studies on chromosomal aberrations and sister chromatid exchanges induced by chemicals. *Proc. Jap. Acad.* **53,** 46–49.

Allen, J. W. & Latt, S. A. (1976a) Analysis of sister chromatid exchange formation *in vivo* in mouse spermatogonia as a new test system for environmental mutagens. *Nature (Lond.)* **260,** 449–451.

Allen, J. W. & Latt, S. A. (1976b) *In vivo* BrdU-33258 Hoechst analysis of DNA replication kinetics and sister chromatid exchange formation in mouse somatic and meiotic cell. *Chromosoma (Berl.)* **58,** 325–340.

Allen, J. W., Shuler, C. F., & Latt, S. A. (1978) Bromodeoxyuridine tablet methodology for *in vivo* studies and DNA synthesis. *Somat. Cell Genet.* **4,** 393–405.

Allen, J. W., Shuler, C. F., Mendes, R. W., and Latt, S. A. (1977) A simplified technique for *in vivo* analysis of sister chromatid exchanges using 5-bromodeoxyuridine tablets. *Cytogenet. Cell Genet.* **18,** 231–237.

Basler, A. (1979) Sister chromatid exchanges *in vivo* in Chinese hamster embryonic liver cells exposed transplacentally to BrdU. *Cytogenet. Cell Genet.* **24,** 193–196.

Bauknecht, T., Vogel, W., Bayer, V., and Wild, D. (1977) Comparative *in vivo* mutagenicity testing by SCE and micronucleus induction in mouse bone marrow. *Hum. Genet.* **35,** 299–307.

Bloom, S. E. (1978) Chick embryos for detecting environmental mutagens. In *Chemical Mutagens: Principles and Methods for Their Detection* Vol. 5. (A. Hollaender and F. J. de Serres, Eds.), Plenum Press, New York, pp. 203–232.

Bloom, S. E. (1979) Detection of sister chromatid exchanges *in vivo* using avian embryos. In *Cytogenetic Assays of Environmental Mutagens* (T. C. Hsu, Ed.) Springer-Verlag, New York.

Bloom, S. E. & Hsu, T. C. (1975) Differential fluorescence of sister chromatids in chicken embryos exposed to 5-bromodeoxyuridine. *Chromosoma* **51,** 261–267.

Bynum, G. D. (1980) personal communications.

Gatti, M., Santini G., Pimpinelli, S., and Olivieri, G. (1979) Lack of spontaneous sister chromatid exchanges in somatic cells of *Drosophila Melanogaster*. *Genetics* **91,** 255–274.

Ikushima, T., and Wolff, S. (1974) Sister chromatid exchanges induced by light-flashes to 5-bromodeoxyuridine and 5-iododeoxyuridine-substituted Chinese hamster chromosomes. *Exp. Cell Res.* **87,** 15–19.

Kanda, N., and Kato, H. (1979) A simple technique for *in vivo* observation of SCE in mouse ascites tumor and spermatogonial cells. *Exp. Cell Res.* **118,** 431–434.

Kato, H. (1974) Spontaneous sister chromatid exchanges detected by BrdU-labelling method. *Nature* **251,** 70–72.

Kato, H. & Sandberg, A. A. (1977) The effect of sera on sister chromatid exchanges *in vitro*. *Exp. Cell Res.* **109,** 445–448.

Kihlman, B. A. & Kornberg, D. (1975) Sister chromatid exchanges in *Vicia faba*. I. Demonstration by a modified fluourescent plus Giemsa technique. *Chromosoma (Berl.)* **51,** 1–10.

Kligerman, A. D. & Bloom, S. E. (1976) Sister chromatid differentiation and exchanges in adult mudminnows *(Umbra limi)* after *in vivo* exposure to 5-bromodeoxyuridine. *Chromosoma* **56,** 101–109.

Kram, D., Bynum, G. D., Senula, G. C., & Schneider, E. L. (1979a) *In utero* sister chromatid exchanges analysis for detection of transplacental mutagens. *Nature* **279,** 531.

Kram, D., Schneider, E. L., Senula, G. C. & Nakanishi, Y. (1979b) Spontaneous and mitomycin-C induced sister chromatid exchanges: comparison of *in vivo* and *in vitro* systems. *Mutat. Res* **60,** 339–347.

Marquardt, H. & Bayer, U. (1977) The induction *in vivo* of sister chromatid exchanges in the bone marrow of the Chinese hamster I. The sensitivity of the system (methyl methanesulphonate). *Mutat. Res.* **56,** 169–176.

Mattevi, M. & Salzano, F. (1975) Senescence and human chromosome changes. *Hum. Genet.* **27,** 1–12.

Monticone, R. E. & Schneider, E. L. (1979) Induction of sister chromatid exchanges in human cells by fluorescent light. *Mutat. Res.* **59,** 329–3?7.

Nakanishi, Y. & Schneider, E. L. (1979) *In vivo* sister chromatid exchange: A sensitive measure of DNA damage. *Mutat. Res.* **60,** 329–337.

Nakanishi, Y., Kram, D., & Schneider, E. L. (1979) Aging and sister chromatid exchange IV. Reduced frequencies of mutagen-induced sister chromatid exchanges in vivo in mouse bone marrow cells with aging. *Cytogenet. Cell Genet.* **24,** 61–67.

Perry, P. & Evans, H. J. (1975) Cytological detection of mutagen-carcinogen exposure by sister chromatid exchange. *Nature* **258,** 121–124.

Polani, P. E., Crolla, J. A., Seller, M. J. & Moir, F. (1979) Meiotic crossing over exchange in the female mouse visualized by BrdU substitution. *Nature* **278,** 348–349.

Popescu, C., Turnbull, D., & DiPaolo, J. A. (1977) Sister chromatid exchange and chromosome aberration analysis with the use of several carcinogens and noncarcinogens: Brief communication. *Natl. Cancer Inst.* **59,** 289–293.

Price, G., Modak, S., & Makinodan, T. (1965) Age associated changes in the DNA of mouse tissue. *Science* **171,** 614–616.

Roszinski-Kocher, G. & Rohrborn, G. (1979) Effects of various cyclophosphamide concentrations *in vivo* on sister chromatid exchanges (SCE) and chromosome aberrations of Chinese hamster bone marrow. *Hum. Genet.* **46,** 51–55.

Scheid, W. (1976) Mechanism of differential staining of BrdU-substituted *Vicia faba* chromosomes. *Exp. Cell Res.* **101,** 55–58.

Schneider, E. L., Chaillet, J., & Tice, R. (1976) *In vivo* BrdU labeling of mammalian chromosomes. *Exp. Cell Res.* **100,** 396–399.

Schneider, E. L., Sternberg, H., & Tice, R. R. (1977) *In vivo* analysis of cellular replication. *Proc. Natl. Acad. Sci.U.S.* **74,** 2041–2044.

Schneider, E. L., Nakanishi, Y., Lewis, J. & Sternberg, H. Simultaneous examination of sister chromatid exchanges and cell replication kinetics in tumor and normal cells *in vivo*. *Cancer Res.* In press.

Schvartzman, J. B. and Cortes, F. (1977) Sister chromatid exchange of Allium cepa. *Chromosoma (Berl.).* **62,** 119–131.

Stetka, D. G., and Wolff, S. (1976) Sister chromatid exchange as an assay for genetic damage induced by mutagenic carcinogens. I. *In vivo* test for compounds requiring metabolic activation. *Mutat. Res.* **41,** 333–342.

Taylor, J. H., Woods, P. S., & Hughes, W. L. (1957) The organization and duplication of chromosomes as revealed by autoradiographic studies using tritium-labelled thymidine. *Proc. Natl. Acad. Sci. U.S.* **43,** 122–128.

Tice, R., Chaillet, J., & Schneider, E. L. (1976) Demonstration of spontaneous sister chromatid exchanges *in vivo*. *Exp. Cell Res.* **102,** 426–428.

Tice, R. R., Costa, D. L. & Drew, R. T. (1980) Cytogenetic effects of inhaled benzene in murine bone marrow: Induction of sister chromatid exchanges, chromosomal aberrations, and cellular proliferation inhibition in DBA/2 mice. *Proc. Natl. Acad. Sci. U.S.* **77,** 2148–2152.

Vogel, W. & Bauknecht, T. (1976) Differential chromatid staining by *in vivo* treatment as a mutagenicity test system. *Nature* **260,** 448–449.

Wheeler, K. & Lett, J. (1975) On the possibility that DNA repair is related to age in nondividing cells. *Proc. Natl. Acad. Sci. U.S.* **71,** 1962–1966.

Wienberg, J. (1977) Technique for the differentiation of sister chromatids in somatic cells of *Drosophila melanogaster*. *Mutat. Res.* **44,** 285–286.

Chapter 9

Sister Chromatid Exchanges in Plants

B. A. KIHLMAN AND H. C. ANDERSSON
DEPARTMENT OF GENETICS
UNIVERSITY OF UPPSALA
UPPSALA, SWEDEN

I. Introduction 244
II. Techniques used for the demonstration of sister chromatid exchanges in plants 245
  A. Autoradiographic demonstration in chromosomes labeled with tritiated thymidine 245
  B. Differential staining with the fluorescence-plus-Giemsa (FPG) technique 247
    1. Labeling with 5-bromodeoxyuridine 247
      a. *Vicia faba* 247
      b. *Allium cepa* 249
      c. *Secale cereale* 249
      d. *Hordeum vulgare* 249
    2. Preparation of slides 249
    3. Treatment and staining of slides 250
III. Results obtained in studies with the autoradiographic and FPG techniques in plants 251
  A. Restricted reunion of chromosome subunits 251
  B. Frequency of SCEs and their distribution between and within chromosomes 251
    1. Autoradiographic studies 251
    2. Studies with the FPG technique 252
  C. Are SCEs induced or spontaneous events? 254

The authors' work discussed in this chapter was supported by a grant from the Swedish Natural Science Research Council.

1. Autoradiographic studies 254
2. Studies with the FPG technique 255
D. Isolabeling 255
1. Autoradiographic studies 255
2. Studies with the FPG technique 256
E. Double SCEs smaller than the width of the chromatid 257
F. Single-strand exchanges 257
G. Dependence on the stage of the cell cycle: studies with the FPG technique 258
H. Induction of SCE with chemical and physical agents: studies with the FPG technique 259
1. Agents with S-dependent effects 259
2. Agents with S-independent effects 260
IV. Concluding remarks 262

I. INTRODUCTION

The phenomenon of sister chromatid exchange (SCE) was first demonstrated in plants. In an extensive study on the mitotic behavior of ring chromosomes in maize, Barbara McClintock (1938) observed double-sized dicentric ring chromosomes and suggested that they were formed as a result of crossing over between sister chromatids. Schwartz (1953) arrived at a similar conclusion in a study of the behavior of a large ring chromosome during meiosis in maize. The first real breakthrough in the study of SCE came, however, when J. H. Taylor and his coworkers introduced tritiated thymidine for labeling chromatids in studies with the autoradiographic technique (Taylor et al., 1957). This now classical study, carried out with *Vicia faba* as experimental material, constitutes the first direct visible demonstration of SCEs.

That these early observations were made in plant cells is not surprising, considering that neither the modern techniques of mammalian cytogenetics nor the methods for *in vitro* culturing of mammalian cells were fully developed at that time [the "dark age" of mammalian cytogenetics, before "the hypotonic miracle" (Hsu, 1979)]. Today the great majority of all studies on SCE are carried out in cultured mammalian cells, and it was also in this material that the new dramatic improvements in the methods for detection of SCEs were made. Since these new methods (based on the effect of incorporated 5-bromouracil on the staining properties of the chromosomes) have a far better resolution than the autoradiographic technique, it is now possible to study SCEs with great accuracy. Problems, such as the phenomenon of isolabeling, which could not be explained with the autoradiographic technique, have now been solved.

Although the new techniques have been adapted to plants, the number of published studies with this material are still rather few and the results are mostly of a confirmatory character. But there are some aspects of SCE that have been studied in more detail in plant than in animal cells, partly by coincidence and partly because the larger plant chromosomes are better suited for this type of study. In another chapter of this book, J. H. Taylor describes his pioneering autoradiographic experiments on SCEs. Since most subsequent studies on SCEs in plants with this technique have added little to Taylor's original observations, the greater part of this chapter has been devoted to studies based on differential staining of sister chromatids after incorporation of 5-bromouracil.

II. TECHNIQUES USED FOR THE DEMONSTRATION OF SISTER CHROMATID EXCHANGES IN PLANTS

A. *Autoradiographic Demonstration in Chromosomes Labeled with Tritiated Thymidine*

In the first mitosis after labeling with tritiated thymidine ([^{3}H]dThd), both chromatids of the chromosomes are uniformly labeled, and any SCEs that may have occurred during this first cell cycle cannot be detected. But in the second mitosis after labeling, each chromosome has one labeled and one unlabeled chromatid, and the SCEs that have taken place since the time of labeling appear as switches of label from one sister chromatid to the other (Figure 9.1). Here we describe how SCEs can be demonstrated autoradiographically in root tips of *Vicia faba*, a well-known material for cytogenetic studies on the effects of mutagenic agents.

Root-tip cells of *Vicia faba* contain 12 chromosomes, one metacentric chromosome pair with a secondary (nucleolar) constriction (M chromosome) and five subtelocentric chromosome pairs (S chromosomes). In the metaphase of colchicine-treated cells, the mean lengths of the M and S chromosomes are 17.2 and 7.5 μm, respectively (Schweizer, 1973).

Roots (main roots or 1–2 cm long lateral roots) of seedlings of *Vicia faba* are exposed for 1–4 hours to 1–2 μCi/ml [^{3}H]dThd. If the SCEs are to be scored in diploid cells, the seedlings are transferred to isotopefree solution (e.g., tap water) for a period long enough to allow cells with labeled chromosomes to enter their second mitosis. The roots are exposed for 3 hours to 0.05% colchicine before fixation overnight at 4°C in alcohol (methanol or ethanol)–acetic acid, 3:1.

In studies on SCE it may be desirable to keep the chromatids produced by the two replication cycles together in one (tetraploid) cell. To achieve

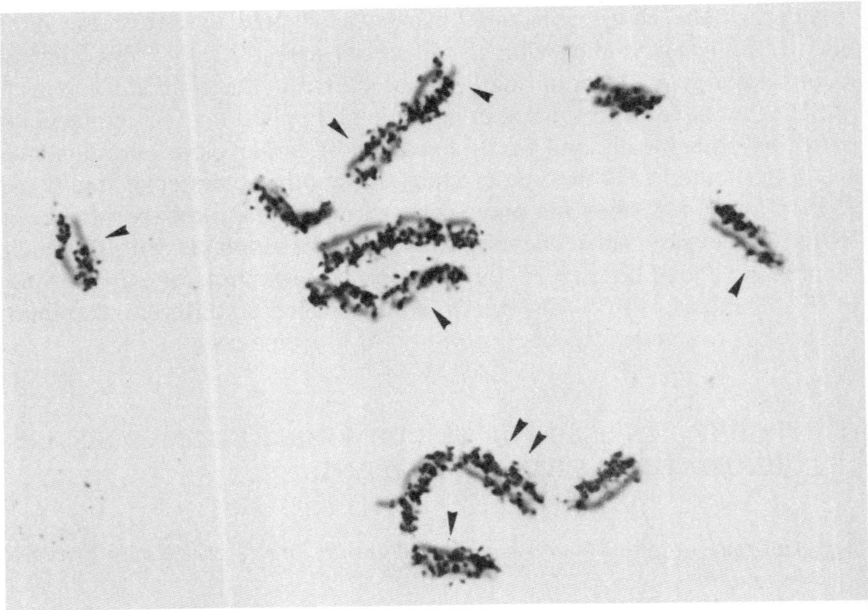

Fig. 9.1. Autoradiograph of root-tip cell of *Vicia faba* in the second metaphase after exposure to [^{3}H]dThd. Arrows indicate SCEs. From Schweizer, 1973, by permission of D. Schweizer.

this, the roots are exposed to a 1–3 hour pulse treatment with 0.025% colchicine at the time when the labeled chromosomes are in their first mitosis after the beginning of the [^{3}H]dThd treatment. A second colchicine treatment is given 3 hours before fixation as described above. Alternatively, tetraploid cells with labeled chromosomes are obtained by exposing the roots to 0.025% colchicine for the entire period between [^{3}H]dThd treatment and fixation.

The fixed roots are hydrolyzed for 8 min in 1 M HC1 at 60°C and are stained with leuco-basic fuchsin (Feulgen stain) for 1 hour. Squash preparations are made of the terminal 1.5 mm portion of the roots on slides coated with a mixture of 10:1 gelatin and chrome alum. Cover slips are removed on Dry Ice and the slides are passed through 100, 85, 70, and 30% ethanol to distilled water. Autoradiographs are made using stripping film or, alternatively, by dipping the slides in nuclear track emulsion. The slides are air dried, exposed for 11–31 days in the dark at 4°C, and developed. After being rinsed in distilled water, the slides are fixed, washed in running water for about ½ hour, air dried, and mounted.

B. *Differential Staining with the Fluorescence-Plus-Giemsa (FPG) Technique*

Following its discovery in cultured mammalian cells (Perry and Wolff, 1974), the FPG technique was soon applied to plant material, first to *Vicia faba* (Kihlman and Kronborg, 1975), then to *Allium cepa* (Schvartzman and Cortés, 1977), *Tradescantia paludosa* (Andersson, unpublished work), *Secale cereale* (Friebe, 1978), and *Hordeum vulgare* (Schubert et al., 1980). Before any satisfactory results could be obtained in plants, several obstacles, not met with in mammalian cells, had to be overcome. Such obstacles included the comparatively poor incorporation of 5-bromodeoxyuridine (BrdUrd), the resistance to squashing provided by the cellulose walls, and the cytoplasmic staining caused by the high RNA content. Our present procedure for differential FPG staining of *Vicia* chromosomes differs considerably from the original version (Kihlman and Kronborg, 1975). The technique also varies somewhat in different laboratories and for different species. When regarded as significant, such local and species-dependent modifications are mentioned.

The details of the FPG technique are described by Latt in another chapter of this book. The technique is based on the fact that the staining properties of chromosomes are altered when the thymine component of their DNA is replaced by 5-bromouracil (BrUra). A chromatid that has incorporated BrUra into both strands of its DNA (a "BB"- chromatid) stains more weakly than a chromatid that has incorporated BrUra into only one strand of its DNA (a "TB"-chromatid), and a TB-chromatid, in turn, does not stain as strongly as a completely unsubstituted "TT"-chromatid. Figure 9.2 shows a differentially stained metaphase cell of *Vicia faba* with TT-TB chromosomes.

1. *Labeling with 5-Bromodeoxyuridine*

(a) VICIA FABA. As is mentioned earlier, plant roots generally do not incorporate sufficient amounts of the thymidine analogue BrdUrd to permit a good differential staining of sister chromatids. The incorporation of BrdUrd is considerably increased, however, if the *de novo* synthesis of thymidylic acid is suppressed by the simultaneous addition of the uridine analogue 5-fluorodeoxyuridine (FdUrd). After phosphorylation to 5-fluorodeoxyuridylic acid, FdUrd specifically and effectively inhibits the enzyme thymidylate synthetase, which catalyzes the methylation of

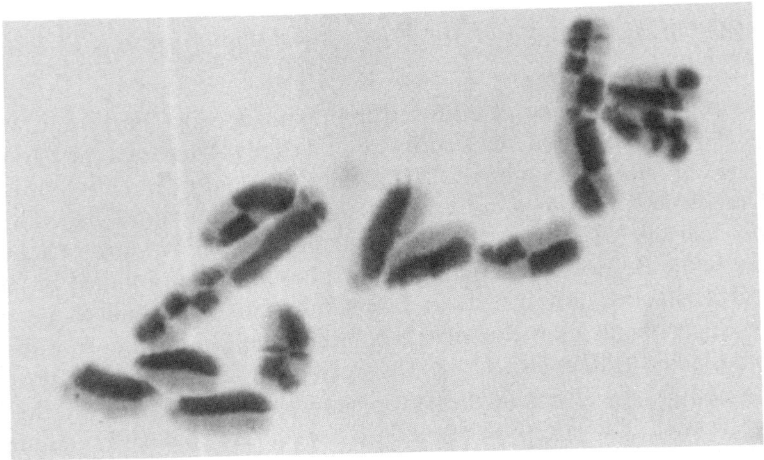

Fig. 9.2. A differentially stained metaphase cell of *Vicia faba* with chromosomes of the TT-TB constitution. From Kihlman and Sturelid, 1978.

uridylic acid to thymidylic acid. After having tried different combinations of BrdUrd and FdUrd, we decided on the following mixture: 10^{-4} M BrdUrd, 10^{-7} M FdUrd, and 5×10^{-6} M uridine (Urd). Uridine is added to the treatment solution to counteract any possible adverse effects of FdUrd on RNA synthesis. The nucleosides were dissolved in tap water.

Lateral roots of *Vicia faba* seedlings are exposed to this nucleoside mixture (BrdUrd-solution) for a period of 16 hours (one round of DNA replication). During the second round of DNA replication, the roots are grown either in the presence of thymidine (dThd) or in a fresh BrdUrd solution. In the former case, the sister chromatids obtain the TT-TB constitution, and in the latter they obtain the TB-BB constitution. The majority of our experiments are carried out according to the first of these alternatives. The dThd solution contains 10^{-4} M dThd and 5×10^{-6} M Urd dissolved in tap water. The relative irreversibility of the inhibitory action of FdUrd on the activity of thymidylate synthetase makes it necessary to grow the roots in the presence of dThd during the second cell cycle. In the absence of dThd, the mitotic cycle is strongly reduced.

After the change of solution at 16 hours, the roots are left in the dThd or fresh BrdUrd solution for a period of 19 hours. The roots are exposed to 0.05% colchicine the last 3 hours before fixation overnight in cold methanol–glacial acetic acid, 3:1. During the whole treatment period at 20°C, care is taken not to expose the roots to light. Approximately the same protocol was used to produce chromatids of the TT-TB constitution in root tips of *Tradescantia paludosa*.

(b) ALLIUM CEPA. In the experiments of Schvartzman and coworkers (Schvartzman and Cortés, 1977; Schvartzman et al., 1979b), roots grown from bulbs of the common onion, *Allium cepa*, were used as experimental material. The treatments were initiated when the roots were 15–20 mm long. All experiments were carried out at 25°C. To obtain chromosomes with chromatids of the TT-TB constitution, roots were grown for 12 hours in 10^{-4} M BrdUrd, 5×10^{-9} M FdUrd, and 10^{-6} M Urd (BrdUrd solution) and then for 14 hours in 10^{-4} M dThd and 10^{-6} M Urd. Chromosomes with chromatids of the TB-BB constitution were obtained by growing the roots for 25 hours in the presence of BrdUrd, the BrdUrd solution being renewed after 10 hours of exposure.

(c) SECALE CEREALE. In the experiments of Friebe (1978) differential staining of rye chromosomes was obtained by using chromosomes of the TB-BB constitution. After germination at 25°C for 48 hours, the roots were exposed in the dark for 24 hours at 25°C to a solution consisting of 10^{-4} M BrdUrd, 5×10^{-8} M FdUrd, and 10^{-6} M Urd dissolved in tap water.

(d) HORDEUM VULGARE. In the experiments of Schubert et al. (1980) differential staining of barley chromosomes was obtained by using chromosomes of the TT-TB constitution. Roots of germinated seeds were exposed in the dark at 24°C for 12 hours to 5×10^{-4} M BrdUrd, 5×10^{-8} M FdUrd, and 10^{-6} M Urd and then for another 12 hours to 2.5×10^{-5} M dThd and 10^{-6} M Urd.

2. *Preparation of Slides*

Root tips require maceration for squashing. We have obtained good maceration by treatment with pectinase. After fixation overnight in methanol–acetic acid at 4°C, the roots are rinsed in a 0.01 M citric acid–sodium citrate buffer, pH 4.7, and are incubated for 2 hours at 27°C with 0.5% pectinase dissolved in the same buffer. The pectinase treatment is terminated by transferring the roots to distilled water, and squash preparations are made in 45% acetic acid on clean slides, coated with a mixture of gelatin and chrome alum, 10 : 1. The procedure for making the preparations is the same as that described previously for Feulgen squashes (Kihlman, 1975a). The cover slips are removed on Dry Ice and the preparations are passed by way of 100 (two changes), 85, 70, 50, and 30% ethanol to distilled water.

The methods for enzymatic maceration vary somewhat in different laboratories and for different materials. In addition to pectinase, cellulase and snail (*Helix pomata*) stomach cytase have been used (Schubert et al., 1979, 1980; Friebe, 1978).

3. *Treatment and Staining of Slides*

To reduce the cytoplasmic staining and improve the contrast between sister chromatids, 200 μl of 0.01% RNase in 0.5 × SSC (saline sodium citrate) is placed on the slides. The preparations are covered with a 24 × 40 mm cover slip and are incubated in a moist chamber for 1 hour at 27°C. With a few modifications in time and temperature, the RNase step has been used by most workers successfully applying the FPG technique to plant material.

After the RNase treatment, the preparations are rinsed in 0.5 × SSC and are stained for 20 min with a solution containing 0.5 ppm 33258 Hoechst in 0.5 × SSC. The stained preparations are rinsed and mounted in 0.5 × SSC, the cover slips being sealed with rubber cement. At this stage it may be convenient to check the differentiation by fluorescence microscopy.

In the next step we expose the preparations for 30 min to long-wave UV at a distance of 6 cm from a radiation source consisting of two Philips 40 W TL09 black light fluorescent lamps (the TL09 lamps produce light primarily in the 320–380 nm range). The slides are then stored for 1 day under ordinary light and temperature conditions in the laboratory and for an additional 3 days at 4°C in the dark. After storage, the slides are incubated for 60 min at 55°C (TB-BB chromosomes for 30 min at 65°C) in 0.5 × SSC. The slides are thoroughly rinsed in 0.017 M phosphate buffer, pH 6.8, and are stained for 6 min in a solution containing 3% Giemsa (Gurr's improved R66) in the same buffer. After the slides are stained, the preparations are rinsed, first in running tap water and then in distilled water, and air dried. The dry preparations are dipped in xylene and mounted in Canada balsam.

This is our standard method. Recently, however, we have successfully applied a much simpler and faster procedure developed by F. Palitti in Rome. After the slides are stained with 33258 Hoechst, they are exposed for 30 min to light from a Osram 300 W Ultra-Vitalux sunlamp in Coplin jars containing 0.5 × SSC. The illumination with the Ultra-Vitalux lamp is followed by a 30 min incubation at 55°C in 0.5 × SSC. The slides are then immediately stained as described for the standard procedure.

The largest deviations between our procedure and those adopted in other laboratories are found in the steps following the RNase treatment.

Furthermore, these same steps seem to be under constant modification both in our and in other laboratories (Schubert et al., 1979, 1980; Schvartzman et al., 1979b).

III. RESULTS OBTAINED IN STUDIES WITH THE AUTORADIOGRAPHIC AND FPG TECHNIQUES IN PLANTS

A. *Restricted Reunion of Chromosome Subunits*

The classical experiments of Taylor et al. (1957) demonstrated that there are chromosome subunits, which are distributed semiconservatively between daughter chromatids. In their replication and segregation patterns, these subunits behave as if they were the complementary polynucleotide chains of a single DNA molecule. Additional and even more convincing evidence for an identity between chromosome subunits and the two strands of a DNA double helix was obtained by Taylor (1958) in a study on the patterns of SCE in *Bellevalia romana*. On the basis of the frequencies of SCE in the first and second cell cycle, Taylor was able to conclude that the chromosome subunits, like the strands of a DNA double helix, have a directional polarity that restricts their reunion. Since these experiments are described by Taylor in another chapter of this book, there is no need to go into any detail here. In a subsequent similar study of the frequencies of SCE in the first and second cell cycle in *Vicia faba*, Geard and Peacock (1969) obtained results that were in good agreement with those of Taylor and therefore supported his conclusions.

B. *Frequency of SCEs and Their Distribution Between and Within Chromosomes*

1. *Autoradiographic Studies*

In *Bellevalia romana* Taylor (1958) found that the frequency of SCE was very nearly proportional to chromosome length. Similar observations were made by Peacock (1963) and by Geard and Peacock (1969), who found that the ratio of the SCE frequencies in the S and M chromosomes of *Vicia faba* closely approximated the metaphase length of these chromosomes. According to Geard and Peacock, the mean number of SCEs per S and M chromosome was 1.64 ± 0.04 and 4.05 ± 0.16, respectively.

Considerably lower numbers of SCE per S and M chromosome were found by Schweizer (1973), who further reported that the SCEs were not equally distributed within the chromosomes. Significant excesses, or occasionally deficiencies, of SCE appeared to be associated with regions showing the most obvious heteropycnotic banding after cold treatment. In the M chromosome, a significant excess of SCE was found in the nucleolar and primary constrictions.

2. *Studies with the FPG Technique*

The frequencies of SCE per cell reported for the five plant species so far investigated are shown in Table 9.1. The data for *Vicia faba* and *Tradescantia paludosa* were obtained in our laboratory, those for *Allium cepa* were obtained by Schvartzman and Cortés (1977) and Schvartzman et al. (1979b), and the SCE frequencies in *Secale cereale* were reported by Friebe (1978). The *Hordeum vulgare* experiments were carried out with a reconstructed karyotype (Schubert et al., 1980). The chromosome constitution (TT-TB or TB-BB) has been indicated in the table because the frequency of SCE seems to be dependent on the number of cell cycles during which the cells have been grown in the presence of BrdUrd.

The lowest frequency of SCE, 11.1/cell, was observed in *Secale cereale* (chromosomes of the TB-BB constitution) and the highest, 44.8 for TT-TB chromosomes and 62.4 for TB-BB chromosomes, in *Allium cepa*. Since there is considerable variation in the amount of DNA per cell in the five species and since the frequency of SCE is likely to be related to the DNA content, the frequency of SCE per picogram of DNA is also given in the table. The DNA content per diploid cell for *Allium, Tradescantia*, and *Vicia* are taken from Baetcke et al. (1967), whereas those for *Secale* and *Hordeum* have been calculated from data published by Bennett et al. (1977) and Skult (1968). In the last case, the DNA content was estimated from phosphorus analyses and a value of 9.23% P in DNA. The frequencies of SCE per picogram of DNA are strikingly low for *Vicia* and *Secale* and high for *Hordeum*.

It was reported in an earlier publication from this laboratory (Kihlman and Kronborg, 1975) that in *Vicia faba* the frequency of SCE is approximately proportional to chromosome length. However, the data that have accumulated since 1975 indicate that there are more SCEs in the M chromosomes than would be expected according to a strict length proportionality. Thus analyses of 400 cells have given an M:S ratio of 2.71:1, which differs markedly from the ratio (Michaelis and Rieger, 1959) that would be expected if the SCEs were distributed in proportion to chromosome length, namely, 2.26:1.

We have not studied the distribution of SCEs within the *Vicia* chromosomes, but Schubert et al. (1979) have reported a random distribution in a

TABLE 9.1

Comparison of the Frequencies of SCE Demonstrated by the FPG Technique in Root Tips of Five Different Plant Species

Species	*Vicia faba* ($2n = 12$)		*Allium cepa* ($2n = 16$)		*Tradescantia paludosa* ($2n = 12$)	*Secale cereale* ($2n = 14$)	*Hordeum vulgare* ($2n = 14$)
Chromosome constitution	TT-TB	TB-BB	TT-TB	TB-BB	TT-TB	TB-BB	TT-TB
Number of chromosomes analyzed	4800	720	300	300	240	1400	1400
SCE/cell	21.0	29.1	44.8	62.4	42.3	11.1	20.6
pg DNA/cell (approx. values)	44	44	54	54	54	25	20
SCE/pg DNA	0.48	0.66	0.83	1.16	0.78	0.44	1.03

reconstructed karyotype except for a clustering in the nucleolar organizing region. A clustering of SCEs in the nucleolar constriction was previously found by Schweizer (1973) with the autoradiographic technique. Vosa (1976) has studied the distribution of SCEs in the M chromosome after exposure of cells to BrdUrd for two rounds of DNA replication. SCEs located in the centromeric region, as well as those occurring in the nucleolar constriction, were not scored. Vosa found a clustering of SCEs in the middle of the long arm of the M chromosome.

In *Allium cepa* Schvartzman and Cortés (1977) found a nonrandom distribution of SCEs within the chromosomes. The frequency of SCE was lower than expected on a random basis in regions that were known to be late replicating or to exhibit C-bands (constitutive heterochromatin).

Similar observations were made by Friebe (1978) in the satellite-bearing chromosome of *Secale cereale*. The frequency of SCE was found to be considerably lower in the telomeric heterochromatin than in the euchromatic regions. Particularly high frequencies of SCE were observed by Friebe in euchromatic regions that were adjacent to heterochromatic regions. A similar clustering of SCE in the junction between heterochromatin and euchromatin had been reported to occur in mammalian cells (Wolff, 1977).

In the chromosomes of *Hordeum*, Schubert et al. (1980) found that with one exception, the frequency of SCE was proportional to chromosome length and independent of the distribution of heterochromatin. The exception, the satellite of chromosome 6, contained fewer SCEs than expected on the basis of length proportionality.

C. Are SCEs Induced or Spontaneous Events?

1. Autoradiographic Studies

On the basis of his studies on the frequencies of SCE during the first and second cell cycle after labeling, Taylor (1958) concluded that the frequency of SCE is "probably not significantly affected by the endogeneous radiation from the tritium." On the other hand, autoradiographic studies on the effects of [^{3}H]dThd incorporation on the frequencies of SCE in a human ring chromosome (Brewen and Peacock, 1969) and in chromosomes of rat kangaroo cells (Gibson and Prescott, 1972) have provided evidence suggesting that SCEs are radiation-induced events. Therefore, the possibility cannot be excluded that a portion of the exchanges are induced by the radiation from the tritium-labeled thymidine.

2. *Studies with the FPG Technique*

Unfortunately, the FPG technique appears to be equally unsuitable for the demonstration of spontaneous SCEs, since like [^{3}H]dThd BrdUrd induces the events it is used to detect. Most of the evidence for SCE being induced by BrdUrd has been obtained in mammalian cells (Wolff, 1977). In plants, the fact that the frequency of SCE appears to be higher in TB-BB than in TT-TB chromosomes (see Table 9.1) is suggestive of BrdUrd dependence. The increase in SCE frequency with increasing concentration of FdUrd in the BrdUrd solution (Kihlman and Kronborg, 1975) could well be the result of an increased incorporation of BrdUrd at the higher FdUrd concentration, rather than an effect of FdUrd itself.

The only experiments in plants that have been especially designed to find out if there is any relationship between the frequency of SCE and the number of BrUra-substituted DNA strands have been carried out by Schvartzman et al. (1979a). These authors studied the frequency of SCE in *Allium* root-tip cells that had been exposed to BrdUrd for only the first or for two or three cycles of DNA replication. The cells were analyzed at the second and third division after the addition of BrdUrd. Third division chromosomes contain both "reciprocal" and "nonreciprocal" SCEs, the former being formed during the last replication cycle and the latter either during the first or the second replication cycle. The authors conclude from the ratios of reciprocal to nonreciprocal SCEs observed in both types of experiment that most of the exchanges detected with the FPG-technique are BrdUrd induced.

D. Isolabeling

1. *Autoradiographic Studies*

The phenomenon named isolabeling by Peacock (1963) has been one of the strongest arguments in favor of the idea that the unduplicated chromosome contains at least two parallel DNA double helices. Isolabeling was first observed by Taylor in *Bellevalia* (Taylor, 1958, 1959) and was further investigated in autoradiographic studies by Peacock (1963) and Schweizer (1973).

Taylor found that although as a rule the chromosomes at the second mitosis after labeling had one chromatid labeled and the other unlabeled at any level of the chromosome, rare instances occurred when both chromatids were labeled for part of their length. Three possible explanations of this phenomenon were considered by Taylor: (1) isolabeling could be the result of the exposure of cells to [^{3}H]dThd for one replication cycle and part of another in combination with asynchronous initiation of DNA synthesis in the various chromosomes; (2) because of the resolution limits of the technique, both sister chromatids would appear nearly uniformly labeled over a segment containing several small SCEs; and (3) the DNA subunits are composed of two strands that occasionally fail to segregate together, indicating that chromosomes have a polynemic structure. However, Taylor's chromosome models were based on a uninemic concept of chromosome structure.

Subsequently, Taylor (1962) added a fourth explanation of isolabeling to the three cited above. He suggested that segmental isolabeling could result from interchromosomal exchange among the four chromatids of adjacent sister chromosomes in the tetraploid cells produced by the colchicine treatment.

Results of subsequent autoradiographic studies on isolabeling have usually been interpreted according to one of these suggestions. Using root tips of *Vicia faba*, Peacock (1963) carried out experiments very similar to those of Taylor in his work on *Bellevalia*. He observed frequent instances of isolabeling in second-division cells and concluded that they provided critical evidence in favor of a polynemic concept of chromosome structure. Schweizer (1973) estimated the resolution limits of autoradiographs and concluded that even under optimal conditions the detection of SCEs closer together than 0.4 μm would not be possible. In most cases SCEs closer together than 0.8 μm would not be resolved but appear as 1.6 μm long isolabeled segments. The experimental material in Schweizer's study was *Vicia faba*.

2. *Studies with the FPG Technique*

According to the calculations of Schweizer (1973), five SCEs in the long arm of one of the S chromosomes in *Vicia faba* would be sufficient to make this arm appear isolabeled over its entire length. In an analysis of the SCE frequency in 3000 S chromosomes, we found five or more SCEs in 1.4% of the S chromosomes. Furthermore, in many cases the size of the chromatid segments involved in single-terminal or double-intercalary SCEs was below the resolution limits of the autoradiographic technique (see Figures 9.2 and 9.3). Such segments would have been scored as isolabeled in autoradiographs, particularly when several of the small SCEs occurred close together. In mammalian cells containing chromosomes labeled with tritiated BrdUrd, the combination of FPG and autoradiographic techniques has clearly demonstrated that closely spaced, multiple SCEs cause apparent isolabeling (Wolff and Perry, 1974).

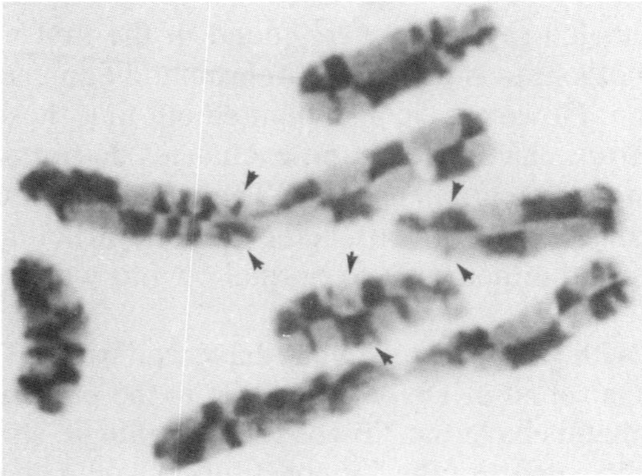

Fig. 9.3. Part of a metaphase cell of *Vicia faba* showing several thiotepa-induced minute double SCEs of different sizes (arrows). From Kihlman, 1975b, by permission of Springer-Verlag, Berlin.

E. Double SCEs Smaller than the Width of the Chromatid

Double SCEs involving material of dimensions smaller than the width of the chromatid have frequently been observed with the FPG technique in plants (Figure 9.3). These minute double SCEs, first reported in *Vicia faba* (Kihlman, 1975b), have been studied in detail in *Allium cepa* by Schvartzman and his coworkers (Schvartzman et al. 1978). They occur both in control and mutagen-treated cells and range in size from a minute dot (at the limit of the resolution of the light microscope) to a narrow line running across almost the entire width of the chromatid.

F. Single-Strand Exchanges

The minute double SCEs are called "sister subchromatid exchanges" by Schvartzman et al. (1978). In our opinion, this term is an unhappy choice, since the term "subchromatid exchange" has been used previously for quite a different phenomenon, namely, the side-arm bridges that can be observed in anaphase after exposure of cells in prophase to ionizing radiation or to other agents with S-independent effects (e.g., Kihlman, 1977).

Taylor (1958, 1959) has pointed out that side-arm bridges in anaphase would be expected if in each sister chromatid only one of the two subunits was involved in the SCE and if the two subunits distal to the exchange were unable to separate and segregate with a subunit of the other chromatid. On the other hand, if such a separation and segregation were possible, single-strand exchanges would give rise to unlabeled segments at the first mitosis after exposure to [^{3}H]dThd. Since neither side-arm bridges nor unlabeled segments were found in the first mitosis, Taylor concluded that both strands of the sister chromatid duplexes are involved in SCE formation. However, he also pointed out that these observations did not demonstrate that single-strand exchanges do not occur, but only that they are not persistent phenomena. They could be first events in the process of exchange formation, but, being unstable structures, they would rapidly be converted either back to normal or into whole chromatid exchanges.

Experiments with the FPG technique have confirmed Taylor's observations. After one round of DNA replication in the presence of BrdUrd, both sister chromatids have the TB constitution. Single-strand exchanges between sister chromatids of the TB constitution would give rise to differentially stained segments of the TT and BB constitution, respectively. No such segments (and no side-arm bridges) have been found in the first mitosis after exposure of root-tip cells of *Vicia faba* to BrdUrd (Kihlman and Kronborg, 1975).

G. *Dependence on the Stage of the Cell Cycle: Studies with the FPG Technique*

It is well-known that lesions induced by alkylating agents in chromosomal DNA do not give rise to chromosomal aberrations until they have been replicated. Thus when cells are exposed to alkylating agents during the G_2-phase of the cell cycle, chromosomal aberrations are not found until the second mitosis after treatment. This S dependence could mean either that the aberrations are formed during S or that the lesions have to be changed by the replication process in order to give rise to aberrations at a later stage of the cell cycle.

Exposure of root-tip cells of *Vicia faba* during G_2 to the alkylating agent thiotepa had no influence on the frequency of SCE in the first mitosis after treatment. On the other hand, when root tips were exposed to thiotepa before or during S, a more than threefold increase in the frequency of SCE was observed (Kihlman, 1975b). These observations suggest that semiconservative DNA replication is required not only for the formation of chromosomal aberrations, but also for the formation of SCEs from thiotepa-induced DNA lesions.

In root-tip cells containing chromosomes with BrUra-substituted DNA, long-wave UV (from Philips 40 W TL09 black light fluorescent lamps) produces chromosomal aberrations independently of chromosome and DNA replication, that is, cells exposed during G_2 contain aberrations during the first mitosis after exposure (Kihlman et al., 1977). Furthermore, long-wave UV is capable of producing a strong enhancement of the frequency of SCE, but only if the cells are irradiated before or during the

S phase (Kihlman et al., 1977). This behavior suggests either that chromosomal aberrations may be formed from lesions that do not give rise to SCE or that the latter always are formed by an S-dependent mechanism.

In root-tip cells of *Allium cepa* Schvartzman et al. (1979b) have analyzed SCEs induced by visible light (emitted from a Philips 60 W incandescent bulb) in chromosomes with BrUra-substituted DNA. The results obtained by these authors clearly demonstrated the BrUra and S dependence of the induction of SCE by visible light. The fact that the production of SCE is an S-dependent process had previously been demonstrated in mammalian cells (Wolff, 1977).

H. *Induction of SCE with Chemical and Physical Agents: Studies with the FPG Technique*

1. *Agents with S-Dependent Effects*

Figure 9.4 shows the frequencies of SCE observed in untreated (control) *Vicia faba* roots and in roots exposed to various chemical substances (inducers). Of the eight substances tested, seven produced a dramatic

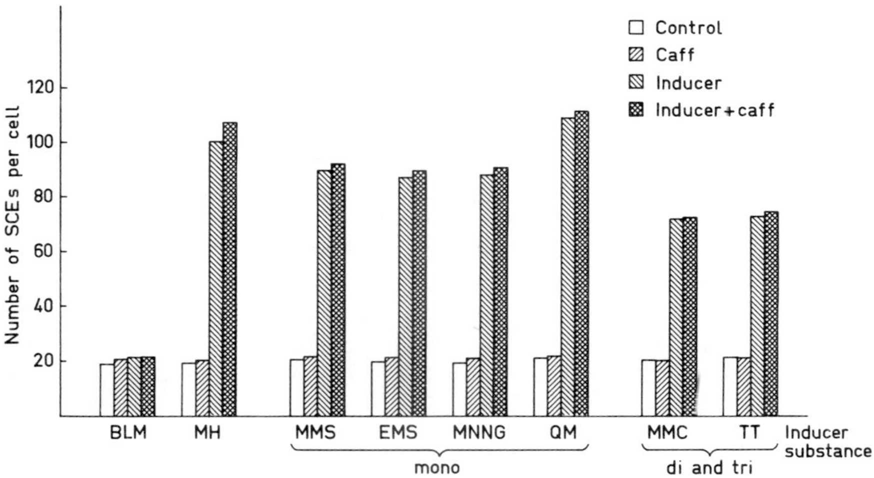

Fig. 9.4. The effects of various mutagenic agents (inducers) with or without posttreatments with caffeine (caff) on the frequencies of SCEs in root tips of *Vicia faba*. BLM = bleomycin; MH = maleic hydrazide; MMS = methyl methanesulfonate; EMS = ethyl methanesulfonate; MNNG = *N*-methyl-*N'*-nitro-*N*-nitrosoguanidine; QM = quinacrine mustard; MMC = mitomycin C; TT = thiotepa. From Kihlman and Sturelid, 1978.

increase in the frequencies of SCE. All these seven substances produced chromosomal aberrations by an S-dependent mechanism. The observation that the frequency of SCE is strongly increased by agents with S-dependent effects is in agreement with similar findings in cultured mammalian cells (Wolff, 1977). The increase in the frequency of SCE is obtained at concentrations lower than those required to produce chromosomal aberrations.

It is further evident in Figure 9.4 that there is little or no effect of posttreatments with caffeine on the induced SCE frequencies. This is in marked contrast to the frequencies of chromosomal aberrations produced by the same agents, which are in all cases dramatically increased by posttreatments with caffeine (Figure 9.5).

2. *Agents with S-Independent Effects*

Figures 9.4 and 9.5 show the effects of eight chemical mutagens. All eight produce chromosomal aberrations, but only seven produce SCEs. The substance that does not have any effect on the frequency of SCE, although it is an efficient producer of chromosomal aberrations, is the antibiotic bleomycin. In contrast to the other seven agents, bleomycin produces chromosomal aberrations independently of DNA and chromosome replication.

A comparatively low capacity to induce SCE appears to be a characteristic common to many agents that produce chromosomal aberrations by an S-independent mechanism. Thus no increase in the frequency of SCE was found after treatments of *Vicia faba* roots with 8-ethoxycaffeine (Kihlman, 1975b). After exposure of bean roots to X rays, a marked increase in the frequency of SCE was obtained only at dosages that produce high frequencies of chromosomal aberrations.

In a comparative study of the effects of X rays under aerobic and anaerobic conditions, an interesting difference was obtained between the results for chromosomal aberrations and those for SCEs (Andersson et al., 1980). For chromosomal aberrations an oxygen enhancement ratio (OER) of 2.4 was obtained. The OER is here defined as the ratio between the dosages that produce the same effect in nitrogen and air. For SCE the much smaller OER of 1.4 was found. When the increase in SCE per cell is plotted against chromosomal aberration frequency for roots irradiated by X rays in air and in nitrogen, it is evident that per aberration more SCEs are produced in nitrogen than in air. This behavior suggests that different types of X-ray-induced DNA lesions are responsible for SCE and chromosomal aberrations.

Paterson and Setlow (1972) have studied the influence of oxygen

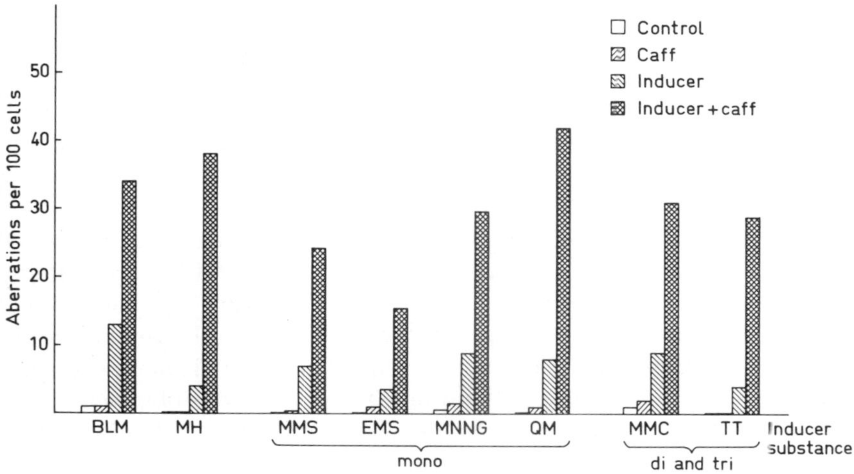

Fig. 9.5. The effects of various mutagenic agents (inducers) with or without posttreatments with caffeine (caff) on the frequencies of chromosomal aberrations in root tips of *Vicia faba*. The inducers are the same as in Figure 9.4. From Kihlman and Sturelid, 1978.

on various types of damage induced by γ-rays in plasmid DNA of *Escherichia coli* minicells. They found that whereas (for a given dose) the frequency of single-strand DNA breaks was about fourfold higher under aerobic than under anoxic conditions, the yield of endonuclease-sensitive sites was only slightly (about 1.3 times) higher in air than in nitrogen. Base damage (or distortions in the DNA double helix) is generally believed to be responsible for the endonuclease-sensitive sites found in γ- and X-ray irradiated DNA. Since high OER values are obtained for strand breaks and chromosomal aberrations and low OER values are obtained for base damage and SCE, it is tempting to conclude that SCEs are caused by base damage rather than by strand breaks whereas both types of damage might, directly or indirectly, give rise to chromosomal aberrations.

Other agents with S-independent effects that have been studied in *Vicia* for their capacity to induce SCE are long-wave UV and the antibiotic streptonigrin. Long-wave UV induces chromosomal aberrations in cells containing chromosomes with BrUra-substituted DNA. The S-independence of the effect is demonstrated by the fact that the chromatid and subchromatid types of aberration are found in the first mitosis after exposure of cells in G_2 and prophase (Kihlman et al., 1977). When cells containing chromosomes of the TB constitution were irradiated with long-wave UV during or before S, a threefold increase in the frequency of SCE was obtained (Kihlman et al., 1977). Thus, in comparison with many other agents with S-independent effects, long-wave UV is an efficient inducer of SCE. The comparatively efficient induction of SCE suggests that the effect of long-wave UV on BrUra-substituted chromosomes is only partly S independent. The occurrence of an S-dependent component in the effect of long-wave UV is further indicated by the fact that both chromatid- and chromosome-type aberrations are obtained when Chinese hamster chromosomes of the TB constitution are exposed to long-wave UV in G_1 (Natarajan et al., 1980).

The antibiotic streptonigrin (SN) is known to produce chromosomal aberrations in an S-independent manner (Kihlman and Odmark, 1965). When root tips of *Vicia faba* were treated with increasing concentrations of SN before or during S, the frequencies of SCE and chromosomal aberrations increased in a parallel manner (Andersson and Kihlman, unpublished). A 3.5-fold increase in the frequency of SCE was obtained at 5×10^{-6} M SN. The results suggest that the effect of SN, like that of long-wave UV on BrUra-substituted chromosomes, is only partly S independent.

The observations described in this section indicate that the DNA lesions giving rise to SCEs are various types of base damage rather than strand breaks. Since the latter apparently can give rise to chromosomal aberrations, the mechanisms for the production of SCEs and chromosomal aberrations must be at least partly different. This conclusion is supported by the observation that posttreatments with caffeine that strongly enhance the frequencies of chromosomal aberrations induced by agents with S-dependent effect have little or no effect on the frequencies of SCE induced by the same agents.

IV. CONCLUDING REMARKS

Studies with the FPG technique in plants, like those carried out with the same technique in mammalian cells, have, on the whole, confirmed Taylor's original observations with the autoradiographic technique and supported his conclusions. Thus the results obtained with the FPG technique strongly support the concept of a uninemic chromosome structure, the subunits of which are identical with the strand of a DNA double helix.

As experimental material, plant root tips have both advantages and disadvantages in comparison with cultured mammalian cells. The advantages are obvious. Root tips of plants such as *Vicia faba* and *Allium cepa* constitute an inexpensive material that is easy to grow and handle and that is available all year round. No expensive equipment is required for carrying out the experiments, nor are sterile conditions necessary. Furthermore, root tips can easily be exposed to known concentrations of chemical substances under controlled environmental conditions.

For cytogenetic studies the material is favorable because of the large number of dividing cells in the root meristem in combination with a comparatively low chromosome number (e.g., *Vicia faba* and *Tradescantia paludosa*, $2n = 12$; *Secale cereale* and *Hordeum vulgare*, $2n = 14$; *Allium cepa*, $2n = 16$).

The large size of the chromosomes of *Vicia faba* and *Allium cepa* makes scoring of SCE easy and accurate. It was no accident that the minute double SCEs were first demonstrated in plants.

Many of the disadvantages of the material are consequences of cells in root tips being surrounded by cell walls that contain cellulose and pectic substances as main constituents. To make squash preparations and to obtain a good separation of the cells in the root meristem, the pectic substances in the middle lamella of the cell wall must be dissolved by hydrolysis with acids or enzymes. Both methods of hydrolysis are liable to affect the structure and morphology of the chromosomes adversely. Furthermore, the cell wall, in combination with the fact that the root meristem is organized into many cell layers, makes it difficult or impossible to use root tips for studies on the effects of radiations with low penetration capacity (e.g., short-wave UV).

A further disadvantage is that in comparison with a culture of mammalian cells, the cells in the root meristem represent a very heterogeneous population with mitotic cycles of different duration and, perhaps, with different sensitivities to mutagenic agents. As a material for detecting agents that may be of a genetic risk to man, root-tip cells have the disadvantage of not allowing for mammalian metabolism.

References

Andersson, H. C., Kihlman, B. A. & Palitti, F. (1981) Production of sister chromatid exchanges by X-rays under aerobic and anaerobic conditions. *Hereditas* **94,** 41–44.

Baetcke, K. P., Sparrow, A. H., Nauman, C. H., & Schwemmer, S. S. (1967) The relationship of DNA content to nuclear and chromosome volumes and to radiosensitivity (LD_{50}) *Proc. Natl. Acad. Sci. U.S.* **58,** 533–539.

Bennett, M. D., Gustafson, J. P., & Smith J. B. (1977). Variation in nuclear DNA in the genus *Secale*. *Chromosoma* **61,** 149–176.

Brewen, J. G. & Peacock, W. J. (1969) The effect of tritiated thymidine on sister-chromatid exchange in a ring chromosome. *Mutat. Res.* **7,** 433–440.

Friebe, B. (1978) Untersuchungen zum Schwesterchromatidenaustausch bei *Secale cereale*. *Microsc. Acta* **81,** 159–165.

Geard, C. R. and Peacock, W. J. (1969) Sister chromatid exchanges in *Vicia faba*. *Mutat. Res.* **7,** 215–223.

Gibson, D. A. & Prescott, D. M. (1972) Induction of sister chromatid exchanges in chromosomes of rat kangaroo cells by tritium incorporation into DNA. *Exp. Cell Res.* **74,** 397–402.

Hsu, T. C. (1979) *Human and Mammalian Cytogenetics. An Historical Perspective,* Springer-Verlag, New York, 186 pp.

Kihlman, B. A. (1975a) Root tips of *Vicia faba* for the study of the induction of chromosomal aberrations. *Mutat. Res.* **31,** 401–412.

Kihlman, B. A. (1975b) Sister chromatid exchanges in *Vicia faba* II. Effects of thiotepa, caffeine and 8-ethoxycaffeine on the frequency of SCEs. *Chromosoma* **51,** 11–18.

Kihlman, B. A. (1977) *Caffeine and Chromosomes,* Elsevier Scientific Publishing Company, Amsterdam, 504 pp.

Kihlman, B. A. & Kronborg, D. (1975) Sister chromatid exchanges in *Vicia faba* I. Demonstration by a modified fluorescent plus giemsa (FPG) technique. *Chromosoma* **51,** 1–10.

Kihlman, B. A. & Odmark, G. (1965) Deoxyribonucleic acid synthesis and the production of chromosomal aberrations by streptonigrin, 8-ethoxycaffeine and 1, 3, 7, 9-tetramethyluric acid. *Mutat. Res.* **2,** 494–505.

Kihlman, B. A. & Sturelid, S. (1978) Effects of caffeine on the frequencies of chromosomal aberrations and sister chromatid exchanges induced by chemical mutagens in root tips of *Vicia faba*. *Hereditas* **88,** 35–41.

Kihlman, B. A., Andersson, H. C., & Natarajan, A. T. (1977) Molecular mechanisms in the production of chromosomal aberrations: studies with the 5-bromodeoxyuridine-labelling method. In *Chromosomes Today,* Vol. 6, (A. de la Chapelle and M. Sorsa, Eds.), Elsevier/North-Holland, Amsterdam, pp. 287–296.

McClintock, B. (1938) The production of homozygous deficient tissues with mutant characteristics by means of the aberrant mitotic behavior of ring-shaped chromosomes. *Genetics* **23,** 315–397.

Michaelis, A. & Rieger, R. (1959) Strukturheterozygotie bei *Vicia faba*. *Der Züchter* **29,** 354–361.

Natarajan, A. T., Kihlman, B. A., & Obe, G. (1980) Use of the 5-bromodeoxyuridine-labelling technique for exploring mechanisms involved in the formation of chromosomal aberrations. II. G1 experiments with Chinese hamster ovary cells. *Mutat. Res.* **73,** 307–317.

Paterson, M. C. & Setlow, R. B. (1972) Endonucleolytic activity from *Micrococcus luteus* that acts on γ-ray-induced damage in plasmid DNA of *Escherichia coli* minicells. *Proc. Natl. Acad. Sci. U.S.* **69,** 2927–2931.

Peacock, W. J. (1963) Chromosome duplication and structure as determined by autoradiography. *Proc. Natl. Acad. Sci. U.S.* **49,** 793–801.

Perry, P. & Wolff, S. (1974) New Giemsa method for the differential staining of sister chromatids. *Nature* **251,** 156–158.

Schubert, I., Sturelid, S., Döbel, P., & Rieger, R. (1979) Intra-chromosomal distribution patterns of mutagen-induced SCEs and chromatid aberrations in reconstructed karyotypes of *Vicia faba*. *Mutat. Res.* **59,** 27–38.

Schubert, I., Künzel, G., Bretschneider, H., Rieger, R., & Nicoloff, H. (1980) Sister chromatid exhanges in barley. *Theor. Appl. Genet.* **56,** 1–4.

Schvartzman, J. B. & Cortés, F. (1977). Sister chromatid exchanges in *Allium cepa*. *Chromosoma* **62,** 119–131.

Schvartzman, J. B., Cortés, F., & López-Sáez, J. F. (1978) Sister subchromatid exchanged segments and chromosome structure. *Exp. Cell Res.* **114,** 443–446.

Schvartzman, J. B., Cortés, F., González-Fernández, A., Gutiérrez, C., & López-Sáez, J. F. (1979a) On the nature of sister-chromatid exchanges in 5-bromodeoxyuridine-substituted chromosomes. *Genetics* **92,** 1251–1264.

Schvartzman, J. B., Postigo, R., & Gutiérrez, C. (1979b) Analysis of visible light-induced sister chromatid exchanges in 5-bromodeoxyuridine-substituted chromosomes. *Chromosoma* **74,** 317–328.

Schwartz, D. (1953) Evidence for sister-strand crossing over in maize. *Genetics* **38,** 251–260.

Schweizer, D. (1973) Vergleichende Untersuchungen zur Längsdifferenzierung der Chromosomen von *Vicia faba* L. *Verhandl. Naturf. Ges. Basel* **83,** 1–75.

Skult, H. (1968) Nucleic acid and protein contents in diploid and polyploid races of barley (*Hordeum vulgare* L.) and timothy (*Phleum pratense* L.). *Acta Acad. Abo.* **28,** 1–12.

Taylor, J. H. (1958) Sister chromatid exchanges in tritium-labeled chromosomes. *Genetics* **43,** 515–529.

Taylor, J. H. (1959) The organization and duplication of genetic material. *Proc. 10th Int. Congr. Genet. Montreal* **1,** 63–78.

Taylor, J. H. (1962) Chromosome reproduction. *Int. Rev. Cytol.* **13,** 39–73.

Taylor, J. H., Woods, P. S., & Hughes, W. L. (1957) The organization and duplication of chromosomes as revealed by autoradiographic studies using tritium-labeled thymidine. *Proc. Natl. Acad. Sci. U.S.* **43,** 122–128.

Vosa, C. G. (1976) Sister chromatid exchanges bias in *Vicia faba* chromosomes. In *Current Chromosome Research* (K. Jones and P. E. Brandham, Eds.), Elsevier/North-Holland, Amsterdam pp. 105–114.

Wolff, S. (1977) Sister chromatid exchange. *Ann.Rev. Genet.* **11,** 183–201.

Wolff, S. & Perry, P. (1974) Differential Giemsa staining of sister chromatids and the study of sister chromatid exchanges without autoradiography. *Chromosoma* **48,** 341–353.

Chapter 10

Sister Chromatid Exchanges in Drosophila

MAURIZIO GATTI
CENTRO DI GENETICA EVOLUZIONISTICA DEL C.N.R.
ISTITUTO DI GENETICA
CITTÀ UNIVERSITARIA, ROMA

I. Introduction 268
II. Methods 269
A. Sister chromatid differentiation by BrdUrd-Hoechst 33258 procedures 269
B. Scoring of SCE in ring chromosomes 271
III. SCE yields at various BrdUrd concentrations and dependence on DNA content 272
IV. SCEs in ring chromosomes 275
V. SCEs in meiotic and mutagen-sensitive mutants 283
VI. Concluding remarks 290

I wish to thank D. L. Lindsley for sharing his unpublished work, S. Pimpinelli and B. S. Baker for critical reading of the manuscript, and G. Belloni for microphotographs. This work was supported by Euratom grants No. 136-74-7 BIOI and No. BI-E-400-I and by a grant from Progetto Finalizzato Promozione Qualità dell' Ambiente.

I. INTRODUCTION

The question of whether sister chromatid exchange (SCE) occurs in meiotic recombination was first investigated in *Drosophila melanogaster*. L. V. Morgan (1933) examined the offspring of females that were heterozygous for *R(1)1** and either a normal rod X or *In(1)dl-49*† and found similar ratios of ring-and rod-bearing progeny. Since SCE occurring in a ring chromosome would generate double-sized dicentric rings resulting in anaphase bridges at the second meiotic division, and thus in a deficiency of ring-bearing progeny, Morgan concluded that sister strand crossing over does not occur in *Drosophila* meiosis. The same conclusion was reached by Novitski (1955), who repeated these experiments using *In(1)AB*†, and by Sandler (1957), who studied crossing over in reversed compound ring X chromosomes. That SCE does not participate in meiotic recombination was also inferred by the analysis of half-tetrads recovered in the progeny of attached-X females of *D. melanogaster* (Beadle and Emerson, 1935; Weinstein, 1936).

However, it should be recognized that these conclusions were based on the assumption that SCE and nonsister strand exchange are equivalent meiotic events. Thus there is the possibility that some SCEs occur independently of crossing over. To test this possibility D. L. Lindsley (unpublished work) analyzed the recombination products in various ring chromosome constitutions on the assumption that SCE is an independent phenomenon superimposed on crossing over. He found that the data were inconsistent with this assumption and in agreement with the notion that SCE does not occur in *Drosophila* meiosis.

The occurrence of SCEs in somatic tissues of *D. melanogaster* was investigated cytologically by examining the mitotic behavior of stable and unstable ring X chromosomes (Braver and Blount, 1949; Hinton, 1955; Brosseau, 1966). However, different ring configurations and different frequencies of anaphase bridges were reported in these studies. Thus they did not permit firm conclusions to be reached as to the spontaneous occurrence of SCE. As noted by Hinton, many of the discrepancies could be attributed to the limitations of the cytological techniques that were available.

In the past few years a series of cytological techniques have been developed in our laboratory that allow excellent metaphase preparations of *Drosophila* chromosomes to be obtained. These techniques have been

---

**R(1)1* is a ring X chromosome. †*In(1)dl-49* and *In(1)AB* are multiply inverted X chromosomes. All chromosomes are described in detail by Lindsley and Grell (1968).

extensively employed in studies on induced chromosome aberrations (Gatti et al., 1974a, b; Gatti et al., 1975; Pimpinelli et al., 1976a; 1977) and on chromosome banding (Gatti et al., 1976; Pimpinelli et al., 1976b). This chapter describes the application of these techniques to the study of sister chromatid exchanges in somatic cells of *D. melanogaster*. The occurrence of SCE has been investigated using two different approaches: (1) differentiation of sister chromatids with BrdUrd labeling techniques and (2) scoring of dicentric rings in stocks carrying monocentric ring chromosomes (Gatti et al., 1979). The results of these studies indicate that in *D. melanogaster* SCEs are not spontaneous events but are induced by the BrdUrd used to reveal them.

In another section of this chapter experiments are reported on sister chromatid exchanges in some meiotic-and mutagen-sensitive mutants of *D. melanogaster* that affect DNA repair processes and/or increase the frequency of spontaneous chromosome aberrations. These mutants can be exploited to investigate the relationships among SCE, chromosome aberrations, and meiotic recombination (Gatti et al., 1980).

II. METHODS

A. *Sister Chromatid Differentiation by BrdUrd–Hoechst 33258 Procedures*

Neural ganglia of *D. melanogaster* can be cultured *in vitro* in saline (NaCl 0.7%) supplemented with 20% fetal calf serum (Pimpinelli et al., 1975; Gatti et al., 1976). Under these conditions the cell cycle of neuroblast cells is of approximately the same length (8 hours) as it is *in vivo* (Gatti et al., 1974b; Pimpinelli et al., 1976a). Thus to obtain metaphases that have incorporated BrdUrd for two successive S phases, neural ganglia are incubated for 13 hours in the presence of the analogue according to the procedure given below (Gatti et al., 1979).

Actively crawling third instar larvae are dissected under a stereoscope in a drop of saline (NaCl 0.7%). After dissection the neural ganglia are washed twice in sterile saline to reduce bacterial and fungal contamination. They are then transferred to Petri dishes (Falcon 35 × 10 mm) containing 2 ml of sterile saline supplemented with 20% fetal calf serum (Gibco). After addition of BrdUrd to this medium, the ganglia are incubated for 13 hours in the dark at 25°C. After 11 hours of treatment, colchicine at a final concentration of 10^{-5}M is added to collect metaphases.

Following incubation with BrdUrd, the ganglia are transferred to a hypotonic solution of 0.5% sodium citrate where they are kept for 10 min at 25°C. They are then fixed for about 10 sec in a mixture of acetic acid, methanol, and distilled water (5.5 : 5.5 : 1) and are squashed in 45% acetic acid under a siliconized cover slip. The cover slip is removed with a razor blade after it is frozen for 10 min on Dry Ice and the slides are air dried at room temperature.

To differentiate sister chromatids the slides are stained for 10 min in 5μg/ml of Hoechst 33258 dissolved in a solution of 0.15 M NaCl, 0.03 M KC1, and 0.01 M phosphate adjusted to pH 7, washed twice in the same buffer, and air dried. They are then mounted in 0.16 M sodium phosphate and 0.04 M sodium citrate (pH 7), and, after the excess mounting buffer is blotted off, the cover slip is sealed with rubber cement.

The slides processed in this way are examined under a Zeiss fluorescence microscope equipped with a 200 W mercury light source for incident illumination and with a combination of filters* transmitting light between 380 and 440 nm, that is excitation filters UG5+BG3, dichroic mirror FT 460, and barrier filter LP 475. In most instances metaphase chromosomes exhibit a clear differentiation of sister chromatids without any further treatment. If necessary, sister chromatid differentiation can be improved by exposing the slides either to sunlight or to a fluorescent bulb. Under these experimental conditions the chromosome fluorescence is rather stable and the absence of fading permits the scoring of preparations under the fluorescence microscope (Figure 10.1).

A direct correlation is observed between the degree of differentiation of sister chromatids and the concentration of BrdUrd with which ganglia are treated. At a concentration of 1μg/m1, although a low degree of sister chromatid differentiation appears to be present in most of the metaphases, only 15% of these are unequivocally differentiated. The fraction of well-differentiated metaphases reaches about 40% after treatment with 3 μg/m1, while at concentrations of 6, 9, and 27 μg/ml most metaphases exhibit a clear differentiation of sister chromatids. The optimal concentration of BrdUrd for obtaining a high yield of well-differentiated metaphases occurs between 6 and 12 μg/ml.

To obtain good fluorescence-plus-Giemsa (FPG) preparations (Figure 10.1) Hoechst-stained chromosomes, after a short exposure to light, are stained wtih 4% Giemsa (Merck) without any intermediate steps. The degree of differentiation of the preparations stained with Giemsa always corresponds to that observed in fluorescence.

Other FPG techniques for sister chromatid differentiation of neuroblast metaphases and *Drosophila* cells in culture have been independently de-

*A wrong combination of filters was erroneously reported in previous papers (Gatti et al., 1976; Gatti et al., 1979).

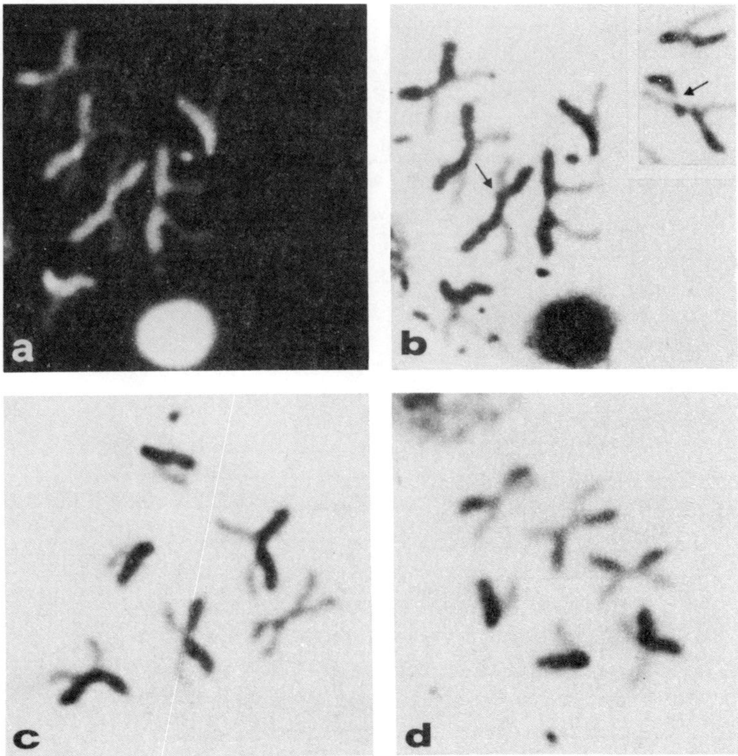

Fig. 10.1. Sister chromatid differentiation in *Drosophila melanogaster*. (*a*) and (*b*) A neuroblast M2 metaphase sequentially stained with Hoechst 33258 and Giemsa. Note an SCE at the euchromatic–heterochromatic junction of an autosome (arrow). A clearer example of an SCE involving these junctions is shown in the partial metaphase of *b*. Euchromatic–heterochromatic junctions have been found to be the preferential site of localization of SCEs in *D. melanogaster* (Gatti et al., 1979; Faccio Dolfini, 1978). (*c*) and (*d*) Neuroblast M3 metaphases. The cell in *d* exhibits a clear three-way differentiation of M3 chromosomes (Miller et al., 1976).

veloped by Wienberg (1977) and Faccio Dolfini (1978). In addition, an FPG technique has been used recently by Tsuji and Tobari (1979) to differentiate ganglion cells of larvae exposed *in vivo* to high concentrations of BrdUrd.

B. *Scoring of SCE in Ring Chromosomes*

SCEs occurring in monocentric ring chromosomes produce double-sized dicentric rings that can be scored as such at metaphase or as bridges at anaphase. However, a more reliable scoring can be obtained by examining c-anaphases where sister centromeres are split and ring sister chromatids are not overlapped.

To increase the yield of c-anaphase figures for scoring, the following procedure was developed (Gatti et al., 1979). After dissection the neural ganglia are incubated for 2 hours in saline (0.7% NaCl) containing 10^{-5} M colchicine. They are then treated with a hypotonic solution of 0.5% sodium citrate for 25 minutes at 30°C, fixed, and squashed in 2% acetic orcein according to our usual procedure (Gatti et al., 1974b).

III. SCE YIELDS AT VARIOUS BrdUrd CONCENTRATIONS AND DEPENDENCE ON DNA CONTENT

Since BrdUrd can damage DNA and induce chromosome aberrations (Hsu and Somers, 1961; Dewey and Humphrey, 1965) it is not clear whether SCEs visualized with BrdUrd—Hoechst 33258 procedures are spontaneous events or are induced by the BrdUrd used to reveal them. One approach to this question is to study the yield of SCEs at increasing concentrations of BrdUrd. The dose-response curves obtained by Kato (1974, 1977a) in both DON Chinese hamster cell lines and in human fibroblasts showed a plateau at low concentrations of BrdUrd followed by a linear increase in the frequency of SCEs at higher concentrations. Similar results were obtained by Tice et al.(1976), who scored SCEs in bone marrow cell preparations from rats injected with increasing concentrations of BrdUrd. These authors concluded that at least some SCEs are not induced by BrdUrd and hence there is a spontaneous level of SCEs. However, Wolff and Perry (1974), using Chinese hamster ovary cells (CHO), obtained a rapid rise in SCE frequency at low concentrations of BrdUrd, followed by a plateau at higher concentrations. Since their dose-response curve could be extrapolated to zero, they concluded that there was insufficient evidence for a background level of spontaneous SCEs. A dose-response curve quite similar to that of Wolff and Perry was obtained by Gibson and Prescott (1972) in autoradiographic experiments carried out on rat kangaroo cells treated with very low doses of tritiated thymidine. These experiments led them to suggest that SCEs are induced by incorporated tritium and that no exchanges occur spontaneously in unlabeled chromosomes.

To determine the relationship between BrdUrd concentration and SCE frequency in *D. melanogaster*, neural ganglia were treated with increasing concentration of BrdUrd (Gatti et al., 1979). The results of two independent experiments, presented in Figure 10.2 and pooled in Table 10.1,

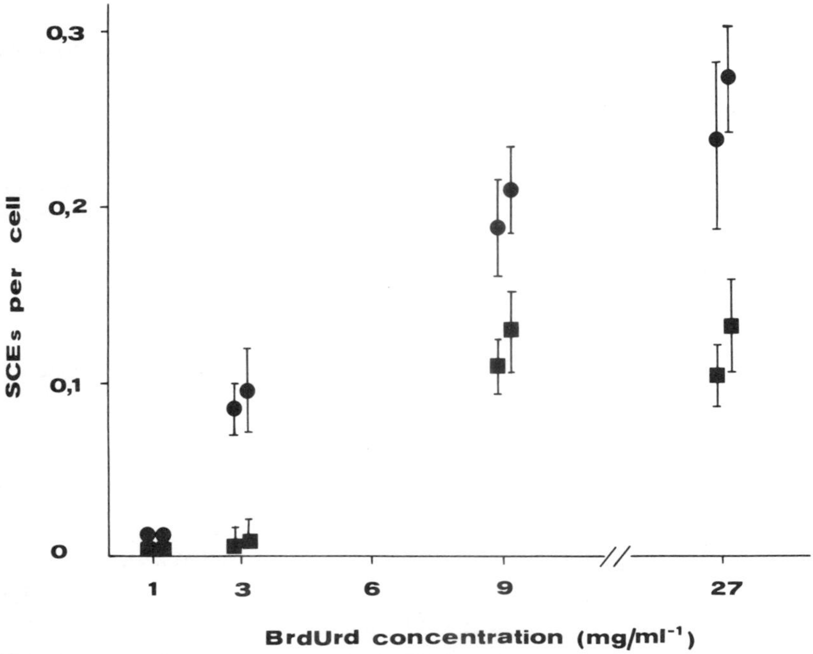

Fig. 10.2. Effect of BrdUrd concentration on the frequency (± s.e.m.) of sister-chromatid exchanges in somatic cells of *D. melanogaster*. (●) Females; (■) males.

show that the neuroblast metaphases of both sexes treated with 1 μg/m1 of BrdUrd do not exhibit spontaneous SCEs. The SCEs occur only at higher concentrations of BrdUrd and the frequency is higher in females than in males. This implies that in wild-type *D. melanogaster* the background level of spontaneous SCE is zero, and they are induced by BrdUrd with a different frequency in the two sexes.

Previous studies have shown that *D. melanogaster* females exhibit a greater frequency of spontaneous chromosome aberrations (Gatti et al., 1974b), as well as X-ray- (Gatti et al., 1974a, b), MMS- (Gatti et al., 1975), and caffeine-induced aberrations (De Marco and Cozzi, 1980). Moreover, a greater frequency of aberrations in females than in males has been found recently in four out of five mutants of *D. melanogaster* that produce high levels of spontaneous chromosome aberrations (Gatti, 1979). To explain these findings, it has been suggested that females differ from males in expressing in their somatic cells some of the enzymes involved in recombination. This, in agreement with the misrepair hypothesis for the origin of chromosome rearrangements (Evans, 1967), would permit a more efficient transformation of DNA lesions into both chromosome aberrations and SCEs (Gatti et al., 1974a, b; Gatti et al., 1979).

TABLE 10.1
Frequency of Sister Chromatid Exchanges in Neuroblast Chromosomes of* Drosophila melanogaster *(data from Gatti et al., 1979)

BrdUrd concentration (μg/ml)	Sex	Number of cells scored	Number of SCEs scored	SCEs per cell
1	♀	183	—	—
	♂	178	—	—
3	♀	512	46	0.09
	♂	338	4	0.01
9	♀	464	94	0.20
	♂	672	76	0.11
27	♀	304	82	0.27
	♂	424	48	0.11

A series of studies carried out on several organisms have shown that the number of SCEs is positively correlated with the length of metaphase chromosomes (reviewed by Kato, 1977a and Wolff, 1977). Moreover, the comparative examination of the frequency of SCEs in various animal and plant species has revealed a good correlation between DNA content per cell and rate of SCE (Kato, 1977a; Wolff, 1977). In fact, when the number of SCEs per metaphase is related to the amount of DNA per cell nucleus, comparable SCE/DNA content ratios are obtained (Table 10.2).

The lack of spontaneous SCE, observed in cells treated with 1 μg/ml of BrdUrd, cannot be explained on the basis of the low DNA content per diploid nucleus of *D. melanogaster*. Although the *Drosophila* genome contains about one-twentieth the DNA of a mammalian cell (0.72 pg/4C, see Table 10.2), the 183 female and 178 male metaphases examined (Table 10.1) contain a total of 260 pg of DNA. Therefore, even if they had the lowest rate of SCE per picogram of DNA so far reported (0.14 SCE/pg DNA, see Table 10.2), they would have shown about 36 SCEs. Moreover, at the higher concentrations of BrdUrd, the *Drosophila* neuroblasts also exhibit an SCE rate per picogram of DNA that is considerably lower than that observed in various *in vivo* and *in vitro* systems (Table 10.2).

A very low frequency of SCEs in *Drosophila* neuroblast cells was first reported by Wienberg (1977), who cultured ganglia for 15 hours in a medium containing 15 μg/ml of BrdUrd. He observed a frequency of 0.005 SCEs/chromosome, corresponding to 0.03 SCEs/cell and to a rate of 0.042 SCEs per picogram of DNA. Since Wienberg scored SCEs only in the euchromatin because of the poor differentiation of heterochromatin in his preparations, and since in *Drosophila* about three-quarters of the SCEs occur in the heterochromatic regions (Gatti et al., 1979), his rate of 0.042 SCEs per picogram of DNA is in excellent agreement with the rates reported in Table 10.2.

Taken as a whole the studies on SCE in *Drosophila* indicate that in this organism SCEs are not spontaneous events in either mitotic or meiotic cells. This behavior suggests two possible explanations: (1) the lack of spontaneous SCEs could be a common feature of all organisms. *Drosophila*, however, would differ from the other organisms in its peculiar resistance to BrdUrd that permits a discernible differentiation of sister chromatids to be achieved without a concomitant induction of SCEs. (2) *Drosophila* could actually differ from other organisms in its lack of a spontaneous background level of SCEs.

Schonberg (1980) suggested that special consideration be given to the second possibility. He argued that the lack of spontaneous SCEs in *D. melanogaster* is a necessity imposed by the somatic pairing of homologous chromosomes. Assuming that SCE and mitotic recombination arise through the same molecular mechanisms, he suggested that organisms such as *Drosophila*, in which homologous chromatids are strictly paired, would have evolved a low somatic recombinogenic activity to avoid mitotic recombination that could lead to the homozygosity of deleterious genes in many somatic lineages. On the other hand, organisms such as mammals or plants lacking somatic pairing could tolerate relatively high somatic recombinogenic activities that would result in the production of some SCEs but not in mitotic recombination.

Although Schonberg's hypothesis is very appealing I feel that it should be considered with caution. In fact, several observations (see Gatti et al., 1980 and Section V of this chapter) indicate that some of the steps involved in the formation of symmetrical chromatid interchanges—interpreted as the cytological picture of mitotic crossing over—do not participate in the formation of SCEs.

IV. SCEs IN RING CHROMOSOMES

Ring chromosomes offer the unique possibility of evaluating the frequency of SCEs in the absence of any exogenous agents for sister chromatid differentiation. Indeed, the frequency of SCEs can be inferred from the rate of dicentric-ring formation in cells carrying monocentric-ring chromosomes (Figure 10.3 and 10.4).

Because of the double chromatid bridges formed at anaphase, dicentric rings are not transmittable from one cell generation to the next. Con-

TABLE 10.2

***Rates of SCEs per Picogram of DNA in Various Animal and Plant Systems*[a] in vivo *and* in vitro *After Two Rounds of DNA Replication in the Presence of Different Concentrations of BrdUrd*[b]**

Species and systems	DNA content per metaphase (4C)	SCEs/pg DNA at different BrdUrd concentrations			
		≤ 1 μg/ml	2.5–3.5 μg/ml	8–12 μg/ml	25–35 μg/ml
Human lymphocytes *in vitro*	11.68 (24)	0.90 (7)	0.35 (3) 0.61 (4) 0.68–0.96 (5) 0.69 (6)	1.07 (1) 0.55 (8) 0.68 (10)	1.50 (1) 0.59 (2) 1.95 (9)
Various lines of human fibroblasts *in vitro*	11.68 (24)	0.28–0.36 (12)	0.77–1.01 (11) 0.80 (4) 0.27–0.36 (12)	0.45–0.63 (12)	
Chinese hamster ovary cells *in vitro*	10.75 (25)	0.73–1.09 (13)	1.40 (13) 1.13 (14)		
Chinese hamster cells line D6 *in vitro*	10.75 (25)	0.21 (15)	0.26 (15)	0.46 (15)	
Primary cell cultures *in vitro* of 23 mammalian species belonging to 11 orders	6.8–16.5 (25)	0.32–0.82 (16)			
Rat bone marrow cells *in vivo*	10.63 (25)	0.14 (17)			

Vicia faba root tips	86.23	(26)							0.24	(18)[c]
Allium cepa root tips	105.64	(26)							0.83	(19)
									0.42	(19)[c]
D. melanogaster ganglion cells ♀♀	0.72	(27)	0–0.025[d]	(20)	0.13	(20)	0.26	(20)	0.35	(20)
D. melanogaster ganglion cells ♂♂			0	(20)	0.028	(20)	0.17	(20)	0.18	(20)
Mouse bone marrow cells *in vivo*	10.75	(25)	0.37	(21)	Multiple injections of BrdUrd					
Mouse spermatogonia *in vivo*	10.75	(25)	0.17	(22)	Multiple injections of BrdUrd					
Chicken embryonic cells *in vivo*	5.00	(28)	0.15	(23)	25 μg BrdUrd/embryo					

[a]In all systems in the upper part of the table except one (rat bone marrow cells) the concentration of BrdUrd to which the cells were exposed is known. However, in this system the SCEs were scored at the minimum BrdUrd concentration that permitted a discernible differentiation of sister chromatids (Tice et al., 1976). It has been assumed that this concentration is less than 1 μg/ml as in the *in vitro* systems. In the three systems listed in the bottom of the table the final concentration of BrdUrd to which the cells were exposed is not known.

[b](1) Latt, 1974; (2) Chaganti et al., 1974; (3) Beek and Obe, 1975; (4) Hayashi and Schmid, 1975; (5) Latt et al., 1975; (6) Sperling et al., 1975: (7) Tice et al., 1975; (8) Crossen et al., 1977; (9) Lambert et al., 1976; (10) Morgan and Crossen, 1977; (11) Wolff et al., 1975; (12) Kato and Stich, 1976; (13) Wolff and Perry, 1974; (14) Perry and Evans, 1975; (15) Kato, 1974; (16) Kato, 1977a; (17) Tice et al., 1976; (18) Kihlman and Kronborg, 1975; (19) Schvartzman and Cortes, 1977; (20) Gatti et al., 1979; (21) Vogel and Bauknecht, 1976; (22) Allen and Latt, 1976; (23) Bloom and Hsu, 1975; (24) Mirsky and Ris, 1951; (25) the DNA content per metaphase has been calculated from the densitometric measurements relative to human cells reported by Kato, 1977a; (26) the DNA contents of *Vicia faba* and *Allium cepa* metaphases were calculated from the average amount of DNA per root tip cell (McLeish and Sunderland, 1961) and from the relative proportion of 2C, 4C, and 8C nuclei in the same material (Sunderland and McLeish, 1961); (27) Rasch et al., 1971; (28) Ris and Mirsky, 1949.

[c]Cells exposed to BrdUrd for only one round of DNA replication.

[d]Frequency of SCEs inferred from the rate of dicentric ring formation of *C(1)TR 94-2*.

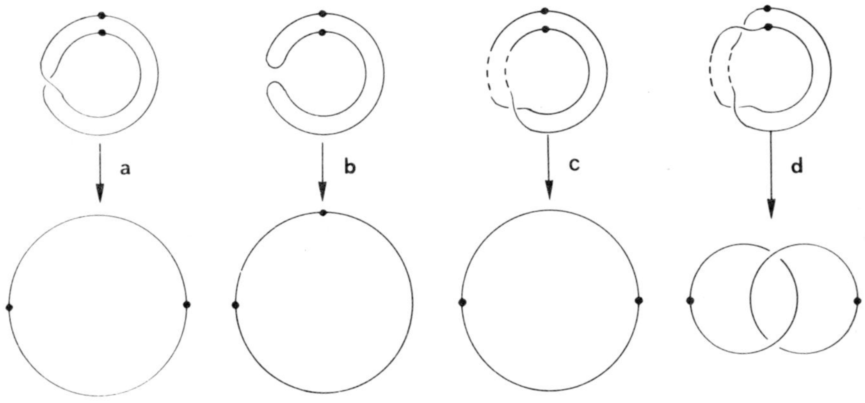

Fig. 10.3. Origin of dicentric rings. (*a*) An SCE, whatever its localization along the ring, produces a double dicentric ring with equal intercentromere distances (cf. Uhl, 1965 and Wolff et al., 1976). (*b*) An isochromatid deletion with sister union produces a double dicentric ring with variable intercentromere distances. A dicentric ring can originate if two chromatids broken at the same locus undergo a restitution of the break after torsion (Bauer, 1942). (*c*) A twisting through 180° produces a symmetrical dicentric ring, while (*d*) a twisting through 360° produces two interlocked single rings. Dicentric rings might be generated by the rotation of the plane of split (Moebius strip phenomenon) after errors occurring during replication of DNA (McClintock, 1938). If we assume that the twisting units are chromatid subunits instead of entire chromatids, *c* gives an up-to-date version of this hypothesis. With a unineme chromosome model, this hypothesis would imply that switches in polarity of chromosomal DNA can occur that would produce dicentric rings after the replication of DNA. It should be noted that such switches in polarity have been found to occur in mammalian chromosomes (Wolff et al., 1976). The events depicted in *a* and *c* are equivalent in that they produce double dicentric rings with equal intercentromere distances and, if they occurred in BrdUrd-labeled rod chromosomes, they would produce cytologically detectable SCEs. Thus they can be considered as hypothetical mechanisms of formation of SCEs. Two SCEs within a ring produce either two free monocentric rings or, alternatively, two interlocked rings if the second SCE does not counteract the first (this case is not depicted here).

sequently, the dicentric rings scorable at metaphase must have been formed during that mitotic cell cycle and not in earlier cycles. Thus in contrast with the [^{3}H] thymidine and most BrdUrd labeling methods that reveal the SCEs formed during two successive cell cycles, the analysis of dicentric-ring formation provides a measure of the rate of SCEs occurring in a single cell cycle.

The cytological analysis of the mitotic behavior of ring chromosomes was pioneered by McClintock (1938), who observed clear figures of double-sized dicentric rings in somatic cells of maize heterozygous for a ring chromosome. She interpreted these chromosome configurations as the cytological consequence of sister-strand exchanges occurring spontaneously in monocentric ring chromosomes. Later these observations

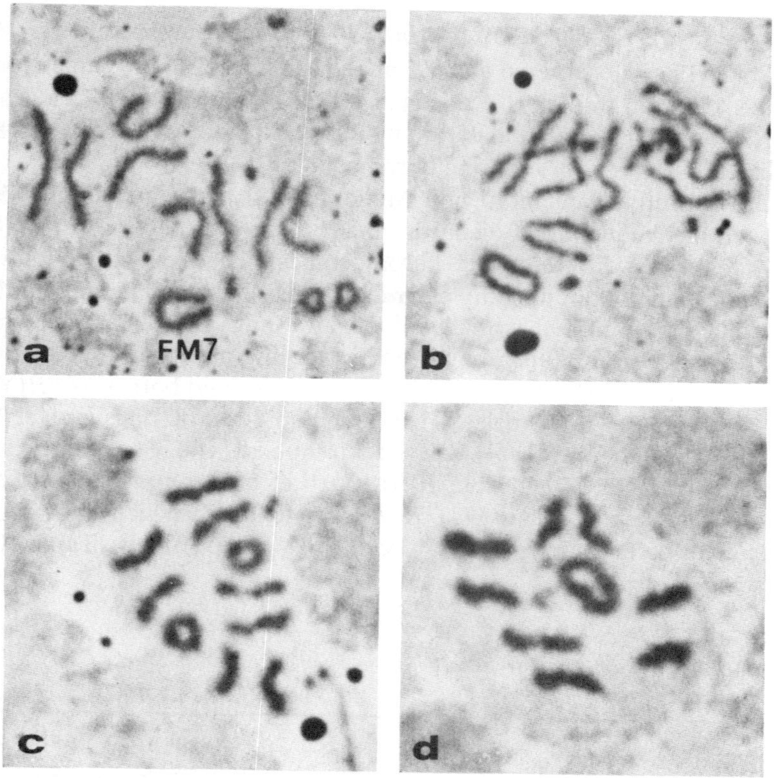

Fig. 10.4. Neuroblast c-anaphases showing monocentric and dicentric rings. (*a*) and (*b*) *R(1)2/FM7*. (*c*) and (*d*) *C(1)TR 94-2/0*.

were substantiated by the study of the frequencies of bridges present in meiotic anaphase I and II of maize heterozygous for a ring chromosome, so leading Schwartz (1953) to conclude that in maize SCE occurs in meiosis as well.

In *D. melanogaster* several ring chromosomes have been synthesized and the mitotic and meiotic behavior of some of them has been analyzed in detail (for a review, see Leigh, 1976). Early cytological investigations were carried out by examining neuroblast chromosome preparations of larvae heterozygous for stable and unstable ring X chromosomes. Braver and Blount (1949) and Brosseau (1966) found interlocking rings and double-sized dicentric rings, while Hinton (1955) observed clear examples of dicentric rings but not of interlocking rings. Moreover, in these three investigations different frequencies of anaphase bridges were recorded. Braver and Blount found 22 and 12% anaphase bridge configurations in

*R(1)2, In-w^{vC}/In(1)dl-49** and *R(1)2/In(1)dl-49** females, respectively, while Brosseau (1966) observed 6 and 3% anaphase bridges in the same type of females. Hinton found a frequency of 10.1% cells with bridges in *R(1)2,In-w^{vC}/In(1)dl-49* females, but only 1.6% in cells carrying a stabilized derivative of this ring chromosome.

More recently Yamamoto and Miklos (1978) and Gatti et al. (1979) analyzed the frequency of dicentric ring formation in several *Drosophila* stocks carrying different ring chromosomes. They examined only c-anaphases and did not find any clear example of interlocking rings. Moreover, they observed that virtually all dicentric rings had equal intercentromere distances and thus they were probably produced by an SCE. Yamamoto and Miklos examined *R(1)2*/Y* males and both X·Y^s/*R*(Y)*L** males and $\overline{XX}$/*R*(Y)*L** females. They found no dicentric rings in over 1000 *R(1)2*/Y cells, whereas they observed a frequency of 0.3% dicentric rings in cells carrying *R*(Y)*L* (Table 10.3). Gatti et al. examined *R(1)2/FM7** heterozygous females and *C(1)TR 94-2/0** females, which showed frequencies of dicentric rings of 0.58 and 0.18%, respectively (Table 10.3).

TABLE 10.3
Frequency of Spontaneous Dicentric Rings in Four Drosophila *Stocks Carrying Monocentric Ring Chromosomes*

Stock	Number of cells scored	Number of dicentric rings	Frequency of dicentric rings (%)	Inferred[c] frequency of SCEs (%)
R(1)2/FM7[a]	3282	19	0.58	11.6
C(1)TR 94-2/0[a]	2820	5	0.18	1.8
R(1)2/Y[b]	1000	0	0.00	0.0
R(Y)*L*[b]	1663	5	0.30	5.4

[a]Data from Gatti et al. (1979).

[b]Data from Yamamoto and Miklos (1978).

[c]The frequency of SCEs per genome for two cell cycles was calculated by multiplying twice the rate of dicentric rings by the ratio of the dimension of the entire genome to that of the ring chromosome.

**R(1)2* is a ring X chromosome; *R(1)2,In-w^{vC}* is a radiation-induced derivative of *R(1)2*; *C(1)TR 94-2* is a compound-ring chromosome synthesized from a tandem attached X chromosome; *R*(Y)*L* is a ring chromosome formed by the long arm of the Y chromosome. *In(1)dl-49* and *FM7* are multiply inverted X chromosomes. All these chromosomes, except *FM7* (Merriam, 1968), are described by Lindsley and Grell (1968). $\overline{XX}$ is an attached-X chromosome; X·Y^s is an X chromosome with Y^s appended as a second arm.

These data indicate that each individual ring chromosome has its own rate of dicentric-ring formation that is largely independent of size. For example, although the *R(1)2* studied by Yamamoto and Miklos (1978) and the *R(1)2* studied by Gatti et al. (1979) both derive from the original *R(1)2* synthesized by Schultz and Catcheside (1938), they seem to have very different rates of dicentric-ring formation.

Moreover, the *R(1)2* examined by Gatti et al. (1979) has a significantly higher frequency of dicentric-ring formation than *C(1)TR 94-2*. Since *C(1)TR 94-2* is double the size of *R(1)2* and contains almost the same proportion of heterochromatin as *R(1)2* (Gatti et al., 1979), it should be about twice as susceptible to any event that converts it into a dicentric ring. Thus, when rates of dicentric-ring formation are adjusted for the relative dimension of the two chromosomes, *R(1)2* has a spontaneous rate of dicentric ring formation 5.6 times as high as that of *C(1)TR 94-2*.

These *R(1)2* and *C(1)TR 94-2* chromosomes were also tested for their sensitivity to BrdUrd-induced SCEs (Gatti et al., 1979). After exposure to 6 μg/m1 of BrdUrd for 7 hours (about one cell cycle) *R(1)2* and *C(1)TR 94-2* exhibit frequencies of dicentrics of 1.96 and 1.74%, respectively (Table 10.4). Thus, if size is taken into account, *R(1)2* also appears twice as susceptible as *C(1)TR 94-2* to the induction of dicentrics by BrdUrd.

To see whether the different behavior of these ring chromosomes reflected different background levels of SCEs in the two stocks *R(1)2/FM7* and *C(1)TR 94-2/0*, Gatti et al. (1979) examined these stocks for the frequency of SCEs in BrdUrd-labeled rod chromosomes and found that they have comparable levels of SCEs (Table 10.4). However, if the frequency of SCEs observed in M2 rod chromosomes is compared with that inferred from the rate of dicentric-ring formation, these frequencies are

TABLE 10.4
***Frequency of BrdUrd-Induced (6 μg/ml for 7 h) Dicentric Rings and Its Comparison With the Frequency of SCEs in BrdUrd-Labeled (6 μg/ml for 13 h) M2 Metaphases in two* Drosophila *Stocks Carrying Monocentric Ring Chromosomes*[a]**

Stock	Frequency of dicentric rings (%)	Inferred[b] frequency of SCEs (%)	Frequency of SCEs in M2 chromosomes (%)
R(1)2/FM7	1.96	39.2	11.3
C(1)TR 94-2/0	1.74	17.4	15.0

[a]Data from Gatti et al. (1979).

[b]Frequency of SCEs per genome for two cell cycles inferred from the rate of dicentric ring formation.

very similar in *C(1)TR 94-2*, but in *R(1)2* the inferred frequency is about three times the observed frequency (Table 10.4).

Taken as a whole these studies in *Drosophila* indicate that different ring chromosomes have different rates of both spontaneous and BrdUrd-induced dicentrics and that these rates are largely independent of ring size. Even rings having a common origin, such as the *R(1)2* derivatives, appear to differ markedly from one another in the rate of dicentric formation, which is in accord with the observation that *Drosophila* ring chromosomes tend to evolve into a series of new chromosomes often having a greater stability (for review, see Leigh, 1976).

One interpretation of these results is that the frequency of SCEs in a given ring chromosome is determined by factors, in addition to size, intrinsic to its structure. Although these factors are currently unknown, the results on ring chromosomes can be tentatively formalized by assuming the existence of three classes of ring chromosomes: (1) rings structurally less sensitive than rod chromosomes to both spontaneous and BrdUrd-induced SCEs; (2) rings having the same sensitivity to SCEs as rod chromosomes both in the presence and in the absence of BrdUrd; and (3) rings structurally more susceptible than rod chromosomes to both spontaneous and BrdUrd-induced SCEs. However, it should be pointed out that at concentrations of BrdUrd at which a saturation effect occurs, the differences among these three types of rings in the rate of SCEs could be greatly reduced.

According to this rationale, the *R(1)2* studied by Gatti et al. (1979) belongs to the third category of rings because it exhibits spontaneous and BrdUrd-induced levels of SCEs much higher than those of rod chromosomes. By contrast, *C(1)TR 94-2* behaves almost exactly like rod chromosomes, showing, in both the presence and the absence of BrdUrd, a rate of dicentric ring formation that corresponds to the SCE level observed in rod chromosomes. From its spontaneous rate of dicentric formation, a rate of 0.009 SCEs per cell cycle can be inferred. Such a frequency of SCEs is compatible with the absence of SCEs found in 183 female metaphases labeled with 1μg/ml of BrdUrd (Table 10.1) and is very similar to the frequency of spontaneous chromosome aberrations, which in *Drosophila* females is 0.006/cell (Gatti et al., 1974b). Therefore, these data on *C(1)TR 94-2* constitute further support for considering SCE a phenomenon that does not occur spontaneously in *D. melanogaster* but that must be regarded, like spontaneous chromosome aberrations, to be the cytological manifestation of errors occurring during DNA metabolism.

A large amount of literature is available on the mitotic behavior of human ring chromosomes. It is worth noting that the situation for these chromosomes is quite similar to that existing in *Drosophila*. Indeed, there is evidence that the mitotic stability of different human ring chromosomes varies markedly independently of ring size (reviewed by Kistenmacher and Punnett, 1970). Particularly interesting in this context is the work of Brewen and Peacock (1969), who studied the occurrence of dicentric rings in the leucocytes of a human male heterozygous for a ring chromosome in both the presence and the absence of tritiated thymidine. They analyzed only c-anaphases and found a spontaneous frequency of dicentric ring formation of 11.9%. Since they examined a ring that was approximately the length of an average human chromosome, this rate of dicentric formation corresponds to a frequency of about 0.24 SCEs per chromosome per two cell cycles and hence to a rate of 1.05 SCEs/pg DNA, which is comparable to the rates observed in BrdUrd-treated human chromosomes (see Table 10.2).

The frequency of dicentric rings in tritium-labeled c-anaphases was 30.2 and 26.0% in M1 and M2 cells, respectively. Thus incorporated tritiated thymidine significantly increased the yield of SCEs in ring chromosomes, inducing a similar rate of SCEs in either the first or the second cell division cycle. However, the frequency of SCEs in M2 rod chromosomes was 27.7% and hence about half the frequency inferred from the rate of dicentric-ring formation. Thus this human ring chromosome appears to behave just like the *R(1)2* of *D. melanogaster*. Therefore, it is possible that this human ring was structurally more susceptible than the corresponding rod chromosome to SCEs and did not provide a correct measurement rate of SCEs in man.

Nevertheless, I feel that once tested for their sensitivity to SCEs (according to the above rationale), some of the numerous human ring chromosomes could be advantageously employed to establish whether or not SCEs occur spontaneously in man.

V. SCEs IN MEIOTIC AND MUTAGEN-SENSITIVE MUTANTS

The molecular mechanisms underlying the formation of SCEs and their relationships with known repair processes are still poorly understood. Largely unknown also are the relationships among SCE, chromosome aberrations, and meiotic recombination. For example, although it has been well established that chromatid interchanges, SCE, and meiotic recombination involve physical exchanges between whole chromatids, it is not known whether these processes occur by the same molecular mechanisms.

Sister chromatid exchanges, chromosome aberrations, and recombinant chromosomes are the end products of a complex series of DNA metabolic events. Consequently, one approach to elucidating the molecular mechanisms underlying these processes is genetic dissection by mutants affecting different aspects of DNA metabolism.

This approach has led to considerable progress with several rare inherited human diseases that affect DNA metabolism and produce elevated frequencies of spontaneous chromosome aberrations (reviewed by Evans, Chapter 7). Cytological studies in Bloom's syndrome, Fanconi's anemia, and ataxia telangiectasia have shown that these conditions are characterized by different chromosomal effects (German, 1972; German et al., 1974; Schroeder and German, 1974). In Bloom's syndrome there is a general increase in all types of chromosome aberrations with a characteristic high incidence of symmetrical (X type) interchanges involving homologous chromosomes at homologous sites. Fanconi's anemia and ataxia telangiectasia also increase all types of aberrations but do not specifically produce symmetrical exchanges between homologous chromosomes. In addition, while Fanconi's anemia and ataxia telangiectasia have normal levels of SCEs, in Bloom's syndrome there is a dramatic increase in the frequency of SCEs (Chaganti et al., 1974; Latt et al., 1975; Galloway and Evans, 1975). These results thus suggest that some chromosome aberrations and SCEs arise by mechanisms that are, at least in part, different.

Although the human mutations affecting DNA repair processes and/or chromosome stability are of great intrinsic interest, they cannot be used for a systematic dissection of the mechanisms of chromosome breakage and rejoining. Indeed only a few of these mutations are available and new human genes involved in DNA metabolism are discovered only rarely. *D. melanogaster* does not present such limitations because in this organism a large number of mutants have been discovered that affect DNA metabolism and often cause chromosome instability (for reviews see Baker et al., 1976a, b; Baker et al., 1980; Boyd et al., 1980; Smith et al., 1980; Würgler and Graf (1980). Moreover, the easy genetic manipulation of *D. melanogaster* permits the collection of additional mutants and the study of the interaction among mutations by construction of multiply mutant genomes. *Drosophila* has the further advantage that the chromosomal effects of mutations can be compared with those on both meiotic and mitotic recombination.

Two classes of mutations, meiotic mutants (mutants affecting recombination and/or chromosome segregation during meiosis; symbol *mei*) and mutagen-sensitive mutants (mutants hypersensitive to killing by mutagens; symbol *mus*) have been shown to be rich sources of defects in processes necessary for maintaining the integrity of nonmutagenized mitotic chromosomes (Baker et al., 1976a, b; Baker et al., 1978; Baker and Smith, 1979; Gatti, 1979). Some of these mutants also have been assayed for their sensitivity to a variety of mutagens and carcinogens and for the normality of several DNA repair pathways (reviewed by Boyd et al., 1980 and by Smith et al., 1980). Table 10.5 summarizes the properties of selected X-linked mutants that have been examined genetically (Baker et al., 1978; Baker and Smith, 1979) and cytologically (Gatti, 1979) for their effects on chromosome stability. The genetic analysis of the effects of these mutants on mitotic chromosomes was performed by examining the frequency, size and morphology of somatic spots in flies homozygous for a particular mutant and simultaneously heterozygous for one or more recessive cuticular markers.

Both genetical and cytological tests showed that *mei-9* and *mei-41*, among meiotic mutants, and *mus-102*, *mus-105*, and *mus-109*, among mutagen-sensitive mutants, exhibit elevated frequencies of chromosome breakage, whereas *mei-218*, *mei-251*, *mus-103*, *mus-104*, and *mus-106* do not affect the frequency of either somatic spots or spontaneous chromosome aberrations. In two mutants, *mei-352* and *mus-101*, increased chromosome instability was detected genetically but not cytologically. Thus these mutants affect chromosome stability by way of mechanisms that do not have clear cytological consequences in neuroblast cells of third instar larvae.

These genetic and cytological studies (reviewed by Baker et al., 1980) demonstrate that many functions utilized in meiotic recombination and/or in the repair of mutagen damage are also required for maintaining the integrity of nonmutagenized somatic chromosomes. These results are interpreted as indicating that the products of *mei-9*$^+$, *mei-41*$^+$, *mus-102*$^+$, *mus-105*$^+$, and *mus-109*$^+$ function in the removal of either spontaneous or mutagen-induced lesions in chromosomal DNA. If such lesions, which may be either primary damages in DNA or intermediates in their metabolism, are not properly metabolized, they could initiate a chain of events leading to the formation of chromosome aberrations.

The cytological analysis of mutant alleles at the *mei-9*, *mei-41*, *mus-102*, *mus-105*, and *mus-109* loci showed that they produce substantial increases in spontaneous chromosome aberrations (Gatti, 1979). In all these mutants only chromatid and isochromatid breaks were observed, there being almost no chromatid and chromosome exchanges. However, since in irradiated wild-type cells with a frequency of breakage of 12%, an exchange frequency of only 0.2% is detected (Gatti et al., 1974a), and since the highest frequency of breakage recorded in these mutants is 14%, this finding does not necessarily imply a defect in exchange formation.

Although chromosome breakage is the predominant pattern of aberrations in all these mutants, very different proportions of chromatid and

TABLE 10.5
***Summary of Properties of Selected X-Linked Meiotic and Mutagen-Sensitive Mutants*[a,b]**

Locus	Map position	Recombination defective[c]	Increased gonial recombination in males	Exhibits sensitivity to				Defective in		
				MMS	HN2	UV	γ or X ray	Repair of single-strand breaks	Excision repair	Post-replication repair
mei-218	0.57	++	−	−	−	−	−	−	−	−
mei-251	Unlocated	+	−			−	−			
mei-352	Unlocated	+	−			−	−			
mei-41	54.2	++	−	++	++	++	++	−	−	++cas
mei-9	6.5	++	−	++	++	++	++	−	++	−
mus-101	44.8	(++)		++	++	+	++	−		+cas
mus-102	0.5	(+)		++	−	−	+	−		−
mus-103	36–44	(−)		+	−	−	−	−	−	−
mus-104	53	(−)		++	++	+	+	−		+cis
mus-105	14.8	(−)		++	−	−	+	−		−
mus-106	36–44	♀ sterile		+	−	−	+			
mus-109	30.2	♀ sterile		++	+	+	++			

[a]The studies on these mutants have been reviewed by Baker et al., 1976a,b; Boyd et al., 1980; Smith et al., 1980; Würgler and Graf, 1980. Data on gonial recombination from Lutken and Baker, 1979.

[b]++: exhibits this characteristic; +: exhibits this characteristic weakly; −: does not exhibit characteristic; blank: not tested. MMS: methyl methanesulfonate; HN2: nitrogen mustard; cas: caffeine sensitive; cis: caffeine insensitive.

[c]Parentheses, recombination status inferred from the frequency of nondisjunction (see Baker et al., 1976a).

isochromatid deletions are produced by mutants at different loci (Gatti, 1979). In mutant alleles at *mei-41, mus-102, mus-105*, and *mus-109* a third to a half of all the breaks are isochromatid deletions, whereas in mutant alleles at *mei-9* most breaks are chromatid deletions.

To further investigate the effect of these meiotic and mutagen-sensitive mutants on chromosome breakage and rejoining, the induction of chromosome aberrations by X rays has been studied (Gatti et al., 1980). The experiments have focused on the repair- and recombination-defective mutants at the *mei-9* and *mei-41* loci. Mutant alleles at these loci were irradiated in S phase with 220 R and the frequencies and types of induced aberrations were determined at the subsequent metaphase. Mutants at both *mei-9* and *mei-41* are about 10 times as sensitive as nonmutant controls to the induction of chromosome aberrations. Here again in *mei-9*-bearing cells most aberrations are chromatid deletions, whereas in the presence of *mei-41* mutants similar frequencies of chromatid and isochromatid deletions are observed. These findings have been taken as evidence for different molecular mechanisms underlying the formation of chromatid and isochromatid deletions. Indeed, they strongly suggest that the lesions (spontaneous and X-ray-induced) that are substrates of *mei-9*$^+$ and *mei-41*$^+$ have very different probabilities, if unrepaired, of producing chromatid and isochromatid breaks.

Mutant alleles at *mei-9* and *mei-41* also differ strikingly in their effects on the induction of chromatid interchanges by X rays (Gatti et al., 1980). In *mei-9*-bearing cells almost no chromatid interchanges are present (0.02%) when the frequency of total aberrations is 1.0–1.3/cell. In *mei-41* cells irradiated with the same dose of X rays total aberrations are 1.1–1.2/cell, but there are about 10% chromatid interchanges. However, in nonmutant control cells of two stocks irradiated with 1320 R 18 and 19% chromatid interchanges are observed when total aberrations are 0.71 and 0.76/cell, respectively. Thus these data indicate that the *mei-9*$^+$ product is necessary for chromatid rejoining. Moreover, since the incidence of interchanges is expected to increase approximately as the square of the frequency of breaks, these data also suggest that the *mei-41*$^+$ function is utilized in a step necessary for the generation of interchanges.

Another way in which *mei-41* mutants affect the formation of chromatid interchanges is by altering the relative proportions of symmetrical (X-type) and asymmetrical (U-type) exchanges (Gatti et al., 1980). In *mei-41*- and *mei-41*^{195}-bearing cells irradiated in the S phase, symmetrical interchanges constituted 12.4 and 6.8% of total exchanges, respectively, whereas in wild-type control cells irradiated in the same phase 30% of exchanges were symmetrical. In *mei-9* alleles only four exchanges were detected and two of these were symmetrical. These observations suggest therefore that the *mei-41*$^+$ product is preferentially utilized in the generation of symmetrical exchanges, while the function specified by *mei-9*$^+$ participates in the formation of both symmetrical and asymmetrical interchanges. This behavior implies that symmetrical interchanges arise by way of molecular mechanisms different from those underlying the formation of asymmetrical interchanges.

To investigate whether the products of *mei-9*$^+$ and *mei-41*$^+$ are required for the formation of SCEs, as well as meiotic recombination and the production of interchanges, the levels of SCEs in these mutants have been determined (Gatti et al., 1980). Neuroblast cells of *mei-41* males exposed for 13 hours to 9 μg/ml of BrdUrd exhibit the same frequency of SCEs as either *FM7* or Oregon-R males used as controls (Table 10.6). On the other hand, in cells carrying the mutant alleles at *mei-9* the frequency of SCEs is about 70% of the control rate. Thus while *mei-41* mutants have normal levels of SCEs, the *mei-9* alleles appear to be slightly defective in the formation of SCEs.

By scoring BrdUrd-substituted chromosomes it also has been found that *mei-9*-bearing cells exposed to BrdUrd have the same frequency of chromosome aberrations as observed in untreated cells, whereas mutants at *mei-41* exhibit about twice as many aberrations as untreated cells (Table 10.6). Moreover, most chromatid deletions scored in *mei-41* in-

TABLE 10.6
Sister Chromatid Exchanges and Chromosome Aberrations in Meiotic Mutants Exposed to BrdUrd (9 μg/ml) for Two Rounds of DNA Replication[a]

Genotype	SCE: No. of cells scored	SCE: No. of SCEs	SCE: %	Spontaneous chromosome aberrations in untreated neuroblasts (%)	Chromosome aberrations in the presence of BrdUrd (%)
Oregon-R ♂	672	76	11.0	0.34	0.59
FM7/Y	678	74	10.91	0.20	0.43
mei-9^b/Y	605	45[b]	7.44	8.47	8.92
mei-9^{ATI}/Y	624	44[c]	7.05	3.37	6.24
mei-41/Y	658	69	10.48	7.36	17.62
mei-41^{195}/Y	837	91	10.87	10.25	21.98

[a]Data from Gatti et al., 1974b; Gatti, 1979; Gatti et al., 1979; Gatti et al., 1980.

[b]Significantly different from the control in χ^2 test with $P < 0.05$.

[c]Significantly different from the control in χ^2 test with $P < 0.025$.

volve the BrdUrd bifilarly substituted chromatid and this gives further support to the specific sensitivity of *mei-41* to BrdUrd.

These data clearly demonstrate that *mei-9* and *mei-41* mutants, which exhibit about a thirtyfold increase in the frequency of spontaneous aberrations, do not show a corresponding increase in the frequency of SCEs. In addition, *mei-41*-bearing cells exhibit a marked sensitivity to the induction of chromosome aberrations by BrdUrd that is not paralleled by an elevated sensitivity to SCEs. Furthermore, mutant alleles at other loci (*mus-102, mus-105,* and *mus-109*) producing high levels of spontaneous chromosome aberrations in somatic cells do not produce a concomitant increase in the frequency of SCEs (Gatti, Pimpinelli, and Baker, unpublished observations). Thus, in view of the results on Fanconi's anemia and ataxia telangiectasia, it can be suggested that the chromosomal lesions responsible for the formation of SCEs are generally different from those that lead to chromosome aberrations. The only genetic defect that produces lesions capable of generating both chromosome aberrations and SCEs is Bloom's syndrome (reviewed by Evans, Chapter 7).

Sister chromatid exchanges, symmetrical interchanges, and meiotic crossing over all involve X-type physical exchanges of whole chromatids and hence might be expected to be generated by similar, if not identical, sequences of events. Nevertheless, our studies on *mei-9* and *mei-41* mutants strongly suggest that these three processes do not occur by the same molecular mechanisms. Mutants at *mei-41* markedly affect meiotic recombination by decreasing the frequency of exchange events (50% of wild type) and altering their distribution along the chromosomes (Baker and Carpenter, 1972). They also affect the formation of X-ray-induced symmetrical interchanges but have no detectable effects on SCE. Mutations at *mei-9* drastically reduce the frequency of meiotic recombination (8% of wild type) without altering the position of exchanges (Baker and Carpenter, 1972; Carpenter and Sandler, 1974) and suppress the formation of X-ray-induced interchanges, but produce only a slight decrease in the frequency of SCEs (70% of wild type). Thus it can be suggested that *mei-9*$^+$ and *mei-41*$^+$ identify steps shared by meiotic recombination and the formation of symmetrical interchanges that do not participate in the generation of most SCEs.

The slight reduction in the SCE rate observed in *mei-9*-bearing cells could be interpreted as indicating that *mei-9*$^+$ has only a peripheral role in the formation of SCEs. Alternatively, if SCEs arise by two mechanisms, one operating at the replication sites of DNA and the other acting postreplicationally in a fashion similar to meiotic crossing over (Kato, 1977a, b), then mutations at *mei-9* could inhibit the formation of those SCEs that arise through the latter mechanism.

Finally, these studies provide some insight into the relationships between the known DNA repair processes and SCE. *Mei-41*-bearing cells, which are defective in a caffeine-sensitive pathway of postreplication repair (Boyd and Setlow, 1976), have a normal level of SCEs. A normal frequency of SCEs has been also observed in *mus-104* cells (Gatti, Pimpinelli, and Baker, unpublished observations), which are defective in a caffeine-insensitive pathway of postreplication repair (Boyd and Setlow, 1976). These data agree very well with the observation that the variant forms of xeroderma pigmentosum (XP) defective in postreplication repair have normal levels of SCEs (Wolff et al., 1975; Kato and Stich, 1976) and support the notion that the formation of SCEs is not a cytological manifestation of postreplication repair.

Mei-9-bearing cells, which are defective in an incision step of excision repair (Boyd et al., 1976), do exhibit a slight reduction in the level of SCEs. By contrast, the XP cells defective in the first steps of excision repair have normal levels of spontaneous SCEs and are very sensitive to the induction of SCEs by a variety of chemical mutagens and carcinogens (Wolff et al., 1975; Wolff et al., 1977). However, this is not the only parameter by which *mei-9* cells differ from excision-defective XP cells. In these XP cells, unlike in *mei-9* neuroblasts, no increase in the frequency of spontaneous chromosome aberrations is found (Wolff et al., 1975; Wolff et al., 1977). Moreover, in contrast with excision-defective XP cells, *mei-9* mutants are extremely sensitive to ionizing radiations and MMS (for reviews see Cleaver and Bootsma, 1975; Baker et al., 1976a). Finally, *mei-9* mutants drastically reduce meiotic recombination (Baker and Carpenter, 1972), while XP patients are probably not defective in this process, since they exhibit normal levels of chiasmata in primary spermatocytes (Hultén et al., 1974). Taken together these data indicate that *mei-9*$^+$ specifies a function different from those defective in patients with xeroderma pigmentosum.

VI. CONCLUDING REMARKS

In *D. melanogaster* SCEs are not spontaneous events in either mitotic or meiotic cells. However, SCEs are observed when *Drosophila* neuroblasts are treated with relatively high concentrations of BrdUrd. Moreover, the response of *Drosophila* cells to mutagens and carcinogens is similar to that of mammalian cells. For example, methylmethanesulfonate substantially increases the frequency of SCEs in *Drosophila* neuroblasts, and mitomycin C can produce an "uncountable" number of SCEs (Pimpinelli and Gatti, unpublished observations).

What makes *Drosophila* one of the most useful organisms for studies on SCE is the possibility of employing mutants at loci that govern the metabolism of chromosomal DNA. The past few years have witnessed a veritable explosion in the discovery of such mutations, and additional mutants are currently being isolated in several laboratories. These mutants should not be used only to elucidate the relationships among SCEs, chromosome aberrations, and meiotic recombination, but also to correlate these processes with gene mutation. However, so far only a few of the extant *mei-* and *mus-* mutants have been tested for their effects on spontaneous mutation and no firm conclusions as to whether they affect this process have been reached (Smith, 1973; Carpenter, unpubl., cited in Baker et al., 1976a; Graf et al., 1979; Würgler and Graf, 1980).

REFERENCES

Allen, J. W. & Latt, S. A. (1976) Analysis of sister chromatid exchange formation in mouse spermatogonia as a new test system for environmental mutagens. *Nature* **260**, 449–451.

Baker, B. S., Boyd, J., Carpenter, A. T. C., Green, M. M., Nguyen, T. G., Ripoll, P., & Smith, P. D. (1976a) Genetic control of meiotic recombination and somatic DNA metabolism in *Drosophila melanogaster*. *Proc. Natl. Acad. Sci. U.S.* **73**, 4140–4144.

Baker, B. S. & Carpenter, A. T. C. (1972) Genetic analysis of sex chromosomal meiotic mutants in *Drosophila melanogaster*. *Genetics* **71**, 255–286.

Baker, B. S., Carpenter, A. T. C., Esposito, M., Esposito, R., & Sandler, L. (1976b) Genetic control of meiosis. *Ann. Rev. Genet.* **10**, 53–134.

Baker, B. S., Carpenter, A. T. C. & Ripoll, P. (1978) The utilization during mitotic cell division of loci controlling meiotic recombination and disjunction in *Drosophila melanogaster*. *Genetics* **90**, 531–578.

Baker, B. S., Gatti, M., Carpenter, A. T. C., Pimpinelli, S., & Smith, D. A. (1980) Effects of recombination-deficient and repair-deficient loci on meiotic and mitotic chromosome behaviour in *Drosophila melanogaster*. In *DNA Repair and Mutagenesis in Eukaryotes*. (W. M. G. Generoso, M. D. Shelby, & F. J. de Serres, Eds) Plenum Press, New York and London, pp. 189–208.

Baker, B. S. & Smith, D. A. (1979) The effects of mutagen-sensitive mutants of *Drosophila melanogaster* in non-mutagenized cells. *Genetics* **92**, 833–847.

Bauer, H. (1942) Röntgenauslösung von chromosomenmutationen bei *Drosophila melanogaster*. II. Die käufigkeit des primären bruchereignisses nach untersuchungen am ring-*X*-chromosom. *Chromosoma* **2**, 407–458.

Beadle, G. W. & Emerson, S. (1935) Further studies of crossing over in attached-X chromosomes of *Drosophila melanogaster*. *Genetics* **20**, 192–206.

Beek, B. & Obe, G. (1975) The human leukocyte test system. VI. The use of sister chromatid exchanges as possible indicators of mutagenic activities. *Humangenetik* **29**, 127–134.

Bloom, S. E. & Hsu, T. C. (1975) Differential fluorescence of sister chromatids in chicken embryos exposed to 5-bromodeoxyuridine. *Chromosoma* **51**, 261–267.

Boyd, J. B. & Setlow, R. B. (1976) Characterization of post replication repair in mutagen-sensitive strains of *Drosophila melanogaster*. *Genetics* **84,** 507–526.

Boyd, J. B., Golino, M. D., & Setlow, R. B. (1976) The *mei-9*[a] mutant of *Drosophila melanogaster* increases mutagen sensitivity and decreases excision repair. *Genetics* **84,** 527–544.

Boyd, J. B., Harris, P. V., Osgood, C. J. & Smith, K. E. (1980) Biochemical characterization of repair-deficient mutants of *Drosophila*. In *DNA Repair and Mutagenesis in Eukaryotes*. (W. M. G. Generoso, M. D. Shelby, F. J. de Serres, Eds) Plenum Press, New York and London, pp. 209–221.

Braver, G. & Blount, J. L. (1949) Somatic eliminations of ring chromosomes in *Drosophila melanogaster*. *(Abstr.) Genetics* **35,** 98.

Brewen, J. C. & Peacock, W. J. (1969) The effect of tritiated thymidine on sister chromatid exchange in a ring chromosome. *Mutat. Res.* **7,** 433–440.

Brosseau, G. E. (1966) Some aspects of ring chromosome behaviour in *Drosophila*. *Dros. Inf. Serv.* **41,** 97–99.

Carpenter, A. T. C. & Sandler, L. (1974) On recombination-defective meiotic mutants in *Drosophila melanogaster*. *Genetics* **76,** 453–475.

Chaganti, R. S. K., Schonberg, S., & German, J. (1974) A many-fold increase in sister chromatid exchanges in Bloom's syndrome lymphocytes. *Proc. Natl. Acad. Sci. U.S.* **21,** 4508–4512.

Cleaver, J. E. & Bootsma, D. (1975) Xeroderma pigmentosum: biochemical and genetic characteristics. *Annu. Rev. Genet.* **9,** 19–38.

Crossen, P. E., Drets, M. E., Arrighi, F. E., & Johnson, D. A. (1977) Analysis of the frequency and distribution of sister chromatid exchanges in cultured human lymphocytes. *Hum. Genet.* **35,** 345–352.

De Marco, A. & Cozzi, R. (1980) Chromosomal aberrations induced by caffeine in somatic ganglia of *Drosophila melanogaster*. *Mutat. Res.* **69,** 55–69.

Dewey, W. C. & Humphrey, R. M. (1965) Increase in radiosensitivity to ionizing radiation related to replacement of thymidine in mammalian cells with 5-bromodeoxyuridine. *Radiat. Res.* **26,** 538–553.

Evans, H. J. (1967) Repair and recovery at chromosome and cellular levels: similarities and differences. *Brookhaven Symp. Biol.* **20,** 111–131.

Evans, H. J. (1981) Sister chromatid exchanges in Human diseases, this volume.

Faccio Dolfini, S. (1978) Sister chromatid exchanges in *Drosophila melanogaster* cell lines *in vitro*. *Chromosoma* **69,** 339–347.

Galloway, S. M. & Evans, H. J. (1975) Sister chromatid exchange in human chromosomes from normal individuals and patients with ataxia telangiectasia. *Cytogenet. Cell Genet.* **15,** 17–29.

Gatti, M. (1979) Genetic control of chromosome breakage and rejoining in *Drosophila melanogaster*: spontaneous chromosome aberrations in X-linked mutants defective in DNA metabolism. *Proc. Natl. Acad. Sci. U.S.* **76,** 1377–1381.

Gatti, M., Pimpinelli, S., & Baker, B. S. (1980) Relationships between chromatid interchanges, sister chromatid exchanges and meiotic recombination in *Drosophila melanogaster*. *Proc. Natl. Acad. Sci. U.S.* **77,** 1575–1579.

Gatti, M., Tanzarella, C., & Olivieri, G. (1974a) Variation with sex of irradiation-induced chromosome damage in somatic cells of *Drosophila melanogaster. Nature* **247,** 151–152.

Gatti, M., Tanzarella, C., & Olivieri, G. (1974b) Analysis of the chromosome aberrations induced by x-rays in somatic cells of *Drosophila melanogaster. Genetics* **77,** 701–719.

Gatti, M., Pimpinelli, S., De Marco, A., & Tanzarella, C. (1975) Chemical induction of chromosome aberrations in somatic cells of *Drosophila melanogaster. Mutat. Res.* **33,** 201–212.

Gatti, M., Pimpinelli, S., & Santini, G. (1976) Characterization of *Drosophila* heterochromatin. I. Staining and decondensation with Hoechst 33258 and Quinacrine. *Chromosoma* **57,** 351–375.

Gatti, M., Santini, G., Pimpinelli, S., & Olivieri, G. (1979) Lack of spontaneous sister chromatid exchanges in somatic cells of *Drosophila melanogaster. Genetics* **91,** 255–274.

Gatti, M., Santini, G., Pimpinelli, S., & Olivieri, G. (1980) Lack of spontaneous sister chromatid exchanges in somatic cells of *Drosophila melanogaster.* A reply (letter to the Editor). *Genetics* **94,** No. 2.

German, J. (1972) Genes which increase chromosomal instability in somatic cells and predispose to cancer. In *Progress in Medical Genetics.* Vol. VIII (A. G. Steinberg, A. G. Bearn, Eds.) Grune & Stratton, New York, pp. 61–101.

German, J., Crippa, L. P. & Bloom, D. (1974) Bloom's syndrome. III. Analysis of the chromosome aberration characteristic of this disorder. *Chromosoma* **48,** 361–366.

Gibson, D. A. & Prescott, D. M. (1972) Induction of sister chromatid exchanges in chromosomes of rat kangaroo cells by tritium incorporated into DNA. *Exp. Cell Res.* **74,** 397–402.

Graf, V., Green, M. M. & Würgler, F. E. (1979) Mutagen-sensitive mutants in *Drosophila melanogaster.* Effects on premutational damage. *Mutat. Res.* **63,** 101–112.

Hayashi, K. & Schmid, W. (1975) The rate of sister chromatid exchanges parallel to spontaneous chromosome breakage in Fanconi's anemia and to Trenimon-induced aberrations in human lymphocytes and fibroblasts. *Humangenetik* **29,** 201–206.

Hinton, C. W. (1955) The behaviour of an unstable ring chromosome of *Drosophila melanogaster. Genetics* **40,** 951–961.

Hsu, T. C. & Somers, C. E. (1961) Effect of 5-bromodeoxyuridine on mammalian chromosomes. *Proc. Natl. Acad. Sci. U.S.* **47,** 396–403.

Hulten, M., de Weerd-Kastelein, E. A., Bootsma, D., Solari, A. J., Skakkebaek, N. E. & Swanbeck, G. (1974) Normal chiasma formation in a male with xeroderma pigmentosum. *Hereditas* **78,** 117–124.

Kato, H. (1974) Spontaneous sister chromatid exchanges detected by a BUdR-labelling method. *Nature* **251,** 70–72.

Kato, H. (1977a) Spontaneous and induced sister chromatid exchanges as revealed by the BUdR-labelling method. *Int. Rev. Cytol.* **49,** 55–93.

Kato, H. (1977b) Mechanisms for sister chromatid exchanges and their relation to the production of chromosomal aberrations. *Chromosoma* **59,** 179–191.

Kato, H. & Stich, F. H. (1976) Sister chromatid exchanges in aging and repair-deficient human fibroblasts. *Nature* **260,** 447–448.

Kihlman, B. A. & Kronborg, D. (1975) Sister chromatid exchanges in *Vicia faba*. I. Demonstration by a modified fluorescent plus Giemsa (FPG) technique. *Chromosoma* **51,** 1–10.

Kistenmacher, M. L. & Punnett, H. H. (1970) Comparative behaviour of ring chromosomes. *Am. J. Human. Genet.* **22,** 304–318.

Lambert, B., Hansson, K., Lindsten, J., Sten, M., & Werelius, B. (1976) Bromodeoxyuridine-induced sister chromatid exchanges in human lymphocytes. *Hereditas* **83,** 163–174.

Latt, S. A. (1974) Sister chromatid exchanges, indices of human chromosome damage and repair: detection by fluorescence and induction by mitomycin C. *Proc. Natl. Acad. Sci. U.S.* **71,** 3162–3166.

Latt, S. A., Stetten, G., Juergens, L. A., Buchanan, G. R., & Gerald, P. S. (1975) Induction by alkylating agents of sister chromatid exchanges and chromatid breaks in Fanconi's anemia. *Proc. Natl. Acad. Sci. U.S.* **72,** 4066–4077.

Leigh, B. (1976) Ring chromosomes and radiation-induced chromosome loss. In *The Genetics and Biology of Drosophila*, Vol. 1b (M. Ashburner & E. Novitski, Eds). Academic Press, New York and London, pp. 505–528.

Lindsley, D. L. & Grell, E. H. (1968) Genetic variation in *Drosophila melanogaster*. Carnegie Institute Washington Publication No. 627.

Lutken, T. & Baker, B. S. (1979) The effects of recombination defective meiotic mutants in *Drosophila melanogaster* on gonial recombination in males. *Mutat. Res.* **61,** 221–227.

McClintock, B. (1938) The production of homozygous deficient tissues with mutant characteristics by means of the aberrant mitotic behaviour of ring-shaped chromosomes. *Genetics* **23,** 315–346.

McLeish, J. & Sunderland, N. (1961) Measurement of deoxyribonucleic acid (DNA) in higher plants by feulgen photometry and chemical methods. *Exp. Cell Res.* **24,** 527–540.

Merriam, J. R. (1968) FM7: first multiple seven. *Drosophila Inf. Serv.* **43,** 64.

Miller, R. C., Aranson, M. M., & Nichols, W. W. (1976) Effects of treatment on differential staining of BUdR labelled metaphase chromosomes: three-way differentiation of M3 chromosomes. *Chromosoma* **55,** 1–11.

Mirsky, A. E. & Ris, H. (1951) The deoxyribonucleic acid content of animal cells and its evolutionary significance. *J. Gen. Physiol.* **34,** 451–462.

Morgan, L. V. (1933) A closed X chromosome in *Drosophila melanogaster*. *Genetics* **18,** 250–283.

Morgan, W. F. & Crossen, P. E. (1977) The incidence of sister chromatid exchanges in cultured human lymphocytes. *Mutat. Res.* **42,** 305–312.

Novitski, E. (1955) Genetic measures of centromere activity in *Drosophila melanogaster*. In *Symposium on Genetic Recombination, 1954. J. Cell Comp. Physiol.* **45,** Suppl. 2, 151–169.

Perry, P. & Evans, H. J., (1975) Cytological detection of mutagen-carcinogen exposure by sister chromatid exchanges. *Nature* **258,** 121–125.

Pimpinelli, S., Gatti, M., & De Marco, A. (1975) Evidence for heterogeneity in heterochromatin of *Drosophila melanogaster*. *Nature* **256,** 335–337.

Pimpinelli, S., Pignone, D., Gatti, M., & Olivieri, G. (1976a) X-ray induction of chromatid interchanges in somatic cells of *Drosophila melanogaster:* variation through the cell cycle of the pattern of rejoining. *Mutat. Res.* **35,** 101–110.

Pimpinelli, S., Santini, G., & Gatti, M. (1976b) Characterization of Drosophila heterochromatin. II. C-and N-banding. *Chromsoma* **57,** 377–386.

Pimpinelli, S., Pignone, D., Santini, G., Gatti, M., & Olivieri, G. (1977) Mutagen specificity in the induction of chromosomal aberrations in somatic cells of *Drosophila melanogaster. Genetics* **85,** 249–257.

Rasch, E. M., Barr, H. J., & Rasch, R. W. (1971) The DNA content of sperm of *Drosophila melanogaster. Chromosoma* **33,** 1–18.

Ris, H. & Mirsky, A. E. (1949) Quantitative cytochemical determination of deoxyribonucleic acid with the feulgen nucleal reaction. *J. Gen. Physiol.* **33,** 125–146.

Sandler, L. (1957) The meiotic behaviour of reversed compound ring X chromosomes in *Drosophila melanogaster. Genetics* **42,** 764–782.

Schonberg, S. (1980) Lack of spontaneous sister chromatid exchanges in somatic cells of *Drosophila melanogaster* (letter to the Editor). *Genetics* **94,** No. 2.

Schroeder, T. M. & German, J. (1974) Bloom's syndrome and Fanconi's anemia: demonstration of two distinctive patterns of chromosome disruption and rearrangement. *Humangenetik* **25,** 299–306.

Schultz, J. & Catcheside, D. G. (1938) The nature of closed X chromosomes in *Drosophila melanogaster. J. Genet.* **35,** 315–320.

Schvartzman, J. B. & Cortés, F. (1977) Sister chromatid exchanges in *Allium cepa. Chromosoma* **62,** 119–131.

Schwartz, D. (1953) Evidence for sister-strand crossing over in maize. *Genetics* **38,** 251–260.

Smith, P. D. (1973) Mutagen sensitivity of *Drosophila melanogaster.* I. Isolation and preliminary characterization of a methyl methansulfonate-sensitive strain. *Mutat. Res.* **20,** 215–220.

Smith, P. D., Snyder, R. D., & Dusenbery, R. L. (1980) Isolation and characterization of repair-deficient mutants of *Drosophila melanogaster.* In DNA repair and mutagenesis in eukaryotes. (W. M. G. Generoso, M. D. Shelby, & F. J. de Serres, Eds), Plenum Press, New York and London, pp. 175–188.

Sperling, K., Wegner, R. D., Riehm, H., & Obe, G. (1975) Frequency and distribution of sister chromatid exchanges in a case of Fanconi's anemia. *Humangenetik* **27,** 227–230.

Sunderland, N. & McLeish, J. (1961) Nucleic acid content and concentration in root cells of higher plants. *Exp. Cell Res.* **24,** 541–554.

Tice, R., Chaillet, J., & Schneider, E. L. (1975) Evidence derived from sister chromatid exchanges of restricted rejoining of chromatid subunits. *Nature* **256,** 642–644.

Tice, R., Chaillet, J., & Schneider, E. L. (1976) Demonstration of spontaneous sister chromatid exchanges *in vivo. Exp. Cell Res.* **102,** 426–429.

Tsuji, H. & Tobari, I. (1979) Detection of sister chromatid exchanges *in vivo* in somatic cells of *Drosophila melanogaster. Mutat. Res.* **62,** 389–392.

Uhl, C. H. (1965) A model of sister-strand crossing over. *Nature* **206,** 1003–1005.

Vogel, W. & Bauknecht, T. (1976) Differential chromatid staining by *in vivo* treatment as a mutagenicity test system. *Nature* **260,** 448–449.

Weinstein, A. (1936) The theory of multiple-strand crossing over. *Genetics* **21,** 155–199.

Wienberg, J. (1977) BrdU-Giemsa-technique for the differentiation of sister chromatids in somatic cells of *Drosophila melanogaster. Mutat. Res.* **44,** 283–286.

Wolff, S. (1977) Sister chromatid exchange. *Annu. Rev. Genet.* **11,** 183–201.

Wolff, S. & Perry, P. (1974) Differential Giemsa staining of sister chromatids and the study of sister chromatid exchanges without autoradiography. *Chromosoma* **48,** 341–353.

Wolff, S., Bodycote, J., Thomas, G. H. & Cleaver, J. E. (1975) Sister chromatid exchange in xeroderma pigmentosum cells that are defective in DNA excision repair or post-replication repair. *Genetics* **81,** 349–355.

Wolff, S., Lindsley, D. L. & Peacock, W. J. (1976) Cytological evidence for switches in polarity of chromosomal DNA. *Proc. Natl. Acad. Sci. U.S.* **73,** 877–881.

Wolff, S., Rodin, B. & Cleaver, J. E. (1977) Sister chromatid exchanges induced by mutagenic carcinogens in normal and xeroderma pigmentosum cells. *Nature* **265,** 347–349.

Würgler, F. E. & Graf, U. (1980) Mutation induction in repair-deficient strains of *Drosophila.* In DNA repair and mutagenesis in eukaryotes (W. M. G. Generoso, M. D. Shelby, & F. J. de Serres, Eds), Plenum Press, New York and London, pp. 223–240.

Yamamoto, M. & Miklos, G. L. G. (1978) Genetic studies on heterochromatin in *Drosophila melanogaster* and their implications for the function of satellite DNA. *Chromosoma* **66,** 71–98.

Index

Abbreviations, chemical, 103
Aberration, chromosome, 41, 42-44, 45-51
relation to SCE, 54
see also Chromosomal aberrations; Chromosome aberrations; Lesions
Acetaldehyde, 214
Acetone, 100
Acetoxyactylaminofluorene (AAAF), 50, 100
Acridine orange, 21-22, 100, 102
Acriflavine, 99
Acrocentric chromosome, SCEs induced in, 52-53
Actinomycin D, 32, 102, 168, 170, 171
Adducts, SCEs from, 54
Adenosine, 18
Adenovirus, 205
Adriamycin (AM), 102, 193, 209
Aerobic conditions, DNA breaks in, 261
Aflatoxin B, 100
Age and SCE incidence, 185
Aging cell populations, 239
Alcohol, 165, 212, 214
N-7 Alkylguanine, 48
Allergic conditions, 167
Allethrin, 99
Allium cepa, 247, 249, 252, 254, 263
Alterations, nature of, 60-61, 63-66
Ames' *Salmonella*/microsome test, 91
2-Aminofluorene, 100
Aminopyrine, 98
Amoebiasis, 168, 169
Anaphase bridges, 279
Andriamycin, 171
Anemia, 183, 187, 191-192, 284
Anesthetics as toxic agent, 98, 175, 211
Angelicin, 50
Aniline HCl, 100
Annato, 98
Anoxic conditions, DNA breaks in, 261
Anthracene, 99
Antipain, 98
Aprt locus, 64
Apurinic endonuclease, 190, 192
Ataxia telangiectasia, 183, 187, 192-194
Atom, heavy, 20
Autologous serum, 161
Autoradiography, 2, 18, 188, 244, 245-246, 251-262, 272
and sister chromatid exchange, 1-15
Azaadine resistance, 65
5-Azacytidine, 100
8-Azaguanine (AG), 64, 75
Azaserine, 100
Azathioprine, 99

Banding, 11-12, 30
Barbital, 99
Basal SCE frequency, 157
Base damage, 261, 262
Baseline frequencies comparison, 29, 62
Bellevalia romana, 3, 4, 10, 251, 255
Benzanthracene, 98
Benzene, 98
Benzo(*a*)pyrene, 76-77, 100
Beta-mercaptoethanol, 25

Bisbenzimidazole dye, 20
Bisulphan, 170, 210
Bleomycin, 46, 91, 98, 100, 102, 170, 260
  in vitro treatment with, 94
Blood donors, smoking, 162
Blood storage, effects on SCEs, 152-153
Bloom's lymphocytes, 93, 97
Bloom's syndrome, 32, 183, 184, 188-190, 191, 284
B lymphocytes, 158-159, 187
Bonds, disulfide, 25
Bone-marrow cells, 106, 120-121, 134, 136, 188, 191
BrdUrd, *see* Bromodeoxyuridine
Breaks in DNA:
  double-strand, 45
  single-strand, 46
  see also DNA
Bridges, half-chromatid, 2
Bromodeoxyuridine (BrdUrd), 18, 50
  administration, 231-233
  base pair rations, 69
  in cell assay, 59, 68-69
  concentration, 68-69, 77, 155-156, 186, 271
  continuous infusion, 232
  and DNA substitution, 69
  does-response, 237
  dye techniques, 88
  -Giemsa techniques, 22
  in human toxicity tests, 151
  induction of SCE with, 68, 77
  labeling, 120-121, 132-136, 247-249
  plants, use with, 247
  procedures, 269-271
  substituted chromosomes, scoring of, 288
7-Bromomethylbenz(*a*)anthracene (BMBA), 70, 71, 75, 100, 102
5-Bromouracil (BrUra), 244, 245, 247
BrUra, *see* 5-Bromouracil (BrUra)
Butanol, 98
N-*n*-butylamine, 100
Butylbuthanolamine, 99
Butylbuthanolnitrosamine, 100
Butylhydroxyanisol, 99
N-*n*-butylurethane, 98

Caffeine, 91, 97, 98, 100, 101, 206, 260, 262
  in vitro treatment with, 96-97
Cancer, 105, 187
  chemotherapy for, 170-173, 177
    drugs, 171
Caprolactone, 98
Captafol, 100
Captan, 212
Caramel, 99
Carcinogenesis, 105, 137
Carcinogens, 54, 104-106
  screening for, 235-237
Carofur, 100
Catalase, 98
CCNU, treatment, 168, 171
Cell cycle, 3, 45, 88, 110, 121
  kinetics, 189, 238-239
Cells, 47, 48, 49
  aging populations of, 239
  CHO, 43, 50
  cultured, 110-115
  fusion, 195
  killing, 188, 193
  late passage, 239
  mediated method, 109-110
  meiotic, SCEs in, 275
  MER, 48
  mitotic, SCEs in, 275
  mutation assay of, 59, 66-70
  neuroblast, 269
  propagation, 156, 158
  renal carcinoma, 202
  target, and metabolic activation, 106-108
  V79, 49
Cellulase, 250
Centric heterochromatin, 19-20
Chemical agents, abbreviations for, 103
  in vivo induction of SCEs by, 98-101, 122-131
Chemical factory workers, toxic exposure of, 162, 174
Chemicals, occupational exposure to, 177
Chemotherapy, 170-173, 177
Chicken embryonic cells, 234, 277
Chinese hamster, 19, 32, 60
  and bleomycin, 94, 95
  cells, 89, 95, 109, 110
  chromosomes, 262
  D-6, 92

fibroblasts, 235
in vitro treatment with mitomycin C, 92
Chinese hamster ovary (CHO) cells, 32, 66, 89, 90, 111, 117, 137, 189, 272, 276
and bleomycin, 94
in vitro treatment with mitomycin C, 92
repair-deficient mutants, 78-79
Chlorambucil, 101, 170
Chloramphenicol, 98
Chlorodiazepoxide, 99
Chloropromazine, 100
Chloropropamide, 100
Chloroprothixene, 100
Choroidal sclerosis, 200
Chromatid aberrations, 43, 44, 109
breaks, 91
isochromatid, 285
crossover exchange, 238
exchange and SCEs, 1-15, 41, 51, 54
interchanges in, 287
labeling, 8, 234
sister:
differentiation, 270
substitution of, 19
subunits, 8, 11, 278
Chromomycinone dyes, 22
Chromosomal aberrations, 115, 193, 197, 208, 258, 260, 283
breakage, 118
gross, 184
spontaneous, 263
viral-induced, 205
see also Chromosome aberrations
Chromosomal DNA, metabolism of, 291
instability, 188, 284
lesions, 289
proteins, 23
sensitivity to BrdUrd-induced SCEs, 281
Chromosome aberrations, 41, 42-44, 45-51, 88, 89, 90, 109
induction of, 42
metronidazole caused, 169
repair, 42
see also Chromosomal aberrations
Chromosome banding, 30
endoreduplicated, 28, 51
endoreplicated, 13
homologous, pairing of, 275
interphase, 31
metaphase, 18, 31
prometaphase, 31
recombinant, 283-284
replication and sister chromatid exchange, 9
reproduction of, 3-6
ring, 51, 244, 275-283
subunits, restricted reunion of, 251
Cigarette smoke, 186
condensates, 101, 164, 213
Cigarette smoking, 212, 213
cis-Pt(II) diamine dichloride, 77
Clastogens, 29, 235
Closing enzymes, *see* Topoisomerases
Cochineal, 98
Cockayne's syndrome (CS), 183, 198, 202
Coffee, 165
Colcemid, 98
Colchicine, 4, 7, 10, 11, 14, 269
Cold/flu, 206
Contraceptive, *see* Oral contraceptive
Copper, 98
Cross-links, 49, 72, 191
Crossing, meiotic, 289
-over, unequal, 30
CsCl banding technique, 11-12
gradients, 31-32
C-type oncoviruses, 206
Culture cell system, 112-114
Culture time, 154, 155
Curcumin, 99
Cyclohexene oxide, 101
Cycloheximide, 98, 99, 100
Cyclophosphamide, 43, 77, 171, 210, 236
Cysteamine, 25, 101
Cysteine, 98, 100
Cytogenetic analyses, 235
Cytogenetic damage as SCEs, 54
as chromosome aberrations, 54
Cytogenetics, mammalian, 244
Cytometry, flow, 22
Cytosine, 18, 99
Cytosine arabinoside, 99, 101, 102, 171, 210
Cytostatics, 176

Cytostatic treatment, 170

Data interpretation limitations, 59, 69-70
Daughter chromosomes, 5
Daunomycin, 102
7-Deazanebularin, 18
Decarbamoyl mitomycin C (DCMMC), 50, 70, 71, 75
Decondensation, chromatin, 19
Delayed release charcoal-BrdUrd, 233
Deoxycytidine, 77
Dexon, 100
6-Diamidino-2-phenylindole (DAPI), 20-21
Diazepam, 98
n-Dibutylamine, 100
Dibutylhydroxytoluene, 98
n-Dibutylnitrosamine, 100
Dibutyl phthalate, 100
Dicarbamol MMC, 191
Dicentric rings, 43
  chromosomes, 244
  double-sized, 51-52, 278
  formation, 280, 281
  frequency of, 237
Dichlorvos, 101, 212
3-Dideoxythymidine triphosphate, 99
Diepoxybutane (DEB), 101, 191
Diethylnitrosamine, 102
Diethylstibestrol, 98
Differentiation, sister chromatid, 25
Dihydrodiols of PAH, 100
Dimethylamine GC1, 100
p-Dimethylaminobenzenediazo sodium sulfonate (DAPA), 100
Dimethylbenz(*a*)anthracene, 100, 197
Dimethylhydrazine, 100
Dimpylate, 98
Diphenyl, 100
Diplochromosomes, 27
Dipyrone, 98
Direct mutagens, 88, 89, 90
  SECs as indicator, 87, 89-104
  SEC short-term test for, 91
  types of, 89-91
  see also Indirect mutagens
Disease, human, and SCEs, 183-215
  state, defined, 184
  states, acquired, 183, 204-214
Disulfide bonds, 25
Di-*tert*-butyl provlavine, 98
Dithiothreitol, 25
DNA, 27
  aerobic breaks in, 261
  binding dyes, 23
  chain growth, 190
  content, 252, 274
  cross-link, 72
  damage, 67, 178
  double-strand breaks in, 45
  elution, 24
  heavy-heavy, 32
  heavy-light, 32
  housekeeping, 203
  interstrand cross-links, 191, 192
  lesions, 60, 72, 73, 74
  ligase, 192
  light-light, 32
  polymerases, 190, 198
  recombinant, 33
  repair, 188
    defects, 183, 187-204
    deficient mutants (CHO), 78-79
    enzymes affecting, 185
    processes, 290
    single-, double-strand, 193
  repeated, 31
  replication, 185, 188
  semiconservative segregation of, 3-6
  single-strand breaks in, 46, 261
  synthesis, 18, 195, 202
  UV damaged, 194
  Watson-Crick model of, 6
Dose response, 59, 70-76, 272
Double-strand breaks, 45, 261
Down's syndrome (DS), 183, 197-198
Drosophila, 12, 237
  americana, 12
  chromatid labeling of, 234
  hydei, 12
  melanogaster, 12, 111, 268, 272
  metaphase preparations of, 268-269
  SCEs in, 267-291
  virilis, 12
DTIC, 168, 171
Dyes, bisbenzimadazole, 20
  chromomycinone, 22
Dyskeratosis congenita, Scoggins type (DC), 183, 200-201

Elastatinal, 98
Embryo fibroblasts, 92
Endonuclease-sensitive sites, 261
Endoreduplicated chromosomes, 28, 51
Endoreplicated chromosomes, 13
Enzyme-deficiency markers, 63-64
  in DNA replication, 185
  nicking and closing, 13
  photoreactivating, 46, 195
Epichlorohydrin, 101, 211, 212
Escherichia coli, 261
Ethanol, 98, 99
Ethenzamide, 101
Ethionamide, 98
8-Ethozycaffeine, 260
Ethylamine, 99
Ethylene oxide, 175-176
Ethyl methanesulfonate (EMS), 50, 70, 71, 75, 100, 102, 190, 191, 206
N-Ethyl-N-nitro-N-nitrosoguanidine (ENNG), 100
Ethylnitrosophenylurea (ENPU), 99
Ethylnitrosourea (ENU), 70, 71, 75, 99, 100
Euchromatic-heterochromatic junction, 271
Euchromatic regions, 31
Eulan U-33, 99
Exchanges, asymmetrical/symmetrical, 287
  expected frequencies of, 8, 10
  single and twin, 8
  sister chromatid, 1-15
Excision repair, 195, 196, 290
Exposure, occupational toxic, 162, 166, 173-174, 183, 211-212
Exposure, therapeutic, to chemical agents, 183, 210-211

Fanconi's anemia, 183, 187, 191-192, 284
Fenitrothion, 101
Fetal calf serum, 161
Fetal rodent chromosomes, BrdUrd labeling of, 236-237
Fibroblasts, 106, 110, 119, 189
  cultured, 188, 194
  embryo, 92, 93
  mitomycin C, 92, 93
  V79-4 (CHO), 137
Fish chromosomes, chromatic labeling of, 234
Flow cytometry, 22
Fluorescence, 19, 20, 28
Fluorescence-plus-Giemsa technique, 51, 151, 152, 247-251, 252, 254, 255, 256, 259-262, 263, 270
Fluorescent brightener, 98, 99
Fluorescent light, 67, 209
5-Fluorodeoxyuridine (FdUrd), 100, 101, 247, 248
5-Fluorouracil, 171, 210
Food additives, chromosomal damage by, 212-214
Formycin, 18
Friedreich's ataxia (FrA), 199-200
Fuelgen staining, 23-24
Fungicides, 211
Furazolidone, 100
Furosemide, 98

Gamma ray, 192, 199, 202
Gardner's syndrome (GS), 183, 201
Genome, cytogenetic damage to, 42
Genotoxic agents, exposure to, 154, 157
Giemsa staining, 19
  see also Fluorescence-plus-Giemsa technique
Glutathione, 98, 101
Glycyrrhizic acid trisodium salt, 99
Gyrases, 14, 185

Hair dye nitrophenylenediamine (NPPD), 236
Hair dyes, synthetic, 212
Half-chromatid bridges, 2
Haloperidol, 98
Halothane, 175
Harlequin (FPG) staining, 43
Harvesting by centrifugation, 151
Heavy atom, 20
Heavy metals, 211
HeLa cell mutant, 65
Helix, DNA, and chromatid size, 6
Helix pomata (snail), 250
Hematoporphyrin, 98
Hemophilia, 184
Hepatocytes, 110
Herpes simplex, 205
Heterochromatin, 19-20, 31
Heteropycnosis, 18
Hodgkin's disease, 207

Hoechst 33242, 20
Hoechst 33258, 19, 100, 151, 250, 269
Holliday model, 33
Horedum vulgare, 247, 249, 252, 254, 263
Hprt locus, 64
Hprt protein, 64-65
Human disease and SCEs, 183-215
Human fibroblasts, 96, 276
Human lymphocytes, 43, 92-93, 94, 96, 97, 156, 177
Human populations, SCE as toxic monitor for, 149-178
Human ring chromosome, 254, 282-283
Hutchison-Gilford progeria syndrome (HG), 198
Hybridization experiments for Bloom's syndrome, 189
Hydergine, 98
Hydralazine, 99
Hydrazine, 100, 101
1-Hydrazinoethanol (HEH), 101
Hydrogen chloride, 98
Hydrogen peroxide, 78
Hydrolysis, 263
Hydroquinone, 100
N-Hydroxy-2-acetylaminofluorene, 100
3,4-Hydroxybenzo(*a*)pyrene, 99, 100
Hydroxybenzoic acid ethyl ester, 99
p-Hydroxybenzoic acid isopropyl ester, 98, 99
Hydroxylation, 109
Hydroxyurea, 98
Hydrozylamine, 99
Hypotonic solution, 269

ICR-191, 70, 71, 75
Imipramine, 98
Indian muntjac and caffeine, 97
Indirect mutagens, 88, 107, 120
Indomethacin, 98
Induction, SCE, 68, 98-104, 122-131
Infectious hepatitis, 206
Ingestion, chemical exposure by, 183, 212-214
Inhalation, chemical exposure by, 183, 212-214
Inherited predispositions to SCE, 183, 187-204
Injections, multiple, SCE detection by, 231
Insecticides, 211
Intercalation, DNA lesions and, 74
Intergenic mutations, 42
Interlocking rings, 280
Interphase chromosomes, 31
Interstitial deletions, 43
Interstrand cross-links, DNA, 191, 192
Intragenic mutations, 42
In utero SCE analyses, 236-237
Inversions, 43
In vitro treatment, 92-93, 94, 96-97
In vivo/in vitro combination technique, 233
  methods, SCE detection, 229-240
  SCE analyses, 230
  SCE test, 119-136
Irradiation, 151. *See also* X-irradiation
Isoascorbic acid Na salt, 98
Isolabeling, 244, 255-257
Isolocus chromatid aberrations, 44
Isoniazid, 101
Isonicotinic acid hydrazide, 191

Junctions, heterochromatic-euchromatin, 31

Kangaroo rat, 31
K-strophantin, 98

Labeling, 8, 120-121, 132-136, 247-249
Laboratory workers, toxic exposure of, 173-174
Lacchaic acid, 101
Lactose operon repressor, 24
Lateral asymmetry, 20
Late-replicating (X chromosome), 19
Lesions, 41, 45, 46-51, 60, 72
Leukemia, 175, 197, 207, 208
Levodopa, 100
Ligases, 185, 192
Light, photoreactivating, 46. *See also* Ultraviolet light (UV)
Lycurim, 102
Lymphocytes, 89, 106, 107, 115-117, 121, 132, 133, 134, 135, 136, 189
  B and T, 156, 187
  human, 21, 117, 118, 119, 156, 187
  SCE frequency, 152-153
  smoking effects on, 164
Lymphomas, 207, 208

Macro culture technique, 153

and micro cultures, differences between, 154
Macrophages, human pulmonary (PAMs), 109
Malathion, 100
Maleic hydrazide, 98
Manganese, 98
M chromosome, 245, 251, 252, 254
Medical treatment, SCE frequencies due to, 167-173
Mefenamic acid, 98
Megaloblastic anemia, 208
Mei-9, 285
Mei-41, 285
Mei-218, 285
Mei-352, 285
Meiosis, mechanism of, 238
Meiotic recombination, 33, 268, 285, 291
Melanoma, 162, 168, 172-173
Melphalan, 102, 168, 171, 172-173
Mercaptoethanol, 98, 100
6-Mercaptoguanosine, 99
6-Mercaptopurine, 98, 100
Metabolic activation, 88, 95
and SCE test, 104-116
Metabolic inhibitors, 95
Metaphase chromosomes, 31, 274
Metaphase preparations, 268-269
Methanol, 98
Methotrexate, 98, 101, 170, 171
8-Methoxoypsoralen, 100, 191
Methylazoxymethanol acetate, 100
7-Methylbenz(*a*)anthracene (MBA), 102
3-Methylcholanthrene (MC), 99, 100
2-Methyl-4-dimethylaminoazobenzene, 99
Methylene blue, 98, 100
Methylmethanesulfonate (MMS), 50, 100, 101, 102, 103, 193, 194, 202, 206, 290
N-Methyl-*N*-motrpspirea (MNU), 100, 101, 103
N-Methyl-N-nitro-N-nitrosoguanidine (MNNG), 100, 102, 103
Methylurea, 98
Metronidazole, 98, 162, 168
4 MeV electrons, 70, 71, 75
MFO, *see* Mixed function oxidases
Mice, 92, 96, 239, 277
Micro culture technique, 153
vs. macro cultures, 154
Micronucleus assay, 235
Microsome-mediated method, 108-109
Misonidazole, 98, 100
Misrepair hypothesis, 273
Mitomycin, 29
Mitomycin C (MMC), 49, 50, 70, 71, 75, 90-91, 92-93, 102, 103, 170, 171, 191, 193, 199, 290
Mitotic behavior, human ring chromosomes, 282-283
Mitotic recombination, 275
Mixed-function oxidases (MFO), 105, 106, 107, 109
Monocentric ring chromosomes, 54, 278
Monozygotic twins, SCE frequencies in, 160
8-MOP, 168, 169, 209
Motor combustion as toxic agent, 176
Mumps, 205
Muntjac, 31
Mus-101, 285
Mus-102, 285
Mus-105, 285
Mus-109, 285
Mutagenic carcinogens, 54
Mutagens:
direct, 87, 89-104
exposure and ill health, 208-214
and indirect, 88, 107, 120
metabolic activation-dependent, 107
screening for, 235-237
Mutants, 203
meiotic, 284
meiotic-sensitive, 269
mutagen-sensitive, 269, 286
phenotypes, 59, 63-66
Mutation alterations, 63
assay, 59, 66-68
intergenic, 42
intragenic, 42
point, 42
and SCE studies, 59, 76-78
single-gene, 59-80
spontaneous, 291

Na-dehydroacetate, 98
Nalidixic acid (Negram), 170
Naphthylflavone, 98
Neocarzinostatine, 99
Neonatal hyperbilirubinaemia, 209

Neoplasms, development of, 183, 187-204
Neoplastic disease, 183, 188, 207-208
Netropsin, 32
Neuroblast cells, 269, 279
Neuroblastoma, 30
Neurofibromatosis, 202
Nicking enzymes, *see* Topoisomerases
Nicotinamide, 101
Nitrofurantoin, 98
Nitrogen mustard, 102, 191
Nitrophenylenediamine (NPPD), 101, 236
4-Nitroquinoline-1-oxide, 48, 50, 99, 100, 101, 102
N-Nitrosodiphenylamine, 100
Nitrosoguanidine (MMNG), 194
Nitrous oxide, 98, 175
Nonautoradiographic models, SCE detection, 17, 18-27
Nonsister strand exchange, 268
Novobiocin, 99
Nursing, occupational toxicity in, 176

O–6-alkylguanine, 48
Occupational toxic exposure, 166, 173-174, 211-212
O^6-ethylation of guanine, 72
Operan, lactose, 24
Operating room personnel, occupational toxic exposure, 174, 175
Oral contraceptives, 166, 167, 186
Organic solvents, 211
Ouabain resistance, 65
2,3,5-Oxide, 99, 100
Oxygen enhancement ratio (OER), 260

Patients, medical, SCE frequencies in, 167-173
Pb-acetate, 98
Pectinase, 249, 250
Perylene, 98
Pesticides, 211
pH treatment, DNA, 25, 27
PHA effect on lymphocyte SCEs, 153
Phenanthrene, 98, 99
Phenol, 99
Phenotypes, mutant, 59, 63-66
Phenylbutazone, 99
Philadelphia chromosome, 207
Phorbol ester (TPA), 76
Photoreactivation, 46-47, 195-196
Phototherapy, 209
Pine-wood tar, 99
Plant chromosomes, chromatid labeling of, 234, 235
Plant root tips, 262-263
Plants, SCEs in, 243-263
Plasticizers, 211
Point mutations, 42
Poisson standard error, 67
Polarity, 11, 27, 251, 278
Poliomyelitis, 209
Polycythemia vera, 186
Polymerases, 185, 193
Polynemic chromosome structure, 256
Polynucleotide strands, 54
Postembryonic development, SCE frequency in, 239
Postreplication repair, 190, 195, 196, 198, 290
Potassium arsenite, 210
Potassium metabisulfite, 99
Potassium sorbate, 100
Potorous tridactylis cells, 46
Proflavin, 99, 100, 102, 236
Progesterone, 98
Promazine, 100
Prometaphase chromosomes, 31
Propane sultone, 100
Propanol, 98
Propidium iodide, 100
Propiolactone, 99, 100
Propyl gallate, 98
Proteins, 12, 23, 24, 64-65
Provlavine (PRO), 70, 71, 75
Psoriasis, 162, 169, 209
Puromycin, 98
PUVA treatment, 168-169
Pyrene, 98, 100
Pyridine, 99
Pyrimidine dimers, 202

Q-negative/positive bands, 30
Quantitative dose-response, 59, 70-76
Quinacrine mustard, 102, 103
Quinoline, 98

Rats, 93, 239, 272, 276
Rauscher leukemia virus, 205
Recombinant DNA techniques, 33
Recombination, meiotic, 268, 285
Refsum's syndrome, 200
Renal cell carcinoma, 202

Repair, 109, 193, 195, 196, 290
Repair deficiency, 50, 78-79, 192, 199, 203
Replication kinetics, 20
Reserpine, 99
Retinoblastoma (Rb), 183, 200
Reverse staining, 25
Ring chromosomes, 2, 43, 51, 244, 268, 280, 282
  classes of, 282
  SCEs in, 275-283
  scoring of SCEs in, 271-272
RNase, 250
RO 10-9359 (retinoid), 100
Rodents, SCE detection in, 231-233
Rose bengal, 98
Rothmund-Thomson syndrome (RT), 183, 201
Roundup, 99
Rous sarcoma virus, 205
Rubber solvent, 98

Saccharin, 98, 99, 100, 212
SCEs, *see* Sister chromatid exchange (SCEs)
S chromosomes, 245, 251, 256
Screening, 235-237
S-dependent agents, 44, 91, 259-260
  SCE induction, 259
Secale cereale, 247, 249, 254, 263
Segregation, 1, 3-6, 11, 27
Selenium, 100
Single-gene mutation, 59-60
  and SCE, compared, 59, 66-70
Single-strand exchanges, 257-258, 278
Sister chromatid exchanges (SCEs), 1-15, 17, 18, 27-31, 42
  alterations producing, 59, 60-63
  analysis, 29, 230, 236-237
  baseline, 29
  and chromosome aberrations, 41, 44-45, 288
  chromosome structure of, 7-11, 252
  clustering, 254
  as cytogenetic damage, 54
  detection, 17, 31-33, 231-233, 235-239
  and disease states, 183-215
  to DNA, content ratios, 274
  double, 257
  in *Drosophila*, 267-291
  frequency, 159
    basal, 157
    blood storage effects on, 152-153
    and BrdUrd concentration, 155-156, 272
    chromosomal distribution of, 251-254
    by dicentric ring method, 275
    in healthy individuals, 185-186
    human, variations in, 157, 178
    smoking effects on, 159-165
    staining procedure effects on, 157
  history of, 184, 244
  as human toxicity monitor, 149-178
  incidence, 184-185
    and age, 185
    in disease states, 183, 204-214
    inherited predispositions to, 183, 187-204
    in vitro, factors affecting, 117-119
    "spontaneous," 188
  induction, 29, 98-104, 249, 259, 272
  in vivo methods to detect, 229-240
  lesions that produce, 41, 46-51
  localization, 30
  lymphocytic, 154
  mean number per chromosome, 251
  mechanism, 78-79, 283-284
  models, 11-13, 49
  molecular mechanisms of, 283-284
  and mutation, 66-70
    compared, 66, 76-78
    dose responses of, 71, 72
    linear relation between, 61-62, 74
    toxicity dependence in, 71, 72
  in plants, 243-263
    techniques to demonstrate, 245-251
  position of, 7
  rate per picogram DNA, 276-277
  resulting from adducts, unremoved, 54
  in ring chromosomes, 275-283
  scoring of, 271-272
  sensitivity and direct mutagens, 89-91
  single, 8
  and single-gene mutation, 59-80
    baseline frequencies of, 62
    compared, 66-70
    experimental evidence in, 70-78, 188
  spontaneous, 237, 272
  test, 91-104
    for humans, 151-152
    in vivo, 119-136

and metabolic activation system, 108-119
sensitivity of, 89-91
unresolved problems in, 13-15
ultraviolet (UV) light-induced, 47, 70, 71, 75, 190, 198, 201, 202, 209, 250, 258, 261, 262
Sister chromatid differentiation, 25
Sister chromatid segregation, 27-28
Skin cancer, 187
Smallpox, 205
Smoking, 159-165, 212-213
Snail (*Helix pomata*), 250
S-9 (9000 g supernatant fraction), 108, 109, 110, 111
S-9 mix, 108, 110, 115, 116
Sodium benzoate, 100
Sodium bisulfite, 101
Spermatogonia, 28
Spinocerebellar degeneration, 199
Staining, 25, 30, 156, 157, 250-251. *See also* Fluorescence-plus-Giemsa technique
Stevioside, 98
Streptonigrin, 261
Student's t-test, 152
"Subchromatid exchange," 258
Sunlamp, 77
Sunset yellow FCF, 99

Tablet implantation, 233
Terminal deletions, 43
Tetraploid chromosome sets, 11
Theophylline, 101
6-Thioguanine (TG), 64, 75
Thio-TEPA, 100, 102, 170, 191
Third-division cells, 27
[^{3}H] Thymidine, 2, 3, 4, 10, 237, 244, 245-246, 254, 256
Thymidine kinase (*tk*), 63
Thymidylate synthetase, 247-248
Thymine dimers, 47, 54, 73
Tilorone, 98
Tissue response, SCE detection, 237
tk Locus (thymidine kinase), 63
T lymphocytes, 158-159, 187
Tobacco, 165, 212
Topoisomerases, 13, 185
Toxic agents, 149-178
types of, 176
Toxic exposure, occupational, 162, 166, 173-174, 183, 211-212
Toxicity dependence, 71
Toxicity tests, 151-153
Tradescantia paludosa, 247, 248, 263
Translocation, 43
Trenimon, 101, 102
Trichloropropene, 99
Trichomoniasis, 168, 169
Tritiated thymidine, 2, 3, 4, 10, 237, 244, 245-246, 254, 256
Tritium, 5, 254, 283
Tuberous sclerosis, 202
Tumor cell studies, 239
Twin exchanges, 8
Twins, human, SCE frequency in, 157

Ultrasound, 209-210
Ultraviolet light (UV), 46, 47, 70, 71, 75, 190, 194, 196, 198, 201, 202, 209
long wave, 250, 258, 261, 262
Unequal crossing-over, 31
Urethane, 98, 100
Uridine (Urd), 248

V79 cells, 65, 117
Vaccinia virus, 205
Vamidothion, 101
Vicia faba, 3, 97, 244, 245, 247, 251, 256, 261, 263, 277
Vinblastine, 99
Vincristine, 98, 99, 102
Vinyl chloride monomer, 211
Virus infections, 183, 186, 205-207
Visible light, SCE induction by, 259
Vitamin A palmitate, 100
Vitamin C, 100, 101, 212
V79 cells, 65, 117

Wilm's tumor, 202
Workers, and occupational toxicity, 173-176

Xenopus cells, 46
Xeroderma, 93
pigmentosum (XP), 183, 187, 194-197, 290
X-irradiation, 44, 45, 193
X-linked meiotic mutants, 286
XP cells, 194-195, 196, 202
XP variant, 194
X-ray, 192, 194, 200, 208, 260, 287
Xylene, 98
Xylitol, 100

Zeiss fluorescence microscope, 270